INTRODUCTION TO THE THEORY OF LINEAR DIFFERENTIAL EQUATIONS

BY

E. G. C. POOLE

Fellow of New College, Oxford

DOVER PUBLICATIONS, INC.
NEW YORK • NEW YORK

Oui, l'oeuvre sort plus belle
D'une forme au travail
Rebelle,
Vers, marbre, onyx, émail

TH. GAUTIER
Émaux et Camées

Published in the United Kingdom by Constable and Company Limited, 10 Orange Street, London, W.C. 2.

This new Dover edition, first published in 1960, is an unabridged and unaltered republication of the first edition published in 1936 by the Oxford University Press. It is published by special arrangement with the Oxford University Press.

Manufactured in the United States of America

Dover Publications, Inc.
180 Varick Street
New York 14, N.Y.

PREFACE

THE study of differential equations began with Newton and Leibnitz, and most of the elementary methods of solution were discovered in the course of the eighteenth century. Where a problem could not be solved in finite terms, expansions in power-series were tentatively employed by Newton. But the theory was not placed on a satisfactory logical basis until about a century ago, when Cauchy distinguished between analytic and non-analytic systems, and constructed rigorous existence-theorems appropriate to each type.

Ordinary linear equations, with which this book deals, have always attracted particular attention by their comparative tractability and their numerous practical applications. Extensive monographs have been devoted to many separate branches of the theory, such as spherical and cylindrical harmonics, expansions in series of orthogonal functions, oscillation and comparison theorems, the Heaviside calculus, polyhedral, elliptic modular and automorphic functions. While some branches arose out of physical problems, others were created by the progress of the theory of functions and of the theory of groups. Many important ideas were first worked out in connexion with the hypergeometric equation by Euler, Gauss, Kummer, Riemann, or Schwarz, and were then generalized by Fuchs, Klein, Poincaré, and many other writers of the highest distinction.

The present *Introduction* is based on lectures to senior undergraduates at Oxford, and is designed for students who have already taken an elementary course of differential equations, but have not yet specialized in one of the more advanced branches. It is not a compendium of this vast subject (to which no single author could do justice), but a selection of investigations of moderate length and difficulty, illustrating those aspects of it which are most familiar to myself. The first five chapters deal with properties common to wide classes of equations, and the last five are devoted to a more detailed examination of the hypergeometric equation, Laplace's linear equation, and the equations of Lamé and Mathieu. I have not discussed systematically the equations of Legendre and Bessel, as there are so many admirable accounts of them in English suitable for students of every grade. On the other hand, I have thought it well to devote a chapter to equations with constant coefficients. I find that candidates in university examinations have great difficulty in constructing the solution of such equations which takes assigned initial values, even when they can write down the complete primitive. A very slight sketch of Heaviside's method should enable them to make short work of this problem, which is of great practical importance. Again, the theory of simultaneous equations with constant coefficients gives an excellent opportunity of introducing in an easy context the notion of invariant factors, which is of fundamental importance in the Fuchsian theory.

The short bibliography and the footnotes serve both to acknowledge my debt to the authorities and to guide the more ambitious reader. Besides some of the great classical memoirs and the systematic treatises of Forsyth, Heffter, and Schlesinger, the books from which I have learnt most are Klein's lectures on the icosahedron and on the hypergeometric function, the masterly summaries of the general theory in the works of Goursat, Jordan, and Picard, and the studies of particular equations in Whittaker and Watson's *Modern Analysis*. Those who wish to learn more about existence-theorems should consult the recent work of Kamke.

While I am solely responsible for the shortcomings of this book, I gladly avail myself of this opportunity of expressing my profound indebtedness to my former tutor, Mr. C. H. Thompson, whose lectures at Queen's College first aroused my interest in differential equations, and later to Professor A. E. H. Love, who inspired and directed my first efforts at research. My interest in everything connected with conformal representation and Schwarz's equation was greatly stimulated by discussions with Mr. J. Hodgkinson. I have received valuable references and information on particular points from Mr. W. L. Ferrar, Dr. F. B. Pidduck, Professor G. Pólya, Professor G. N. Watson, and Professor E. T. Whittaker, not to mention many others who have courteously sent me offprints of their papers. My pupil, Mr. G. D. N. Worswick, Scholar of New College, has been of great assistance to me in reading the proofs. Last but not least, I desire to thank the Delegates of the Clarendon Press for accepting the book, and the Staff for their unfailing skill and courtesy in printing it.

E. G. C. P.

CONTENTS

I. EXISTENCE THEOREMS. LINEARLY INDEPENDENT SOLUTIONS
- § 1. The Method of Successive Approximations 1
- § 2. Solutions in Power-Series 4
- § 3. Linearly Independent Solutions 6
- § 4. Wronskian Determinants 10
- § 5. The General Linear Equation 12
- EXAMPLES I 15

II. EQUATIONS WITH CONSTANT COEFFICIENTS
- § 6. Heaviside's Solution of Cauchy's Problem . . . 18
- § 7. Operators in D and δ 22
- § 8. Simultaneous Equations. Invariant Factors . . 27
- EXAMPLES II 30

III. SOME FORMAL INVESTIGATIONS
- § 9. Linear Operators 33
- § 10. Adjoint Equations 36
- § 11. Simultaneous Equations with Variable Coefficients . . 39
- EXAMPLES III 41

IV. EQUATIONS WITH UNIFORM ANALYTIC COEFFICIENTS
- § 12. Group of the Equation 45
- § 13. Canonical Transformations 47
- § 14. Hamburger Sets of Solutions 53
- § 15. Fuchs's Conditions for a Regular Singularity . . 55
- EXAMPLES IV 59

V. REGULAR SINGULARITIES
- § 16. Formal Solutions in Power-Series 62
- § 17. Convergence 65
- § 18. Apparent Singularities 68
- § 19. The Method of Frobenius 70
- § 20. The Point at Infinity. Equations of the Fuchsian Class . 74
- EXAMPLES V 79

VI. THE HYPERGEOMETRIC EQUATION
- § 21. Riemann's P-Function 83
- § 22. Kummer's Twenty-four Series 87
- § 23. Group of Riemann's Equation 88

§ 24. Recurrence Formulae. Hypergeometric Polynomials	.	92
§ 25. Quadratic and Cubic Transformations	. . .	97
§ 26. Continuation of the Hypergeometric Series	. .	100
§ 27. Hypergeometric Integrals . . .	. .	104
EXAMPLES VI	. .	109

VII. CONFORMAL REPRESENTATION

§ 28. Schwarz's Problem . . .	. .	118
§ 29. The Reduced Curvilinear Triangle .	. .	122
§ 30. Symmetrical Continuation . .	. .	125
§ 31. Some Special Cases . . .	. .	129
EXAMPLES VII	. .	133

VIII. LAPLACE'S TRANSFORMATION

§ 32. Laplace's Linear Equation . .	. .	136
§ 33. The Confluent Hypergeometric Equation	. .	139
§ 34. Integral Representations of Kummer's Series	.	143
§ 35. Bessel's Equation . . .	. .	147
EXAMPLES VIII	. .	150

IX. LAMÉ'S EQUATION

§ 36. Lamé Functions . . .	. .	154
§ 37. Introduction of Elliptic Functions .	. .	159
§ 38. Oscillation and Comparison Theorems	. .	164
§ 39. The General Equation of Integral Order	. .	167
§ 40. Equations of Picard's Type .	. .	170
EXAMPLES IX	. .	175

X. MATHIEU'S EQUATION

§ 41. Nature and Group of Mathieu's Equation	. .	178
§ 42. The Methods of Lindstedt and Hill .	. .	182
§ 43. Mathieu Functions . . .	. .	187
§ 44. The Methods of Lindemann and Stieltjes	. .	193
EXAMPLES X	. .	197

SHORT BIBLIOGRAPHY 200

INDEX OF NAMES 201

I

EXISTENCE THEOREMS. LINEARLY INDEPENDENT SOLUTIONS

1. The Method of Successive Approximations

Standard Forms. AN ordinary differential equation is said to be in the *canonical form* when the highest derivative of the unknown function is given by an explicit formula

$$y^{(n)} = F(x; y, y^{(1)}, y^{(2)}, \ldots, y^{(n-1)}). \tag{1}$$

This is evidently equivalent to a system of n simultaneous equations of the first order

$$\left. \begin{array}{l} Dy_i = y_{i+1} \quad (i = 1, 2, \ldots, n-1), \\ Dy_n = F(x; y_1, y_2, \ldots, y_n) \quad (D \equiv d/dx). \end{array} \right\} \tag{2}$$

More generally, it may be shown that every well-determined system of simultaneous differential equations is equivalent to a *normal canonical system*

$$Dy_i = F_i(x; y_1, y_2, \ldots, y_n) \quad (i = 1, 2, \ldots, n). \tag{3}$$

Consider, for example, three equations $\Phi_i = 0$ ($i = 1, 2, 3$), in three unknowns (u, v, w) and their derivatives up to $(u^{(\alpha)}, v^{(\beta)}, w^{(\gamma)})$. In a domain where the Jacobian

$$J \equiv \frac{\partial(\Phi_1, \Phi_2, \Phi_3)}{\partial(u^{(\alpha)}, v^{(\beta)}, w^{(\gamma)})} \neq 0, \tag{4}$$

the equations can be solved for the three highest derivatives and written in the canonical form

$$u^{(\alpha)} = F, \qquad v^{(\beta)} = G, \qquad w^{(\gamma)} = H; \tag{5}$$

and, on introducing auxiliary variables as in (1), we obtain a normal canonical system of order $(\alpha + \beta + \gamma)$. If, however, $J \equiv 0$, the three relations connecting $(u^{(\alpha)}, v^{(\beta)}, w^{(\gamma)})$ are not algebraically independent, and can be replaced by two equations involving the highest derivatives, and one where they do not appear, say $\Omega = 0$. We then have three possible cases.

(i) If $\Omega \equiv 0$, the system is *indeterminate*, the three given equations being algebraically equivalent to not more than two.

(ii) If Ω does not vanish identically, but does not involve any of the unknowns or their derivatives, the system is *incompatible*, or else the problem is incorrectly formulated.

(iii) If Ω involves the unknowns and their derivatives of order not higher than $(u^{(\alpha-\kappa)}, v^{(\beta-\kappa)}, w^{(\gamma-\kappa)})$, suppose $u^{(\alpha-\kappa)}$ is actually present. Differentiating κ times the relation $\Omega = 0$, we may eliminate all derivatives of u above $u^{(\alpha-\kappa)}$ from the system, without introducing any new derivatives above $v^{(\beta)}$ or $w^{(\gamma)}$. The apparent order of the system will thus be reduced by κ, and the process may be repeated if necessary until a well-determined system has been obtained.

Real Linear Systems. The most general normal canonical system which is linear in the unknowns may be written in the standard form

$$(S_n) \qquad Dy_i = \sum_{j=1}^{n} a_{ij}(x) y_j + f_i(x) \quad (i = 1, 2, ..., n),$$

the coefficients (a_{ij}) and (f_i) being any functions of x.

We shall assume in the first instance that the independent variable x is essentially real; the coefficients and the unknowns may be complex; but, if they are, we can separate real and imaginary parts and obtain a purely real system of order $2n$; accordingly we may assume them to be real also, without essential loss of generality. The following fundamental theorem is a particular case of Cauchy's general existence theorem for non-linear systems.

THEOREM. *If the coefficients of the real linear system* (S_n) *are continuous in the finite interval* $(x' \leqslant x \leqslant x'')$, *the system is identically satisfied by functions with continuous derivatives*

$$y_i = \phi_i(x) \quad (i = 1, 2, ..., n),$$

which are uniquely determined by the arbitrary initial values

$$\phi_i(\xi) = \eta_i \quad (i = 1, 2, ..., n)$$

at a point $x = \xi$ *of the interval.*

The simplest proof of Cauchy's existence theorem for the most general system is by Picard's method of successive approximations, which is particularly suitable for linear systems. It is closely connected with methods used by Liouville (1837) and Caqué (1864), and with the 'calculus of integral matrices' developed by Peano, Baker, and Schlesinger.†

We take as a first approximation

$$\phi_i^{(0)}(x) \equiv \eta_i \quad (i = 1, 2, ..., n), \tag{6}$$

† L. Schlesinger, *Vorlesungen über lineare Differentialgleichungen* (1908).

and construct further sets of functions according to the rule

$$\phi_i^{(k)}(x) = \eta_i + \int_\xi^x \Big[\sum_{j=1}^n a_{ij}(t)\phi_j^{(k-1)}(t) + f_i(t)\Big]\,dt. \tag{7}$$

We shall show that the required solution is

$$\phi_i(x) = \lim_{k\to\infty} \phi_i^{(k)}(x) \quad (i = 1, 2, ..., n). \tag{8}$$

To examine these limits, let us write them as infinite series

$$\phi_i(x) = \eta_i + \sum_{k=1}^\infty [\phi_i^{(k)}(x) - \phi_i^{(k-1)}(x)] = \eta_i + \sum_{k=1}^\infty U_i^{(k)}(x), \tag{9}$$

where

$$\left. \begin{aligned} U_i^{(1)}(x) &= \int_\xi^x \Big[\sum_{j=1}^n a_{ij}(t)\eta_j + f_i(t)\Big]\,dt, \\ U_i^{(k)}(x) &= \sum_{j=1}^n \int_\xi^x a_{ij}(t) U_j^{(k-1)}(t)\,dt. \end{aligned} \right\} \tag{10}$$

Since (η_i) are constants, and since the coefficients of the system (S_n) are continuous, we can choose positive numbers A and M such that

$$|a_{ij}(x)| \leqslant A, \quad \Big|\sum_{j=1}^n a_{ij}(x)\eta_j + f_i(x)\Big| \leqslant MnA, \tag{11}$$

in the interval $(x' \leqslant x \leqslant x'')$. Now $\{U_i^{(k)}(x)\}$ are defined successively as integrals of continuous functions, and are therefore themselves continuous; and (10) and (11) give

$$|U_i^{(1)}(x)| \leqslant MnA|x-\xi| \quad (x' \leqslant x \leqslant x''). \tag{12}$$

Suppose that, for a certain positive integer k, we have

$$|U_i^{(k)}(x)| \leqslant Mn^k A^k |x-\xi|^k/k!. \tag{13}$$

Then the recurrence formulae (10) give the inequalities

$$\begin{aligned} |U_i^{(k+1)}(x)| &\leqslant \sum_{j=1}^n \Big|\int_\xi^x A \cdot Mn^k A^k |t-\xi|^k\,dt/k!\Big| \\ &\leqslant Mn^{k+1}A^{k+1}|x-\xi|^{k+1}/(k+1)!; \end{aligned} \tag{14}$$

and since (13) is true for $k = 1$, it is true in general, by induction.

We now have uniformly in the entire interval

$$|U_i^{(k)}(x)| \leqslant Mn^k A^k (x''-x')^k/k!; \tag{15}$$

and so the terms of the series (9) are dominated by those of a series of positive constants $M\exp[nA(x''-x')]$, which is convergent. Hence the series $\{\phi_i(x)\}$ converge uniformly in the interval and their sums are continuous. Making $k \to \infty$ in (7), we have

$$\phi_i(x) = \eta_i + \int_\xi^x \Big[\sum_{j=1}^n a_{ij}(t)\phi_j(t) + f_i(t) \Big] dt. \tag{16}$$

By putting $x = \xi$, we verify the initial values $\phi_i(\xi) = \eta_i$; and, because the integrands (16) are continuous, we may differentiate and so verify that $y_i = \phi_i(x)$ is a solution of the system (S_n).

Uniqueness. If there were two sets of functions satisfying all the conditions, their differences $\Phi_i(x) \equiv \phi_i(x) - \phi_i^*(x)$ $(i = 1, 2, ..., n)$ would satisfy the relations

$$\Phi_i(x) = \sum_{j=1}^n \int_\xi^x a_{ij}(t)\Phi_j(t)\, dt. \tag{17}$$

Since $\{\Phi_i(x)\}$ are continuous, we can find an upper bound

$$|\Phi_i(x)| \leqslant B \quad (x' \leqslant x \leqslant x''). \tag{18}$$

By the same argument as in (14) we can then prove by induction that
$$|\Phi_i(x)| \leqslant Bn^k A^k |x-\xi|^k/k! \leqslant Bn^k A^k (x''-x')^k/k! \tag{19}$$

for all positive integers k. By making $k \to \infty$ we have $|\Phi_i(x)| < \epsilon$, uniformly in the interval, where ϵ is an arbitrarily small constant. Hence $\Phi_i(x) \equiv 0$, or the two solutions are identical.

2. Solutions in Power-Series

Analytic Linear Systems. The system (S_n) is said to be *analytic* if the coefficients (a_{ij}) and (f_i) are analytic functions of a real or complex variable x, in the sense of Cauchy. A point $x = \xi$ where every coefficient is holomorphic is called an *ordinary point*; all other points are called *singular*. At an ordinary point the coefficients can be expanded as Taylor series:

$$a_{ij}(x) = \sum_{k=0}^\infty a_{ij}^{(k)}(x-\xi)^k, \qquad f_i(x) = \sum_{k=0}^\infty f_i^{(k)}(x-\xi)^k, \tag{1}$$

which are convergent when $|x-\xi| < d(\xi)$, where $d(\xi)$ is the shortest distance between $x = \xi$ and the nearest singular point.

The following existence theorem consists of a particular case of Cauchy's existence theorem for analytic non-linear systems, together with a rider by Fuchs specifying the radius of convergence of the solution, when the system is linear.

THEOREM. *At an ordinary point $x = \xi$ the analytic linear system (S_n) is identically satisfied by n power-series*

$$y_i = \phi_i(x) = \sum_{k=0}^{\infty} c_i^{(k)}(x-\xi)^k \quad (i = 1, 2, ..., n), \tag{2}$$

which are uniquely determined by the initial values $\phi_i(\xi) = \eta_i$. These series are convergent when $|x-\xi| < d(\xi)$, where $d(\xi)$ is the shortest distance between $x = \xi$ and the nearest singular point.

To obtain a formal solution we substitute the power-series (2) in the initial conditions and in the system (S_n), and equate coefficients of like powers of $(x-\xi)$ on both sides. The initial conditions give first

$$c_i^{(0)} = \eta_i \quad (i = 1, 2, ..., n), \tag{3}$$

and then we have

$$(k+1)c_i^{(k+1)} = \sum_{j=1}^{n} \sum_{s=0}^{k} a_{ij}^{(k-s)} c_j^{(s)} + f_i^{(k)} \quad (i = 1, 2, ..., n; k = 0, 1, 2, ...). \tag{4}$$

It is evident, by induction with respect to k, that every coefficient is uniquely determined.

Let x be any point where $|x-\xi| = r < d(\xi) = d$, and let $R = \tfrac{1}{2}(r+d) < d$. The series (1) for the coefficients are absolutely convergent when $|x-\xi| \leqslant R$; so that we can choose a positive number M such that

$$|a_{ij}^{(k)}|R^k, \quad |f_i^{(k)}|R^k < M. \tag{5}$$

The relations (4) then yield the inequalities

$$(k+1)|c_i^{(k+1)}| \leqslant MR^{-k}\Big[\sum_{j=1}^{n}\sum_{s=0}^{k} R^s |c_j^{(s)}| + 1\Big]. \tag{6}$$

If we put $c^{(k)} = \sum_{j=1}^{n} |c_j^{(k)}|$, we obtain from (6) by adding

$$(k+1)c^{(k+1)} \leqslant nMR^{-k}\Big[\sum_{s=0}^{k} R^s c^{(s)} + 1\Big]. \tag{7}$$

We shall clearly obtain a dominant power-series

$$\sum_{k=0}^{\infty} c_i^{(k)}(x-\xi)^k \leqslant \sum_{k=0}^{\infty} c^{(k)}(x-\xi)^k \leqslant \sum_{k=0}^{\infty} C^{(k)}(x-\xi)^k \tag{8}$$

if we put $C^{(0)} = c^{(0)}$, and then

$$(k+1)C^{(k+1)} = nMR^{-k}\Big[\sum_{s=0}^{k} R^s C^{(s)} + 1\Big]. \tag{9}$$

But this gives

$$(k+1)C^{(k+1)} = nMC^{(k)} + R^{-1}kC^{(k)}, \tag{10}$$

and hence
$$\lim_{k\to\infty} \frac{C^{(k+1)}}{C^{(k)}} = \lim_{k\to\infty}\left[\frac{k+nMR}{(k+1)R}\right] = \frac{1}{R}. \tag{11}$$

Hence the dominant series $\sum C^{(k)}(x-\xi)^k$ and, *a fortiori*, the n series (2) are absolutely convergent if $|x-\xi| = r < R < d$. The series $\sum C^{(k)}(x-\xi)^k$ is determined by the differential equation

$$DY = nM\left[1-\left(\frac{x-\xi}{R}\right)\right]^{-1}(Y+1), \tag{12}$$

and the initial value $Y = c^{(0)}$, when $x = \xi$. By separation of the variables we have the solution

$$Y = -1 + (c^{(0)}+1)\left[1-\left(\frac{x-\xi}{R}\right)\right]^{-nMR}, \tag{13}$$

and this explains the binomial recurrence formula (10).

This investigation shows that the *solutions* of a linear analytic system are holomorphic at all finite points of the plane where the *coefficients* are holomorphic. Accordingly the singularities of the solutions are known *a priori* by inspection of the coefficients. This fundamental property was pointed out by Fuchs.

3. Linearly Independent Solutions

Combination of Solutions. If $y_i = \phi_i(x)$ and $y_i = \phi_i^*(x)$ are any two solutions of the linear system (S_n), then $y_i = \phi_i(x) - \phi_i^*(x)$ is a solution of the corresponding *homogeneous system*

$$(S_n^*) \qquad Dy_i = \sum_{j=1}^{n} a_{ij}(x)y_j \quad (i = 1, 2, ..., n).$$

We therefore require only one particular integral of (S_n), together with the complete primitive of (S_n^*).

If we have m solutions of the homogeneous system (S_n^*)
$$y_i = \phi_{ij}(x) \quad (i = 1, 2, ..., n; \; j = 1, 2, ..., m), \tag{1}$$
another solution is
$$y_i = \sum_{j=1}^{m} c_j \phi_{ij}(x) \quad (i = 1, 2, ..., n), \tag{2}$$

where (c_j) are any constants. The solutions (1) are said to be *linearly connected* if constants (not all zero) exist such that

$$\sum_{j=1}^{m} c_j \phi_{ij}(x) \equiv 0 \quad (i = 1, 2, ..., n). \tag{3}$$

THEOREM. *The necessary and sufficient condition that m solutions (1) of the homogeneous system (S_n^*) should be linearly connected is that the rank of the matrix $(\phi_{ij}(x))$ should be less than m.*

The condition is necessary; for, if the relations (3) hold, every m-rowed determinant of the matrix vanishes. Suppose now that the condition is satisfied and that the rank is k ($k < m, k \leqslant n$). We can number the variables and solutions so that the determinant

$$\Delta_k \equiv |\phi_{ij}(x)| \neq 0 \quad (i,j = 1, 2, ..., k). \tag{4}$$

Corresponding to any solution $y_i = \phi_{is}(x)$ of the set, we can find a unique set of multipliers (u_{js}) such that

$$\phi_{is}(x) = \sum_{j=1}^{k} u_{js} \phi_{ij}(x) \quad (i = 1, 2, ..., n;\ s = 1, 2, ..., m), \tag{5}$$

these relations being compatible, because every determinant of $(k+1)$ rows of the matrix vanishes. On introducing this solution in (S_n^*), the terms multiplied by (u_{js}) cancel, leaving

$$\sum_{j=1}^{k} \phi_{ij}(x) D u_{js} = 0 \quad (i = 1, 2, ..., n); \tag{6}$$

and, because $\Delta_k \neq 0$, we must have

$$D u_{js} = 0, \qquad u_{js} = \text{constant}, \tag{7}$$

so that the condition is sufficient.

At an ordinary point $x = \xi$ we can construct n solutions, taking arbitrary initial values such that the determinant $\Delta(\xi) = |\phi_{ij}(\xi)| \neq 0$, and these solutions are necessarily linearly independent. Since $k \leqslant n$, every other solution can be written in the form

$$y_i = \sum_{j=1}^{n} c_j \phi_{ij}(x) \quad (i = 1, 2, ..., n), \tag{8}$$

where (c_j) are constants. Any such system of n linearly independent solutions is called a *fundamental system*.

Jacobi's Determinant. The determinant $\Delta \equiv |\phi_{ij}(x)|$ of any n solutions of the homogeneous system (S_n^*) can be evaluated by means of an auxiliary differential equation. In an interval or domain where $\{a_{ij}(x)\}$ are continuous we have

$$\begin{aligned}
D\Delta &= \sum_{i=1}^{n} \sum_{k=1}^{n} \frac{\partial \Delta}{\partial \phi_{ik}} D\phi_{ik}(x) \\
&= \sum_{i=1}^{n} \sum_{j=1}^{n} \sum_{k=1}^{n} \frac{\partial \Delta}{\partial \phi_{ik}} a_{ij}(x) \phi_{jk}(x) \\
&= \sum_{i=1}^{n} \sum_{j=1}^{n} a_{ij}(x) \left[\sum_{k=1}^{n} \phi_{jk}(x) \frac{\partial \Delta}{\partial \phi_{ik}} \right].
\end{aligned} \tag{9}$$

If we introduce the Kronecker deltas
$$\delta_{ii} = 1, \qquad \delta_{ij} = 0 \quad (i \neq j), \tag{10}$$
and use the property that a determinant with two identical rows vanishes, we may write
$$D\Delta = \sum_{i=1}^{n} \sum_{j=1}^{n} a_{ij}(x) \delta_{ij} \Delta$$
$$= \Big[\sum_{i=1}^{n} a_{ii}(x)\Big] \Delta. \tag{11}$$

This is integrated by separating the variables and gives Jacobi's result
$$\Delta(x) = \Delta(\xi) \exp\Big[\sum_{i=1}^{n} \int_{\xi}^{x} a_{ii}(t)\, dt\Big]. \tag{12}$$

Hence, *either* $\Delta(x) \equiv 0$, and the n solutions are linearly connected; *or else* $\Delta(x) \neq 0$ in the entire interval of continuity, the exponential factor being neither zero nor infinite. Thus the necessary and sufficient condition for a fundamental system is that $\Delta(\xi) \neq 0$ at any one point of the interval.

Method of Variation of Parameters. If a fundamental system of solutions of the homogeneous system (S_n^*) is known, the non-homogeneous system (S_n) is soluble by quadratures. Since $\Delta \equiv |\phi_{ij}(x)| \neq 0$, a set of multipliers (u_j) is uniquely determined by the relations
$$y_i = \sum_{j=1}^{n} u_j \phi_{ij}(x) \quad (i = 1, 2, ..., n). \tag{13}$$

On substituting these expressions in (S_n) and cancelling terms multiplied by (u_j), there remain the relations
$$\sum_{j=1}^{n} \phi_{ij}(x) D u_j = f_i(x) \quad (i = 1, 2, ..., n), \tag{14}$$
which give uniquely
$$Du_j = \sum_{i=1}^{n} \frac{1}{\Delta} \frac{\partial \Delta}{\partial \phi_{ij}} f_i(x). \tag{15}$$

For the *principal solution* $y_i = \Phi_i(x;\xi)$, determined by the conditions $y_i = 0$ at $x = \xi$, we must have also $u_j = 0$ at $x = \xi$; and (15) accordingly gives
$$u_j = \sum_{i=1}^{n} \int_{\xi}^{x} \frac{1}{\Delta(t)} \frac{\partial \Delta(t)}{\partial \phi_{ij}} f_i(t)\, dt; \tag{16}$$

and (13) now becomes, with a change of suffixes,

$$y_k = \Phi_k(x;\xi) = \sum_{i=1}^{n}\sum_{j=1}^{n}\int_{\xi}^{x} \frac{1}{\Delta(t)}\frac{\partial \Delta(t)}{\partial \phi_{ij}}\phi_{kj}(x)f_i(t)\,dt. \tag{17}$$

We can write this in the form

$$\Phi_k(x;\xi) = \sum_{i=1}^{n}\int_{\xi}^{x}\phi_{ki}(x;t)f_i(t)\,dt \quad (k=1,2,\ldots,n), \tag{18}$$

where $\quad \phi_{ki}(x;t) \equiv \sum_{j=1}^{n}\frac{1}{\Delta(t)}\frac{\partial \Delta(t)}{\partial \phi_{ij}}\phi_{kj}(x) \quad (i,k=1,2,\ldots,n). \tag{19}$

If t is regarded as a constant, $y_k = \phi_{ki}(x;t)$ is a solution of the homogeneous system (S_n^*), taking at $x = t$ the values $\phi_{ki}(t;t) = \delta_{ki}$. Since the determinant of these values is unity, the solutions $y_k = \phi_{ki}(x;t)$ $(i,k = 1,2,\ldots,n)$ are a fundamental system, which can be constructed directly from the initial conditions at $x = t$ and then introduced in (18).

The method can also be used to reduce by m the order of the homogeneous system (S_n^*), when m linearly independent solutions are known. Let the variables be numbered so that the determinant $\Delta_m \equiv |\phi_{ij}(x)| \neq 0$ $(i,j = 1,2,\ldots,m)$, and let us put

$$\left.\begin{aligned} y_i &= \sum_{j=1}^{m}u_j\phi_{ij}(x) & (i=1,2,\ldots,m), \\ y_i &= \sum_{j=1}^{m}u_j\phi_{ij}(x)+Y_i & (i=m+1,\ldots,n). \end{aligned}\right\} \tag{20}$$

These relations give (u_j) and (Y_i) uniquely. Substituting, and cancelling the terms in (u_j), we have

$$\left.\begin{aligned} \sum_{j=1}^{m}\phi_{ij}(x)Du_j &= \sum_{k=m+1}^{n}a_{ik}(x)Y_k & (i=1,2,\ldots,m), \\ \sum_{j=1}^{m}\phi_{ij}(x)Du_j+DY_i &= \sum_{k=m+1}^{n}a_{ik}(x)Y_k & (i=m+1,\ldots,n). \end{aligned}\right\} \tag{21}$$

The first set gives, since $\Delta_m \neq 0$,

$$Du_j = \sum_{k=m+1}^{n}\sum_{i=1}^{m}\frac{1}{\Delta_m}\frac{\partial \Delta_m}{\partial \phi_{ij}}a_{ik}(x)Y_k; \tag{22}$$

and when these values are substituted in the second set, we get a homogeneous system of order $(n-m)$ of the type

$$DY_i = \sum_{k=m+1}^{n}A_{ik}(x)Y_k \quad (i=m+1,\ldots,n). \tag{23}$$

If this system can be solved, (u_j) can then be obtained from (22) by quadratures.

4. Wronskian Determinants

Linearly Connected Functions. The functions $\{\phi_i(x)\}$ are said to be *linearly connected* if constants (c_i) not all zero exist such that $\sum c_i \phi_i(x) \equiv 0$. If there are n functions, which are differentiable $(n-1)$ times, we have the relations

$$\sum_{i=1}^{n} c_i D^{j-1}\phi_i(x) = 0 \quad (j = 1, 2, \ldots, n); \tag{1}$$

and on eliminating (c_i) we obtain the determinantal relation

$$W(\phi_1, \phi_2, \ldots, \phi_n) \equiv \begin{vmatrix} \phi_1 & \phi_2 & \cdots & \phi_n \\ D\phi_1 & D\phi_2 & \cdots & D\phi_n \\ \cdots & \cdots & \cdots & \cdots \\ D^{n-1}\phi_1 & D^{n-1}\phi_2 & \cdots & D^{n-1}\phi_n \end{vmatrix} = 0. \tag{2}$$

The vanishing of this expression, which is called a *Wronskian determinant*, is thus a *necessary* condition for the functions to be linearly connected. The converse is true only with qualifications. We suppose that $W(\phi_1, \phi_2, \ldots, \phi_n) \equiv 0$, and that there is no sub-set of $(n-1)$ functions whose Wronskian vanishes identically. For, if there were, we could reason on the sub-set. We can then prove the following

THEOREM. *If n functions $\{\phi_i(x)\}$ are differentiable $(n-1)$ times, and if $W(\phi_1, \phi_2, \ldots, \phi_n) \equiv 0$, but $W(\phi_1, \phi_2, \ldots, \phi_{n-1}) \neq 0$ in a certain interval, then the functions are connected by a linear relation with constant coefficients, which is valid in the entire interval.*

We can determine uniquely multipliers (u_i) such that

$$D^{j-1}\phi_n(x) = \sum_{i=1}^{n-1} u_i D^{j-1}\phi_i(x) \quad (j = 1, 2, \ldots, n). \tag{3}$$

By combining the jth relation with the derivative of the $(j-1)$th, we have

$$\sum_{i=1}^{n-1} Du_i D^{j-1}\phi_i(x) = 0 \quad (j = 1, 2, \ldots, n-1); \tag{4}$$

and because $W(\phi_1, \phi_2, \ldots, \phi_{n-1}) \neq 0$, we must have $Du_i = 0$, and so

$$\phi_n(x) = \sum_{i=1}^{n-1} c_i \phi_i(x) \quad (c_i = \text{constant}). \tag{5}$$

The relation can be extended to any interval where the derivatives exist and where *all* the Wronskians of sub-sets of $(n-1)$ functions nowhere vanish simultaneously. The necessity for the condition was indicated by Peano, and may be simply illustrated. Consider two functions with continuous derivatives defined as follows:

$$\left.\begin{array}{ll} \phi_1(x) \equiv 2x^2, \quad \phi_2(x) \equiv x^2 & (x \geqslant 0); \\ \phi_1(x) \equiv x^2, \quad \phi_2(x) \equiv 2x^2 & (x \leqslant 0). \end{array}\right\} \quad (6)$$

Then $\phi_1(x) = 2\phi_2(x)$ $(x \geqslant 0)$, but $\phi_1(x) = \frac{1}{2}\phi_2(x)$ $(x \leqslant 0)$. The theorem is inapplicable to any interval containing $x = 0$, because $\phi_1(0) = 0 = \phi_2(0)$, so that the subsidiary condition is not satisfied, although $W(\phi_1, \phi_2) \equiv 0$. If, however, the functions are *analytic*, and if $\sum c_i \phi_i(x) \equiv 0$ in any finite interval, the relation holds everywhere, by the principle of analytical continuation.

The General Linear Homogeneous Equation. If the n functions $\{\phi_i(x)\}$ are differentiable n times, and if $W(\phi_1, \phi_2, ..., \phi_n) \neq 0$ in a certain interval or domain, then the expression $y = \sum c_i \phi_i(x)$, where (c_i) are constants, must satisfy the relation

$$\frac{W(\phi_1, \phi_2, ..., \phi_n, y)}{W(\phi_1, \phi_2, ..., \phi_n)} = 0. \quad (7)$$

The form of this relation is unchanged if $\{\phi_i(x)\}$ are replaced by n linearly independent combinations with constant coefficients

$$\Phi_i(x) = \sum_{j=1}^{n} c_{ij} \phi_j(x) \quad (i = 1, 2, ..., n), \quad (8)$$

where the determinant $|c_{ij}| \neq 0$. For, by the rule for multiplying determinants, we have

$$\left.\begin{array}{l} |c_{ij}| W(\phi_1, \phi_2, ..., \phi_n) = W(\Phi_1, \Phi_2, ..., \Phi_n), \\ \begin{vmatrix} c_{ij} & 0 \\ 0 & 1 \end{vmatrix} W(\phi_1, \phi_2, ..., \phi_n, y) = W(\Phi_1, \Phi_2, ..., \Phi_n, y), \end{array}\right\} \quad (9)$$

and hence

$$\frac{W(\phi_1, \phi_2, ..., \phi_n, y)}{W(\phi_1, \phi_2, ..., \phi_n)} = \frac{W(\Phi_1, \Phi_2, ..., \Phi_n, y)}{W(\Phi_1, \Phi_2, ..., \Phi_n)}. \quad (10)$$

Conversely, consider any homogeneous equation

$$(E_n^*) \qquad D^n y + p_1(x) D^{n-1} y + ... + p_n(x) y = 0, \quad (11)$$

and suppose that $\{\phi_i(x)\}$ are n known linearly independent solutions, so that $W(\phi_1, \phi_2, ..., \phi_n) \neq 0$ in a certain interval.

Eliminating $\{p_r(x)\}$ from (11) and the equations satisfied by $\{\phi_i(x)\}$, we have

$$W(\phi_1, \phi_2, ..., \phi_n, y) = 0; \quad (12)$$

and, since we assumed that $W(\phi_1, \phi_2, ..., \phi_n) \neq 0$, Wronski's theorem shows that the most general solution of (11) is of the form $y = \sum_{i=1}^{n} c_i \phi_i(x)$, where (c_i) are constants.

Every solution of (E_n^*) is accordingly expressible in terms of any set of n linearly independent solutions, which are called a *fundamental system*.

5. The General Linear Equation

The Complete Primitive. We shall now specialize some of the preceding results for the particular system

$$\left.\begin{aligned} Dy_i &= y_{i+1} \quad (i = 1, 2, ..., n-1), \\ Dy_n &+ \sum_{i=1}^{n} p_{n-i+1}(x) y_i = f(x), \end{aligned}\right\} \quad (1)$$

which is equivalent to the general linear equation

(E_n) $\qquad D^n y + p_1(x) D^{n-1} y + ... + p_n(x) y = f(x).$

If x is real and all the coefficients are continuous in the interval $(x' \leqslant x \leqslant x'')$, there is a solution with n continuous derivatives, which is uniquely determined by the arbitrary initial values $D^i y = \eta^{(i)}$ $(i = 0, 1, ..., n-1)$ at any point $x = \xi$ of the interval. If x is real or complex and the coefficients are *analytic*, the solution is analytic and has no singularities in the finite part of the plane, except at singularities of the coefficients.

The difference of any two solutions of (E_n), say $y = \phi(x) - \phi^*(x)$, satisfies the corresponding homogeneous equation

(E_n^*) $\qquad D^n y + p_1(x) D^{n-1} y + ... + p_n(x) y = 0.$

We therefore require only one *particular integral* of (E_n), together with a *complementary function* which is the complete primitive of (E_n^*).

The Abel-Liouville Formula. The analogue of Jacobi's determinant for the system (1) is the Wronskian $W(\phi_1, \phi_2, ..., \phi_n)$ of any n solutions of the homogeneous equation (E_n^*). By a double application of the rule that determinants with two rows identical vanish, we have

$$DW = \sum_{j=1}^{n} \left[\sum_{i=1}^{n} \frac{\partial W}{\partial \phi_i^{(j-1)}} \phi_i^{(j)}(x) \right]$$

$$= \sum_{i=1}^{n} \frac{\partial W}{\partial \phi_i^{(n-1)}} \phi_i^{(n)}(x)$$

$$= -\sum_{j=1}^{n} p_j(x)\left[\sum_{i=1}^{n} \frac{\partial W}{\partial \phi_i^{(n-1)}} \phi_i^{(n-j)}(x)\right]$$

$$= -p_1(x)\left[\sum_{i=1}^{n} \frac{\partial W}{\partial \phi_i^{(n-1)}} \phi_i^{(n-1)}(x)\right]$$

$$= -p_1(x)W. \tag{2}$$

Separating the variables and integrating, we have

$$W(x) = W(\xi)\exp\left[-\int_{\xi}^{x} p_1(t)\,dt\right], \tag{3}$$

a result due to Abel for $n = 2$ and to Liouville in general. In a finite domain where $p_1(x)$ is continuous, $W(\phi_1, \phi_2, ..., \phi_n)$ vanishes either identically or not at all. The necessary and sufficient condition that n solutions $\{\phi_i(x)\}$ should form a fundamental system of solutions of (E_n^*) is that their Wronskian should not vanish at one point $x = \xi$, chosen at random in the domain.

Method of Variation of Parameters. To solve the non-homogeneous equation (E_n), knowing a fundamental system of solutions of (E_n^*), we put

$$D^{j-1}y = \sum_{i=1}^{n} u_i\, D^{j-1}\phi_i(x) \quad (j = 1, 2, ..., n). \tag{4}$$

These relations give unique values for (u_i), since $W(\phi_1, \phi_2, ..., \phi_n) \neq 0$. By combining (4) and their derivatives we now find

$$0 = \sum_{i=1}^{n} Du_i\, D^{j-1}\phi_i(x) \quad (j = 1, 2, ..., n-1), \tag{5}$$

and

$$D^n y = \sum_{i=1}^{n} [u_i D^n \phi_i(x) + Du_i\, D^{n-1}\phi_i(x)]. \tag{6}$$

When the expressions (4) and (6) for the unknown and its derivatives are substituted in (E_n), terms multiplied by (u_i) cancel, leaving

$$\sum_{i=1}^{n} Du_i\, D^{n-1}\phi_i(x) = f(x). \tag{7}$$

From (5) and (7) we now get

$$Du_i = f(x)\frac{1}{W(x)}\frac{\partial W(x)}{\partial \phi_i^{(n-1)}} \quad (i = 1, 2, ..., n). \tag{8}$$

If the *principal solution* at $x = \xi$, say $y = \Phi(x;\xi)$, is determined by the conditions $D^{j-1}y = 0$ $(j = 1, 2, ..., n)$ when $x = \xi$, we find that

$u_i = 0$ $(i = 1, 2,..., n)$ when $x = \xi$. Hence

$$u_i = \int_{\xi}^{x} \frac{f(t)}{W(t)} \frac{\partial W(t)}{\partial \phi_i^{(n-1)}} dt, \tag{9}$$

$$\Phi(x;\xi) = \sum_{i=1}^{n} \phi_i(x) \int_{\xi}^{x} \frac{f(t)}{W(t)} \frac{\partial W(t)}{\partial \phi_i^{(n-1)}} dt. \tag{10}$$

If t is a constant, the expression

$$\phi(x;t) \equiv \sum_{i=1}^{n} \frac{\phi_i(x)}{W(t)} \frac{\partial W(t)}{\partial \phi_i^{(n-1)}} \tag{11}$$

is that solution of the homogeneous equation (E_n^*), which is determined by the initial values

$$\left.\begin{array}{l} D^{j-1}y = 0, \quad \text{when } x = t \quad (j = 1, 2,..., n-1); \\ D^{n-1}y = 1, \quad \text{when } x = t. \end{array}\right\} \tag{12}$$

With this definition of $\phi(x;t)$ we obtain a formula of great importance for the principal solution, which is due to Cauchy, namely,

$$\Phi(x;\xi) = \int_{\xi}^{x} \phi(x;t)f(t)\, dt. \tag{13}$$

We can also apply the method of variation of parameters to reduce a homogeneous equation (E_n^*), when we know m independent solutions $(m < n)$; a simpler method will be given later, so that the details will be left as an exercise. If $W(\phi_1, \phi_2,..., \phi_m) \neq 0$, we write

$$\left.\begin{array}{l} D^{j-1}y = \sum_{i=1}^{m} u_i D^{j-1}\phi_i(x) \quad (j = 1, 2,..., m), \\ D^{j-1}y = \sum_{i=1}^{m} u_i D^{j-1}\phi_i(x) + Y_j \quad (j = m+1,..., n), \end{array}\right\} \tag{14}$$

in the normal canonical system (1), where now $f(x) \equiv 0$.

We have $(n-1)$ relations

$$\left.\begin{array}{ll} \sum_{i=1}^{m} Du_i D^{j-1}\phi_i(x) = 0 & (j = 1, 2,..., m-1), \\ \quad = Y_{m+1} & (j = m), \\ \quad = Y_{j+1} - DY_j & (j = m+1,..., n-1), \end{array}\right\} \tag{15}$$

and also the relation

$$D^n y = \sum_{i=1}^{n} [u_i D^n \phi_i(x) + Du_i D^{n-1}\phi_i(x)] + DY_n. \tag{16}$$

From (14), (16), and (E_n^*) we find

$$\sum_{i=1}^{n} Du_i\, D^{n-1}\phi_i(x) + DY_n + \sum_{r=1}^{n-m} p_r(x)Y_{n+1-r} = 0. \tag{17}$$

On eliminating (Du_i) between (15) and (17), we have a canonical system of order $(n-m)$ in (Y_j), which can be replaced by a single homogeneous equation of order $(n-m)$ for Y_{m+1}; while (14) gives

$$Y_{m+1} = \frac{W(\phi_1, \phi_2, \ldots, \phi_m, y)}{W(\phi_1, \phi_2, \ldots, \phi_m)}, \tag{18}$$

so that Y_{m+1} (regarded as an expression in y) is the left-hand side of the equation of order m admitting the given solutions.

EXAMPLES. I

1. Reduce to canonical systems the pairs of equations
$$[D^2 y + D^2 z + y = 1, \quad D^2 y \pm D^2 z \pm z = 0].$$

2. The system
$$Dy = F_1(x, y, z), \quad D^2 y = F_2(x, y, z, Dy, Dz)$$
is in general of the second order. Show by examples that it may be (i) indeterminate, (ii) incompatible, or (iii) of order lower than the second.

3. Show by the method of successive approximations that the equation $Dy = a(x)y$ is satisfied by
$$y = \eta \exp\left[\int_\xi^x a(t)\, dt\right];$$
and that the equation $Dy = a(x)y + f(x)$ is satisfied by
$$y = \int_\xi^x f(s) \exp\left[\int_s^x a(t)\, dt\right] ds.$$

4. Solve from first principles the systems:
  (i) $Dy = xy$; $y = 1$, when $x = 0$.
  (ii) $Dy = ay/(1+x)$; $y = 1$, when $x = 0$.
  (iii) $Dy = z$, $Dz = -y$; $y = A$, $z = B$, when $x = 0$.
  (iv) $Du = v$, $Dv = w$, $Dw = u$; $u = A$, $v = B$, $w = C$, when $x = 0$.

5. If $y_i = \phi_{ij}(x)$ is a set of m solutions ($m \leq n$) of the homogeneous system (S_n^*), and if constants (c_j) not all zero exist such that
$$\sum_{j=1}^{m} c_j \phi_{ij}(\xi) = 0 \quad (i = 1, 2, \ldots, n)$$
for a particular value ξ, then $\sum_{j=1}^{m} c_j \phi_{ij}(x) \equiv 0$.

6. If $(\phi_{ij}(x))$ is a matrix of n^2 arbitrary differentiable functions, whose determinant does not vanish identically, then there is a unique homogeneous system (S_n^*) admitting the n solutions $y_i = \phi_{ij}(x)$ $(i, j = 1, 2, \ldots, n)$. If the determinant vanishes identically, the system (S_n^*) does not exist unless the

functions are connected by relations with *constant* coefficients

$$\sum_{j=1}^{n} c_j \phi_{ij}(x) = 0 \quad (i = 1, 2, ..., n).$$

If the rank of the matrix is k, any $(k+1)$ solutions must be connected by relations with *constant* coefficients. When these conditions are satisfied, the system (S_n^*) is indeterminate.

7. The homogeneous system (S_n^*) cannot admit m linearly independent solutions $y_i = \phi_{ij}(x)$, and also a solution $y_i = \sum_{j=1}^{m} u_j \phi_{ij}(x)$, where (u_i) are *not* constants.

8. If (y_i) is a solution of the homogeneous system (S_n^*) and the determinant $|\alpha_{ij}(x)| \neq 0$, then $Y_j \equiv \sum_{i=1}^{n} \alpha_{ij}(x) y_i$ satisfy another homogeneous system of order n. The expression $Y \equiv \sum_{i=1}^{n} \alpha_i(x) y_i$, with coefficients differentiable n times, satisfies a linear homogeneous equation of order n or less.

9. SYSTEMS WITH CERTAIN KNOWN SOLUTIONS. If the homogeneous system (S_n^*) has $(n-1)$ known solutions $y_i = \phi_{ij}(x)$, which are linearly independent, show that for any other solution

$$\begin{vmatrix} y_1, & y_2, & ..., & y_n \\ \phi_{11}, & \phi_{21}, & ..., & \phi_{n1} \\ \vdots & & & \\ \phi_{1(n-1)}, & \phi_{2(n-1)}, & ..., & \phi_{n(n-1)} \end{vmatrix} = C \exp\left[\sum_{i=1}^{n} \int_{\xi}^{x} a_{ii}(t)\, dt\right].$$

Complete the solution of the system by quadratures, using the method of variation of parameters.

10. The homogeneous equation (E_n^*) has $(n-1)$ known independent solutions. Show that any other solution satisfies the equation of order $(n-1)$

$$W(\phi_1, \phi_2, ..., \phi_{n-1}, y) = C \exp\left[-\int_{\xi}^{x} p_1(t)\, dt\right],$$

which is soluble by quadratures.

11. If $\phi_1(x)$ is a known solution of the homogeneous equation (E_n^*), show that the equation can be simplified by d'Alembert's substitution $y = \phi_1(x) u$; and that it is reduced to an equation of order $(n-1)$ by Fuchs's substitution $y = \phi_1(x) \int z\, dx$.

12. REMOVAL OF A TERM. If the equation (E_n^*) is transformed by writing $y = u \exp\left[-\frac{1}{n} \int p_1(x)\, dx\right]$, the equation satisfied by u has no term in $D^{n-1} u$. If in (E_n^*) we take a new independent variable $z = \psi(x)$, determined by the relation $n(n-1)\psi''(x) + 2p_1(x)\psi'(x) = 0$, the new equation satisfied by y has no term in $d^{n-1} y/dz^{n-1}$.

13. Every solution of the non-homogeneous equation (E_n) satisfies the homogeneous equation of order $(n+1)$

$$f(x)[D^{n+1} y + p_1 D^n y + (p_1' + p_2) D^{n-1} y + ... + p_n' y] -$$
$$- f'(x)[D^n y + p_1 D^{n-1} y + ... + p_n y] = 0.$$

14. Solve $xD^2y-(2x+1)Dy+(x+1)y = 0$, given that the quotient of two particular integrals is x^2.

15. Solve $(x^3+x^2)D^2y+xDy-(x+1)^3y = 0$, given that the product of two particular integrals is constant.

16. CAUCHY'S FORMULA FOR A REPEATED INTEGRAL. Show that the principal solution of $D^n y = f(x)$, determined by the conditions $D^{i-1}y = 0$ $(i = 1, 2, ..., n)$ when $x = \xi$, is
$$y = \int_\xi^x \frac{(x-t)^{n-1}}{(n-1)!} f(t)\, dt.$$

Hence obtain Cauchy's form of the remainder in Taylor's theorem
$$\phi(x) = \phi(\xi) + \frac{(x-\xi)\phi'(\xi)}{1!} + \cdots + \frac{(x-\xi)^{n-1}\phi^{(n-1)}(\xi)}{(n-1)!} + \int_\xi^x \frac{(x-t)^{n-1}\phi^{(n)}(t)}{(n-1)!}\, dt.$$

17. If $F_0(x) \equiv f(x)$, $F_n(x) \equiv \int_0^x F_{n-1}(t)\, dt$, prove by integration by parts that
$$F_n(x) = \int_0^x \frac{(x-t)^{n-1} f(t)}{(n-1)!}\, dt.$$

18. Give another proof of Cauchy's formula by means of the relation
$$\int_0^x dt_1 \int_0^{t_1} dt_2 \cdots \int_0^{t_{n-1}} f(t_n)\, dt_n = \int_0^x dt_n \int_{t_n}^x dt_{n-1} \cdots \int_{t_2}^x f(t_n)\, dt_1.$$

II
EQUATIONS WITH CONSTANT COEFFICIENTS

6. Heaviside's Solution of Cauchy's Problem†

Introduction. The simplest class of linear differential equations are those with constant coefficients, which can be solved with the aid of elementary functions. But the systems occurring in mechanical and electrical problems may be very complicated, and give scope for labour-saving symbolical methods. If we require a particular solution taking assigned initial values, the operational calculus of Heaviside affords the most appropriate shorthand for the method of successive approximations. If we require the complete primitive with the arbitrary constants of integration displayed in the simplest form, but without reference to any particular set of initial values, the most effective method is the symbolical calculus of Boole.

Solution in Power-Series. The homogeneous system with constant coefficients

$$Dy_i = \sum_{j=i}^{n} a_{ij} y_j \quad (i = 1, 2, ..., n) \tag{1}$$

is analytic and free from singularities in the finite part of the plane. In the neighbourhood of a typical point $x = 0$ the solutions can be expanded in power-series, which will converge for all finite values of x. Let us write these in the form

$$y_i = \phi_i(x) = \sum_{k=0}^{\infty} c_i^{(k)} \frac{x^k}{k!} \quad (i = 1, 2, ..., n). \tag{2}$$

Then the solution taking the values $y_i = \eta_i$ at $x = 0$ is given by the relations

$$c_i^{(0)} = \eta_i, \qquad c_i^{(k+1)} = \sum_{j=1}^{n} a_{ij} c_j^{(k)}. \tag{3}$$

Let $(a_{ij}^{(k)})$ be the kth power of the matrix (a_{ij}), and let $(a_{ij}^{(0)})$ denote the unit matrix (δ_{ij}), whose elements are Kronecker deltas. By the multiplication rule we have

$$a_{ij}^{(k+1)} = \sum_{s=1}^{n} a_{is} a_{sj}^{(k)}, \qquad a_{ij}^{(\alpha+\beta)} = \sum_{s=1}^{n} a_{is}^{(\alpha)} a_{sj}^{(\beta)}. \tag{4}$$

---

† J. R. Carson, *Electrical Circuit Theory and the Operational Calculus* (New York, 1926); H. Jeffreys, *Operational Methods in Mathematical Physics* (Cambridge, 1927); E. J. Berg, *Rechnung mit Operatoren* (Munich and Berlin, 1932); P. Humbert, *Le calcul symbolique* (Paris, 1934).

Then the relations (3) give

$$c_i^{(k)} = \sum_{j=1}^{n} a_{ij}^{(k)} \eta_j \quad (i = 1, 2, ..., n; k = 0, 1, ..., \infty); \tag{5}$$

and

$$y_i = \sum_{j=1}^{n} \left[\sum_{k=0}^{\infty} a_{ij}^{(k)} \frac{x^k}{k!} \right] \eta_j$$

$$= \sum_{j=1}^{n} \phi_{ij}(x) \eta_j \quad (i = 1, 2, ..., n), \tag{6}$$

where $y_i = \phi_{ij}(x)$ is the solution with the initial values $\phi_{ij}(0) = \delta_{ij}$ at the origin.

If every element of the matrix (a_{ij}) has the upper bound $|a_{ij}| \leqslant A$, then $|a_{ij}^{(k)}| \leqslant n^{k-1} A^k$ for positive integers k. Hence the series (6) are absolutely convergent for all finite values of x, and at least as rapidly as $\exp(nAx)$. To solve the non-homogeneous system

$$Dy_i = \sum_{j=1}^{n} a_{ij} y_j + f_i(x) \quad (i = 1, 2, ..., n), \tag{7}$$

we observe that $y_i = \phi_{ij}(x-t)$ is the solution of the homogeneous system (1) which takes the initial values $y_i = \delta_{ij}$ at any given point $x = t$. Applying the method of variation of parameters as in § 3, we see that

$$y_i = \sum_{j=1}^{n} \int_0^x \phi_{ij}(x-t) f_j(t) \, dt \quad (i = 1, 2, ..., n) \tag{8}$$

is the principal solution of (7), given by $y_i = 0$ when $x = 0$.

Heaviside Operators. To confirm these results by the method of successive approximations, we introduce the notation

$$\frac{1}{p} f(x) \equiv \int_0^x f(t) \, dt. \tag{9}$$

By Cauchy's formula for a repeated integral, we have, for positive integers m,

$$\frac{1}{p^m} f(x) = \int_0^x dt_1 \int_0^{t_1} dt_2 \ldots \int_0^{t_{m-1}} f(t_m) \, dt_m$$

$$= \int_0^x \frac{(x-t)^{m-1}}{(m-1)!} f(t) \, dt. \tag{10}$$

If $f(x) \equiv 1$, we may omit the operand and write simply

$$\frac{1}{p^m} \equiv \frac{x^m}{m!}. \tag{11}$$

Proceeding exactly as in § 1, we write the solution as a uniformly convergent series $y_i = \eta_i + \sum_{k=1}^{\infty} U_i^{(k)}(x)$, where

$$\left.\begin{aligned} U_i^{(1)}(x) &= \frac{1}{p}\Big[\sum_{j=1}^{n} a_{ij}\eta_j + f_i(x)\Big], \\ U_i^{(k+1)}(x) &= \sum_{j=1}^{n} \frac{a_{ij}}{p} U_j^{(k)}(x). \end{aligned}\right\} \tag{12}$$

Since the operation of integration is commutative with that of multiplication by a constant, we find that (12) are satisfied by

$$U_i^{(k)}(x) = \sum_{j=1}^{n} \Big[\frac{a_{ij}^{(k)}}{p^k}\eta_j + \frac{a_{ij}^{(k-1)}}{p^k} f_j(x)\Big]; \tag{13}$$

and the solution is

$$y_i = \sum_{j=1}^{n} \Big[\sum_{k=0}^{\infty} \frac{a_{ij}^{(k)}}{p^k}\Big]\eta_j + \sum_{j=1}^{n} \Big[\sum_{k=0}^{\infty} \frac{a_{ij}^{(k)}}{p^{k+1}} f_j(x)\Big]. \tag{14}$$

Using the rules (10) and (11) to interpret these operators, we find, exactly as in (6) and (8),

$$\begin{aligned} y_i &= \sum_{j=1}^{n} \Big[\sum_{k=0}^{\infty} a_{ij}^{(k)} \frac{x^k}{k!}\Big]\eta_j + \sum_{j=1}^{n} \int_0^x \Big[\sum_{k=0}^{\infty} a_{ij}^{(k)} \frac{(x-t)^k}{k!}\Big] f_j(t)\, dt \\ &= \sum_{j=1}^{n} \phi_{ij}(x)\eta_j + \sum_{j=1}^{n} \int_0^x \phi_{ij}(x-t) f_j(t)\, dt. \end{aligned} \tag{15}$$

Summation of the Series. Consider the auxiliary series

$$w_j = \sum_{k=0}^{\infty} \frac{c_j^{(k)}}{\lambda^k} \quad (j = 1, 2, ..., n), \tag{16}$$

which are certainly convergent if $|\lambda| > nA$. By reason of (3), these are found to satisfy the relations

$$\lambda(w_j - \eta_j) = \sum_{i=1}^{n} a_{ji} w_i \quad (j = 1, 2, ..., n), \tag{17}$$

which may be written

$$\sum_{i=0}^{n} (\lambda\delta_{ji} - a_{ji}) w_i = \lambda\eta_j \quad (j = 1, 2, ..., n). \tag{18}$$

If $\Delta(\lambda) \equiv |\lambda\delta_{ji} - a_{ji}|$ is the determinant of this system of linear algebraic equations and $\Delta_{ji}(\lambda)$ the minor of the element $[\lambda\delta_{ji} - a_{ji}]$, we have

$$w_i = \sum_{j=1}^{n} \frac{\lambda\Delta_{ji}(\lambda)}{\Delta(\lambda)} \eta_j \quad (i = 1, 2, ..., n). \tag{19}$$

We can now write (14) symbolically in the form

$$y_i = \sum_{j=1}^{n} \frac{\Delta_{ji}(p)}{\Delta(p)} \{p\eta_j + f_j(x)\}. \qquad (20)$$

Since $\Delta(p)$ is of degree n and $\{\Delta_{ji}(p)\}$ are at most of degree $(n-1)$, the operators can be expanded formally in negative integral powers of p and interpreted term by term.

Heaviside's Partial-Fraction Rule. The expansions of the operators may be found by resolving $\{\Delta_{ji}(p)/\Delta(p)\}$ into partial fractions, which are interpreted as follows.

$$\frac{p}{(p-\alpha)^m} = \sum_{r=0}^{\infty} \frac{(m+r-1)!\,\alpha^r}{(m-1)!\,r!\,p^{m+r-1}}$$

$$= \sum_{r=0}^{\infty} \frac{\alpha^r x^{m+r-1}}{(m-1)!\,r!} = \frac{x^{m-1}}{(m-1)!} e^{\alpha x}; \qquad (21)$$

$$\frac{1}{(p-\alpha)^m} f(x) = \sum_{r=0}^{\infty} \frac{(m+r-1)!\,\alpha^r}{(m-1)!\,r!\,p^{m+r}} f(x)$$

$$= \sum_{r=0}^{\infty} \int_0^x \frac{\alpha^r (x-t)^{m+r-1}}{(m-1)!\,r!} f(t)\,dt$$

$$= \int_0^x \frac{(x-t)^{m-1}}{(m-1)!} e^{\alpha(x-t)} f(t)\,dt$$

$$= e^{\alpha x} \frac{1}{p^m} [e^{-\alpha x} f(x)]. \qquad (22)$$

If $\Delta_{n-1}(p)$ is the H.C.F. of all the first minors $\{\Delta_{ji}(p)\}$, and if $\Delta(p)/\Delta_{n-1}(p)$ is divisible by $(p-\alpha)^r$, then the solution involves terms of the types $x^s e^{\alpha x}$ $(s = 0, 1, 2,..., r-1)$.

Non-Commutative Operators. The operator p, of which only negative integral powers have been defined, is not identical with D, nor are the two operators commutative. For we have

$$\left.\begin{aligned}D\frac{1}{p}f(x) &= D\int_0^x f(t)\,dt = f(x), \\ \frac{1}{p}Df(x) &= \int_0^x f'(t)\,dt = f(x) - f(0);\end{aligned}\right\} \qquad (23)$$

$$D^m \frac{1}{p^n} f(x) = \frac{1}{p^{n-m}} f(x) \quad (n > m) \\ = D^{m-n} f(x) \quad (m > n);$$
(24)

$$\frac{1}{p^m} D^m f(x) = f(x) - f(0) - \frac{x}{1!} f'(0) - \ldots - \frac{x^{m-1}}{(m-1)!} f^{(m-1)}(0);$$
(25)

$$\frac{1}{p^n} D^m f(x) = \int_0^x \frac{(x-t)^{n-1}}{(n-1)!} f^{(m)}(t) \, dt.$$
(26)

Summary. The practical rule for solving the system (7) with given initial values is first to integrate between the limits 0 and x, and write

$$y_i = \eta_i + \sum_{j=1}^n \frac{a_{ij}}{p} y_j + \frac{1}{p} f_i(x) \quad (i = 1, 2, \ldots, n).$$
(27)

This is a system of integral equations equivalent to the given differential equations *plus* the given initial conditions. We now solve (27) formally as linear equations in (y_j), as though p were a number. This gives the formulae (20), which are to be interpreted by Heaviside's partial-fraction rule.

7. Operators in D and δ

Linear Operators. If we write

$$F(D)y \equiv \left[\sum_{i=0}^n a_i D^{n-i}\right] y \equiv \sum_{i=0}^n a_i D^{n-i} y,$$
(1)

where (a_i) are constants, the expression $F(D)$ is called a *linear operator with constant coefficients*. Since the differential operator D is commutative with constants, we have

$$F(D)G(D)y = \left[\sum_{i=0}^n a_i D^{n-i}\right]\left[\sum_{j=0}^m b_j D^{m-j}\right] y$$

$$= \sum_{i=0}^n \sum_{j=0}^m a_i b_j D^{m+n-i-j} y$$

$$= \sum_{k=0}^{m+n} \left[\sum_{i=0}^k a_i b_{k-i}\right] D^{m+n-k} y$$

$$= G(D)F(D)y, \text{ by symmetry.}$$
(2)

Hence such operators are commutative with one another.

Again, by Leibnitz's theorem, we have

$$D^m[e^{\alpha x}y] = \sum_{r=0}^{m}\binom{m}{r}D^r e^{\alpha x}D^{m-r}y$$

$$= e^{\alpha x}\sum_{r=0}^{m}\binom{m}{r}\alpha^r D^{m-r}y$$

$$= e^{\alpha x}(D+\alpha)^m y, \qquad (3)$$

and so, for any polynomial in D,

$$F(D)[e^{\alpha x}y] = e^{\alpha x}F(D+\alpha)y. \qquad (4)$$

Homogeneous Equations. The algebraic equation $F(\lambda) = 0$ is called the *characteristic equation* of the differential equation $F(D)y = 0$. If its roots are known, we may write $F(D) \equiv a_0 \prod(D-\lambda_i)$, with the factors in any order. If $\lambda = \lambda_i$ is a root of multiplicity n_i, we may write the equation as

$$F(D)y \equiv F_i(D)[(D-\lambda_i)^{n_i}y] = 0, \qquad (5)$$

and this is certainly satisfied if

$$(D-\lambda_i)^{n_i}y = 0, \qquad (6)$$

or if

$$e^{\lambda_i x}D^{n_i}[e^{-\lambda_i x}y] = 0. \qquad (7)$$

A particular integral with n_i arbitrary constants is therefore $y = e^{\lambda_i x}P^{(n_i-1)}(x)$, where $P^{(n_i-1)}(x)$ is an arbitrary polynomial of degree (n_i-1). Taking all the roots in turn, we have a solution with n arbitrary constants

$$y = \sum_{(i)} e^{\lambda_i x}P_i^{(n_i-1)}(x). \qquad (8)$$

We can prove that this is the complete primitive by evaluating the Wronskian determinant of the n solutions corresponding to the n constants; if this is not zero, the solutions must be linearly independent. In the first place we have

$$W(e^{\lambda_1 x}, e^{\lambda_2 x}, \ldots, e^{\lambda_n x}) \equiv e^{\sum \lambda_i x}\begin{vmatrix} 1 & 1 & . & 1 \\ \lambda_1 & \lambda_2 & . & \lambda_n \\ . & . & . & . \\ \lambda_1^{n-1} & \lambda_2^{n-1} & . & \lambda_n^{n-1} \end{vmatrix}$$

$$\equiv e^{\sum \lambda_i x}\prod_{1\leqslant s<r\leqslant n}(\lambda_r-\lambda_s), \qquad (9)$$

which is an identity in (λ_i) as well as in x. This expression does not vanish if (λ_i) are all unequal.

If we differentiate with respect to λ_2, and then put $\lambda_2 = \lambda_1$, we have

$$W(e^{\lambda_1 x}, xe^{\lambda_1 x}, e^{\lambda_3 x}, \ldots, e^{\lambda_n x}) = 1!\, e^{\sum \lambda_i x}\prod_{r=3}^{n}(\lambda_r-\lambda_1)^2 \cdot \prod_{3\leqslant s<r\leqslant n}(\lambda_r-\lambda_s). \quad (10)$$

If we now differentiate twice with respect to λ_3, and then put $\lambda_3 = \lambda_1$, we have

$$W(e^{\lambda_1 x}, xe^{\lambda_1 x}, x^2 e^{\lambda_1 x}, e^{\lambda_4 x}, \ldots, e^{\lambda_n x})$$
$$= 1! \, 2! \, e^{\sum \lambda_i x} \prod_{r=4}^{n} (\lambda_r - \lambda_1)^3 \cdot \prod_{4 \leq s < r \leq n} (\lambda_r - \lambda_s); \quad (11)$$

and the Wronskian corresponding to any combination of equal roots of the characteristic equation may be evaluated in the same way.

Non-Homogeneous Equations. To solve $F(D)y = f(x)$, we construct the algebraic partial fraction identity

$$\frac{1}{F(\lambda)} \equiv \sum_{(i)} \sum_{r=1}^{n_i} \frac{C_i^{(r)}}{(\lambda - \lambda_i)^r}. \quad (12)$$

This gives an identity between operator polynomials

$$1 \equiv \sum_{(i)} F_i(D) \Big[\sum_{r=1}^{n_i} C_i^{(r)} (D - \lambda_i)^{n_i - r} \Big]$$

or
$$1 \equiv \sum_{(i)} F_i(D) \psi_i(D) \quad \text{say}, \quad (13)$$

where $\{F_i(D)\}$ are defined as in (5). We can accordingly write

$$f(x) \equiv \sum_{(i)} F_i(D) \psi_i(D) f(x) \equiv \sum_{(i)} F_i(D) f_i(x); \quad (14)$$

and the equation $F(D)y = f(x)$ can now be satisfied by putting $y = \sum y_i$, where
$$F(D) y_i = F_i(D) f_i(x), \quad (15)$$

or
$$F_i(D)[(D - \lambda_i)^{n_i} y_i - f_i(x)] = 0. \quad (16)$$

These relations are certainly satisfied if

$$(D - \lambda_i)^{n_i} y_i = f_i(x), \quad (17)$$

and so the problem is reduced to the solution of 'simple equations' of the type (17).

To solve $(D - \lambda)^k y = f(x)$, we write it as

$$D^k[e^{-\lambda x} y] = e^{-\lambda x} f(x). \quad (18)$$

The solution is now given by k integrations, which may be reduced to a single integration by Cauchy's formula, if we ignore constants of integration. The latter yield only terms duplicated in the complementary function. We write symbolically

$$y = \frac{1}{(D - \lambda)^k} f(x) = e^{\lambda x} \frac{1}{D^k}[e^{-\lambda x} f(x)]. \quad (19)$$

The practical rule derived from this discussion is to resolve formally into partial fractions the inverse operator

$$y = \frac{1}{F(D)}f(x) = \sum_{(i)} \sum_{r=1}^{n_i} \frac{C_i^{(r)}}{(D-\lambda_i)^r} f(x), \qquad (20)$$

and to interpret each symbolical fraction by the rule given in (19).

Equations Soluble without Quadratures. For many important applications the right-hand side is of the form

$$f(x) \equiv \sum_j e^{\mu_j x} \psi_j(x), \qquad (21)$$

where (μ_j) may be complex, and where $\{\psi_j(x)\}$ are polynomials. It is sufficient to solve the typical equation

$$F(D)y = e^{\mu x}\psi(x). \qquad (22)$$

If $\psi(x)$ is of degree $(m-1)$, the right-hand side is annihilated by $(D-\mu)^m$; and so the solution of (22) is included in that of

$$(D-\mu)^m F(D)y = 0. \qquad (23)$$

If we write down the complete primitive of (23), and then omit terms annihilated by $F(D)$, we find that there is a particular integral of the form $y = e^{\mu x}\chi(x)$, where $\chi(x)$ is a polynomial of degree $(m-1)$, if $F(\mu) \neq 0$, or of degree $(m+k-1)$, if $\lambda = \mu$ is a k-tuple root of $F(\lambda) = 0$.

We could, of course, find $\chi(x)$ by the method of indeterminate coefficients; but this is unnecessarily tedious. The required particular integral is to satisfy both the equations

$$F(D)y = e^{\mu x}\psi(x), \qquad (D-\mu)^{m+k}y = 0. \qquad (24)$$

If $F(D) \equiv (D-\mu)^k G(D)$, let us construct by the H.C.F. process the identity between polynomials

$$A(D)G(D) + B(D)(D-\mu)^m \equiv 1, \qquad (25)$$

where $A(D)$ is of degree $(m-1)$ in D. If we operate with $A(D)$ on the first of the equations (24) and with $B(D)$ on the second and add, we find

$$(D-\mu)^k y = A(D)e^{\mu x}\psi(x); \qquad (26)$$

so that, if $y = e^{\mu x}\chi(x)$, we get

$$D^k\chi(x) = A(D+\mu)\psi(x). \qquad (27)$$

The operator $A(D+\mu)$ can be found without calculating $B(D)$; for

the identity (25) corresponds to a relation between rational functions:
$$\frac{1}{G(\mu+h)} \equiv A(\mu+h) + \frac{B(\mu+h)h^m}{G(\mu+h)}$$
$$\equiv A_0 + A_1 h + A_2 h^2 + \ldots + A_{m-1} h^{m-1} + \frac{B(\mu+h)h^m}{G(\mu+h)}. \quad (28)$$

Hence $A(\mu+h)$ is identical with the first m terms of the Taylor expansion of $1/G(\mu+h)$ in ascending powers of h. Since $A_0 = 1/G(\mu)$ is neither zero nor infinite, the expression
$$[A_0 + A_1 D + A_2 D^2 + \ldots + A_{m-1} D^{m-1}]\psi(x) \equiv \psi^*(x) \quad (29)$$
is a polynomial of the *same degree* $(m-1)$ as $\psi(x)$. The required polynomial $\chi(x) = D^{-k}\psi^*(x)$ may be written down by inspection, or as the result of k successive integrations. The constants of integration correspond to terms already accounted for in the complementary function. The practical rule for solving (22), where $\psi(x)$ is a polynomial, is to write
$$y = \frac{1}{F(D)}[e^{\mu x}\psi(x)] = e^{\mu x}\frac{1}{F(D+\mu)}\psi(x), \quad (30)$$
and to expand the operator in *ascending* powers of D, omitting all terms which annihilate $\psi(x)$,
$$y = e^{\mu x}\frac{1}{D^k}[A_0 + A_1 D + \ldots + A_{m-1} D^{m-1}]\psi(x)$$
$$= e^{\mu x}\frac{1}{D^k}\psi^*(x) \quad (\text{if } D^m\psi(x) \equiv 0). \quad (31)$$

Euler's Homogeneous Equation. The equation
$$a_0 x^n D^n y + a_1 x^{n-1} D^{n-1} y + \ldots + a_n y = f(x), \quad (32)$$
where (a_i) are constants, can be reduced to an equation with constant coefficients by putting $x = e^t$. If $\delta \equiv d/dt \equiv x\,d/dx$, we can prove by induction the well-known identity
$$x^m D^m y \equiv \delta(\delta-1)\ldots(\delta-m+1)y; \quad (33)$$
and so (32) can be transformed into
$$b_0 \delta^n y + b_1 \delta^{n-1} y + \ldots + b_n y = f(e^t). \quad (34)$$
Corresponding to (4), we have the rule $F(\delta)(x^\alpha y) = x^\alpha F(\delta+\alpha)y$. If $(\delta-\alpha_i)^{n_i}$ is a factor of the operator with constant coefficients $F(\delta)$, then a solution of $F(\delta)y = 0$ is given by $y = x^{\alpha_i}P_i^{(n_i-1)}(\log x)$, where $P_i^{(n_i-1)}(\log x)$ is an arbitrary polynomial of degree (n_i-1) in $\log x$.†

† The reader should investigate directly the Wronskian determinant of the solutions of $F(\delta)y = 0$.

As an exercise in these operators, we observe that the equation $D^n y = f(x)$ may also be written in the form

$$\delta(\delta-1)...(\delta-n+1)y = x^n f(x). \tag{35}$$

By resolving into partial fractions the expression

$$y = \frac{1}{\delta(\delta-1)...(\delta-n+1)} x^n f(x) \tag{36}$$

and interpreting the inverse operators, we obtain another proof of Cauchy's formula for a repeated integral

$$\frac{1}{D^n} f(x) = \int_{\xi}^{x} \frac{(x-s)^{n-1}}{(n-1)!} f(s)\,ds + \text{a polynomial of degree } (n-1). \tag{37}$$

8. Simultaneous Equations. Invariant Factors[†]

Characteristic Equation. Consider the homogeneous system

$$\Phi_i \equiv \sum_{j=1}^{n} F_{ij}(D) y_j = 0 \quad (i = 1, 2, ..., n), \tag{1}$$

where $(F_{ij}(D))$ is a matrix of linear operators with constant coefficients whose determinant $\Delta(D) \equiv |F_{ij}(D)|$ is of degree N. If we add the results of operating on the equations with the minors of the elements of the kth column of the determinant, we eliminate all the unknowns except one, which is found to satisfy the equation

$$\Delta(D) y_k = 0 \quad (k = 1, 2, ..., n). \tag{2}$$

The corresponding algebraic equation $\Delta(\lambda) = 0$ is called the *characteristic equation* of the system. Suppose first that the roots (λ_r) are unequal. Then we must have

$$y_k = \sum_{r=1}^{N} c_{kr} e^{\lambda_r x} \quad (k = 1, 2, ..., n), \tag{3}$$

and, on substituting these expressions in (1), we get

$$\sum_{r=1}^{N} \left[\sum_{j=1}^{n} F_{ij}(\lambda_r) c_{jr} \right] e^{\lambda_r x} = 0 \quad (i = 1, 2, ..., n). \tag{4}$$

The coefficients of $(e^{\lambda_r x})$ must vanish separately, since these functions are linearly independent; and so we have for each root n relations

$$\sum_{j=1}^{n} F_{ij}(\lambda_r) c_{jr} = 0 \quad (i = 1, 2, ..., n). \tag{5}$$

These are compatible and determine uniquely the ratios $(c_{1r}:c_{2r}:...:c_{nr})$; for we have, on the one hand, $\Delta(\lambda_r) = |F_{ij}(\lambda_r)| = 0$; while, on the

[†] C. Jordan, *Cours d'analyse*, 3, 175–9; M. Bôcher, *Higher Algebra*, 262–78; H. W. Turnbull and A. C. Aitken, *Canonical Matrices*, 19–31, 176–8.

other hand, at least one first minor of the determinant is not zero. For suppose that every first minor of $\Delta(\lambda)$ were divisible by $(\lambda-\lambda_r)$; then the reciprocal determinant $\left|\dfrac{\partial \Delta(\lambda)}{\partial F_{ij}}\right| \equiv \Delta^{n-1}(\lambda)$ would be divisible by $(\lambda-\lambda_r)^n$, and so $\lambda = \lambda_r$ could not be merely a simple root of $\Delta(\lambda) = 0$. Thus each root of the characteristic equation gives a solution with one arbitrary constant. Solutions of the type $y_i = c_i e^{\lambda x}$, where the unknowns remain in a fixed ratio to one another, are called in dynamics *normal solutions*.

The reader should now verify, by the method of § 1, that the system (1) is equivalent to a normal canonical system whose order is equal to the degree of $\Delta(D)$. If the determinant vanishes identically, the system may be replaced by a smaller number of equations and so is indeterminate.

The method of indeterminate coefficients is equally applicable when the characteristic equation has equal roots, but the discussion of the linear equations corresponding to (5) becomes very laborious. In certain cases the equation (2) can be simplified; for if $\Delta_{n-1}(D)$ is the H.C.F. of all the first minors of $\Delta(D) \equiv |F_{ij}(D)|$, we may divide those minors by $\Delta_{n-1}(D)$ before we operate on (1), and so we have

$$\left[\frac{\Delta(D)}{\Delta_{n-1}(D)}\right] y_k = 0 \quad (k = 1, 2, ..., n). \tag{6}$$

But the problem can be very much more clearly treated by simplifying the given system (1). The process is a direct application of H. J. S. Smith's canonical reduction of a matrix of polynomials.

Equivalent Systems. We employ two kinds of transformations, both of which are *reversible*. We may replace two equations $(\Phi_\alpha = 0 = \Phi_\beta)$ by an equivalent pair $\{\Phi_\alpha + F(D)\Phi_\beta = 0 = \Phi_\beta\}$, where $F(D)$ is any linear operator with constant coefficients, since either pair implies the other. This process is called *reduction by rows* of the matrix $\{F_{ij}(D)\}$. Or we may replace two variables (y_α, y_β) by

$$\{y'_\alpha \equiv y_\alpha + F(D) y_\beta,\ y'_\beta \equiv y_\beta\},$$

since either pair can be written in terms of the other. This is called *reduction by columns*.

Our object is to reduce the system to be solved to as few equations as possible, and, failing this, to reduce to a minimum the degree of the lowest operator in the matrix. If any operator is a constant other than zero, the accompanying variable can be explicitly written in terms of the others and eliminated from the system. Every such opportunity of elimination and reduction of the system is to be seized. Suppose now that every operator actually involves D. We pick out the operator of lowest degree, and look for another operator in the same column not exactly divisible by it. For example, if $F_{11}(D)$ is the lowest operator and $F_{21} \equiv QF_{11}+F_{21}^*$, where $F_{21}^*(D)$ is of lower degree than $F_{11}(D)$, but not identically zero, we replace $\Phi_2 = 0$ by $[\Phi_2 - Q(D)\Phi_1] = 0$; the new matrix will have at least one operator of lower order than before. If the lowest operator is a factor of every operator in its column, we make all of them identical and obtain a system of the type

$$M(D)y_1 + \sum_{j=2}^{n} G_{ij}(D)y_j = 0 \quad (i = 1, 2,..., n), \tag{7}$$

where no operator is of lower order than $M(D)$.

Suppose now that one of the operators is not exactly divisible by $M(D)$, say $G_{12}(D) \equiv M(D)Q(D)+G_{12}^*(D)$, where $G_{12}^*(D)$ is of lower degree than $M(D)$ but not identically zero. Then the matrix can be reduced by columns on taking $[y_1+Q(D)y_2]$ as a new variable instead of y_1. If, however, every operator in (7) is divisible by $M(D)$, we eliminate y_1 from all equations except the first and write

$$\left. \begin{array}{l} M(D)z_1 = M(D)[y_1+H_2(D)y_2+...+H_n(D)y_n] = 0, \\ \displaystyle\sum_{j=2}^{n} M(D)H_{ij}(D)y_j = 0 \quad (i = 2, 3,..., n). \end{array} \right\} \tag{8}$$

The first equation need not again be disturbed. The remaining set may be reduced in the same manner as before, and we shall finally obtain a system of the type

$$E_1(D)z_1 = 0, \qquad E_2(D)z_2 = 0, \qquad ..., \qquad E_n(D)z_n = 0, \tag{9}$$

where each operator $E_k(D)$ is divisible by the one preceding it. The old variables (y_i) can be expressed in terms of the new variables (z_i) and vice versa. If k of the original unknowns have been eliminated, the number of equations in the system (9) will be only $(n-k)$; or we may suppose that $E_i(D) \equiv 1 \quad (i = 1, 2,..., k)$, so that the first k equations give the explicit formulae for the eliminated variables.

We have now only to write down the complete primitives of the equations (9) for (z_i) and to introduce these expressions in the formulae giving (y_i) in terms of (z_i). The method of indeterminate coefficients is not required.

Invariant Factors. At each step of the reduction, we have merely added to one row or column of the matrix $(F_{ij}(D))$ a certain multiple of another. This leaves invariant the determinant $\Delta(D) \equiv |F_{ij}(D)|$; but much more than this is true.

Let $\Delta_k(D)$ be a polynomial in D (with the highest coefficient unity) defined as the H.C.F. of all k-rowed determinants of the matrix $(F_{ij}(D))$. If $\Delta_k^*(D)$ is the corresponding H.C.F. after an elementary transformation, it is easily seen that every new k-rowed determinant is divisible by $\Delta_k(D)$; hence $\Delta_k^*(D)$ is divisible by $\Delta_k(D)$. But the process is reversible, so that $\Delta_k(D)$ is divisible by $\Delta_k^*(D)$, and therefore $\Delta_k^*(D) \equiv \Delta_k(D)$ $(k = 1, 2, ..., n)$. But these invariants $\{\Delta_k(D)\}$ can be written down by inspection of the reduced canonical matrix

$$\begin{pmatrix} E_1(D) & 0 & \cdots & 0 \\ 0 & E_2(D) & \cdots & 0 \\ \cdot & \cdot & \cdot & \cdot \\ 0 & 0 & \cdots & E_n(D) \end{pmatrix}. \tag{10}$$

Since $E_{i+1}(D)$ is divisible by $E_i(D)$, every k-rowed minor which does not vanish identically is divisible by $\prod_{i=1}^{k} E_i(D)$. If the highest coefficients in $\{E_i(D)\}$ are reduced to unity, we have therefore

$$\Delta_k(D) \equiv E_1(D)E_2(D)...E_k(D) \equiv \Delta_{k-1}(D)E_k(D). \tag{11}$$

Hence $\{E_k(D)\}$ are likewise invariants of the original matrix; these expressions $\{E_k(D)\}$ are called its *invariant factors*. They can be found by rational operations, without solving the characteristic equation $\Delta(\lambda) = 0$.

EXAMPLES. II

1. HEAVISIDE OPERATORS. Verify the formulae

$$\frac{p}{p-in} = e^{inx}, \qquad \frac{p^2}{p^2+n^2} = \cos nx, \qquad \frac{np}{p^2+n^2} = \sin nx,$$

$$\frac{p}{p^2+n^2}f(x) = \int_0^x \cos n(x-t)f(t)\, dt,$$

$$\frac{n}{p^2+n^2}f(x) = \int_0^x \sin n(x-t)f(t)\, dt.$$

2. (i) The solution of $F(D)y = 0$ determined by the initial conditions

$$y = 0, \quad Dy = 0, \quad ..., \quad D^{n-2}y = 0, \quad D^{n-1}y = 1,$$

when $x = 0$, is
$$y = \frac{p}{F(p)}.$$

(ii) The solution of $F(D)y = f(x)$ determined by the initial conditions
$$y = 0, \quad Dy = 0, \quad ..., \quad D^{n-1}y = 0,$$
when $x = 0$, is
$$y = \frac{1}{F(p)} f(x).$$

3. The solution of $F(D)y = 0$ determined by the initial conditions
$$D^i y = \eta^{(i)}, \quad \text{when} \quad x = 0 \quad (i = 0, 1, ..., n-1),$$
is
$$y = P(x) - \frac{1}{F(p)}[F(D)P(x)],$$
where
$$P(x) \equiv \eta^{(0)} + \eta^{(1)}\frac{x}{1!} + \eta^{(2)}\frac{x^2}{2!} + ... + \eta^{(n-1)}\frac{x^{n-1}}{(n-1)!}.$$

4. Verify that
$$\frac{1}{(p^2-1)^n} f(x) = \int_0^x dt \int_0^t f(s) \Phi(x,t,s) \, ds$$
$$= \int_0^x f(s) \, ds \int_s^x \Phi(x,t,s) \, dt,$$
where
$$\Phi(x,t,s) \equiv e^{-x+2t-s} \frac{(x-t)^{n-1}(t-s)^{n-1}}{(n-1)!(n-1)!};$$
and also that
$$\frac{1}{(p^2-1)^n} f(x) = \int_0^x \frac{f(s) e^{s-x}(x-s)^{2n-1} \, ds}{(n-1)!(n-1)!} \int_0^1 e^{2u(x-s)} u^{n-1}(1-u)^{n-1} \, du.$$

5. Express as integrals
$$\frac{1}{(p-\alpha)^m (p-\beta)^n} f(x), \quad \frac{1}{(p^2 - 2p\cos\theta + 1)^n} f(x), \quad \frac{1}{[(p-\alpha)^2 + \beta^2]^n} f(x),$$
$$\frac{1}{(p^{2n}+1)} f(x), \quad \frac{1}{(p^{2n+1}+1)} f(x), \quad \frac{1}{(p^{2n} - 2p^n \cos 2n\theta + 1)} f(x).$$

6. If $R(p)$ is a rational function, whose denominator is of degree not lower than the numerator, show that
$$e^{\alpha x} R(p) = R(p-\alpha) e^{\alpha x} = \frac{R(p-\alpha) p}{(p-\alpha)}.$$
Hence prove by induction that
$$\frac{(2n-1)! \, p}{(p^2+1^2)(p^2+3^2)...\{p^2+(2n-1)^2\}} = \sin^{2n-1} x,$$
$$\frac{(2n)!}{(p^2+2^2)(p^2+4^2)...\{p^2+(2n)^2\}} = \sin^{2n} x.$$

[H. V. Lowry, *Phil. Mag.* (7) **13** (1932), 1033–48, 1144–63.]

7. Verify Lowry's formulae by showing that $y = \sin^{2n-1} x$ satisfies the equation
$$(D^2+1^2)(D^2+3^2)...\{D^2+(2n-1)^2\}y = 0,$$
with the conditions
$$D^i y = 0 \quad (i = 0, 1, ..., 2n-2), \quad D^{2n-1} y = (2n-1)!,$$
when $x = 0$.

8. A sequence of Legendre trigonometric polynomials is defined by the relations
$$P_0(\cos x) = 1, \qquad P_1(\cos x) = \cos x,$$
$$(n+1)P_{n+1}(\cos x) - (2n+1)\cos x\, P_n(\cos x) + nP_{n-1}(\cos x) = 0.$$
Verify the formulae
$$P_{2n+1}(\cos x) = \frac{p^2(p^2+2^2)(p^2+4^2)\ldots\{p^2+(2n)^2\}}{(p^2+1^2)(p^2+3^2)(p^2+5^2)\ldots\{p^2+(2n+1)^2\}},$$
$$P_{2n}(\cos x) = \frac{(p^2+1^2)(p^2+3^2)\ldots\{p^2+(2n-1)^2\}}{(p^2+2^2)(p^2+4^2)\ldots\{p^2+(2n)^2\}}.$$

[B. van der Pol, *Phil. Mag.* (7) **7** (1929), 1153–62; **8** (1929), 861–98.]

9. BOREL'S RELATION. If the Heaviside operators $\{F_i(p)\}$, generating ordinary power-series $\{f_i(x)\}$, are connected by the relation
$$F_1(p) \equiv \frac{1}{p}F_2(p)F_3(p),$$
then
$$f_1(x) = \int_0^x f_2(x-t)f_3(t)\,dt.$$

10. Show that
$$\int_0^x P_{2n}\{\cos(x-t)\}\sin^{2n-1}t\,dt = \frac{\sin^{2n}x}{2n}. \quad [\text{LOWRY.}]$$

11. CAUCHY'S METHOD. If $F(z)$, $G(z)$ are polynomials of degree n, m respectively ($m < n$), and if C is a contour enclosing all the zeros of $F(z)$, show that
$$\frac{G(p)}{F(p)} = \frac{1}{2i\pi}\int_C e^{xz}\frac{G(z)}{F(z)}\,dz,$$

$$\frac{1}{F(p)}f(x) = \frac{1}{2i\pi}\int_0^x f(\xi)\,d\xi\int_C e^{(x-\xi)z}\frac{dz}{F(z)}.$$

[A. L. Cauchy, *Œuvres* (2) **6**, 252–5, **7**, 40–54, 255–66; C. Hermite, *Bulletin des sc. math.* (2) **3** (1879), 311–25.]

12. Solve the simultaneous equations

(1) $\left.\begin{array}{l}D^2u = 2u-v-w\\D^2v = -u+2v-w\\D^2w = -u-v+2w\end{array}\right\};\qquad$ (ii) $\left.\begin{array}{l}D^2u = 2u+v+w\\D^2v = u+2v+w\\D^2w = u+v+2w\end{array}\right\}.$

III

SOME FORMAL INVESTIGATIONS

9. Linear Operators

Operators with Variable Coefficients. If we write

$$F(x,D)y \equiv \left[\sum_{r=0}^{n} p_r(x)D^{n-r}\right]y \equiv \sum_{r=0}^{n} p_r(x)D^{n-r}y, \tag{1}$$

the expression $F(x,D)$ is called an *ordinary linear differential operator*. If $\{\phi_i(x)\}$ are functions of x, differentiable sufficiently often, and (c_i) are constants, we have

$$\left.\begin{array}{l} D^m(\phi_1+\phi_2) \equiv D^m\phi_1+D^m\phi_2, \\ D^m(\sum c_i\phi_i) \equiv \sum c_i D^m\phi_i. \end{array}\right\} \tag{2}$$

By combining such relations, we have for any operator $F \equiv F(x,D)$ of the type (1)

$$\left.\begin{array}{l} F(\phi_1+\phi_2) \equiv F\phi_1+F\phi_2, \\ F(\sum c_i\phi_i) \equiv \sum c_i F\phi_i. \end{array}\right\} \tag{3}$$

But it is *not* in general true that $FG\phi$ is the same as $GF\phi$, for distinct linear operators F, G.

Division and Factors. Operator polynomials in D have many analogies with ordinary polynomials in a variable. Consider two operators

$$\left.\begin{array}{l} F \equiv p_0(x)D^n+p_1(x)D^{n-1}+\ldots+p_n(x), \\ G \equiv q_0(x)D^m+q_1(x)D^{m-1}+\ldots+q_m(x), \end{array}\right\} \tag{4}$$

where $m \leqslant n$. Let us form the expression

$$Fy - [A_0(x)D^{n-m}+A_1(x)D^{n-m-1}+\ldots+A_{n-m}(x)]Gy, \tag{5}$$

and let us determine successively

$$\left.\begin{array}{l} A_0 = p_0/q_0, \\ A_1 = \{p_1 - A_0 q_1 - (n-m)A_0 Dq_0\}/q_0, \\ \cdot\quad\cdot\quad\cdot\quad\cdot\quad\cdot\quad\cdot\quad\cdot \end{array}\right\} \tag{6}$$

so that the expression (5) shall have no term in $(D^n y, D^{n-1}y,\ldots,D^m y)$. We see that $\{A_i(x)\}$ are uniquely determined, and we have a relation

$$Fy = QGy + Ry, \tag{7}$$

where $Q \equiv \sum A_i D^{n-m-i}$, and R is an operator of order not exceeding $(m-1)$. These are analogous to the ordinary quotient and remainder of two polynomials.

If we know m ($< n$) linearly independent solutions of $Fy = 0$, let the equation satisfied by them be

$$Gy \equiv \frac{W(\phi_1, \phi_2, ..., \phi_m, y)}{W(\phi_1, \phi_2, ..., \phi_m)} = 0. \tag{8}$$

If we construct the identity (7) and substitute for y the solutions $(\phi_1, \phi_2, ..., \phi_m)$, we have

$$R\phi_i \equiv B_0 D^{m-1}\phi_i + ... + B_{m-1}\phi_i \equiv 0 \quad (i = 1, 2, ..., m); \tag{9}$$

and since $W(\phi_1, \phi_2, ..., \phi_m) \neq 0$, these give

$$B_i \equiv 0 \quad (i = 0, 1, ..., m-1).$$

Hence the relation becomes

$$Fy \equiv QGy, \tag{10}$$

so that G is an *inner* or *right-hand factor* of the operator F. We can thus break up the equation into the pair

$$Qz = 0, \qquad Gy = z. \tag{11}$$

The first equation is of order $(n-m)$; and, when it is solved, the second is soluble by quadratures, by the method of variation of parameters.

Highest Common Factor. To discover whether two given equations have any common solutions, we use a process of Brassine analogous to the extraction of the H.C.F. of two polynomials. Let (F_1, F_2) be two given operators, of which the former is of the higher order; we construct a sequence of operators (F_i) of steadily decreasing order $n_1 > n_2 > n_3 > ... > n_k$, in accordance with the scheme

$$\left.\begin{aligned} F_1 &= Q_1 F_2 + F_3, \\ F_2 &= Q_2 F_3 + F_4, \\ &\cdot \quad \cdot \quad \cdot \quad \cdot \\ F_{k-1} &= Q_{k-1} F_k \quad (F_{k+1} \equiv 0). \end{aligned}\right\} \tag{12}$$

After a finite number of steps we must have $F_{k+1} \equiv 0$; and the last operator which does not vanish identically is an inner factor of all the preceding ones, since

$$F_{k-1} = Q_{k-1} F_k, \qquad F_{k-2} = (Q_{k-2} Q_{k-1} + 1) F_k, ... \tag{13}$$

But, on the other hand, each of these operators can be written as $F_i = (A_i F_1 + B_i F_2)$; for we have

$$F_3 = F_1 - Q_1 F_2, \qquad F_4 = -Q_2 F_1 + (1 + Q_2 Q_1) F_2, ... \tag{14}$$

From the identity

$$F_k y = A_k F_1 y + B_k F_2 y, \tag{15}$$

we see that every solution common to $(F_1 y = 0, F_2 y = 0)$ must also satisfy $F_k y = 0$. Hence F_k is the *highest common inner factor* of (F_1, F_2).

Consider now two *non-homogeneous* equations

$$F_1 y = f_1(x), \qquad F_2 y = f_2(x). \tag{16}$$

This system is in all respects equivalent to the pair

$$F_2 y = f_2(x), \qquad F_3 y = f_1(x) - Q_1 f_2(x) = f_3(x), \text{ say,} \tag{17}$$

where $F_3 = (F_1 - Q_1 F_2)$ as in (12); for we can deduce either pair from the other. Thus the original system can be replaced successively by the pairs

$$F_i y = f_i(x), \qquad F_{i+1} y = f_{i+1}(x) \quad (i = 2, 3, ..., k). \tag{18}$$

But, since F_{k+1} is identically zero, the last equation necessitates that $f_{k+1}(x)$ should be identically zero, for otherwise the equations are incompatible. This means that

$$f_{k+1}(x) \equiv A_{k+1} f_1(x) + B_{k+1} f_2(x) \equiv 0, \tag{19}$$

is a necessary condition for the compatibility of the system. The condition is sufficient; for, in the kth equivalent pair (18), one equation is merely $0 = 0$, so that the system is effectively equivalent to the single condition

$$F_k y = f_k(x). \tag{20}$$

Least Common Multiple. Let the operators (F_1, F_2) have no common inner factor; and let us expand as linear homogeneous functions of $(D^i y)$ the $(n_1 + n_2 + 2)$ expressions

$$D^i F_1 y \quad (i = 0, 1, ..., n_2), \qquad D^j F_2 y \quad (j = 0, 1, ..., n_1). \tag{21}$$

These $(n_1 + n_2 + 2)$ linear forms, homogeneous in $(n_1 + n_2 + 1)$ quantities $(y, Dy, ..., D^{n_1+n_2} y)$, must be connected by an identity

$$My \equiv \sum_{i=0}^{n_2} \alpha_i(x) D^i F_1 y \equiv \sum_{j=0}^{n_1} \beta_j(x) D^j F_2 y, \tag{22}$$

or
$$My \equiv \Psi_1 F_1 y \equiv \Psi_2 F_2 y, \tag{23}$$

where (Ψ_1, Ψ_2) are operators of order (n_2, n_1) respectively.

The equation $My = 0$ is satisfied by any solution of *either* $F_1 y = 0$ or $F_2 y = 0$; so that the operator M is analogous to the *least common multiple* of two polynomials. The relation (23) is equivalent to the one expressing that the last remainder in the H.C.F. process is $F_{k+1} \equiv A_{k+1} F_1 + B_{k+1} F_2 \equiv 0$.

Again, if the H.C.F. of (F_1, F_2) is of order $n_k = \nu > 0$, we have
$$F_1 y \equiv G_1(F_k y) \equiv G_1 z, \\ F_2 y \equiv G_2(F_k y) \equiv G_2 z, \quad \rbrace \tag{24}$$
where the operators (G_1, G_2) are of order $(n_1-\nu, n_2-\nu)$. By the same method, we now obtain an equation $My = 0$, of order only $(n_1+n_2-\nu)$, which is satisfied by every solution of either equation. Conversely, if we have an identity (23), where (Ψ_1', Ψ_2') are only of order $(n_2-\nu, n_1-\nu)$, the given equations have ν common solutions. For let $\{\phi_i(x)\}$ be n_1 linearly independent solutions of $F_1 y = 0$. We have then
$$\Psi_2'[F_2 \phi_i(x)] = 0 \quad (i = 1, 2, ..., n_1), \tag{25}$$
and so $\{F_2 \phi_i(x)\}$ are all solutions of an equation of order $(n_1-\nu)$. Since not more than $(n_1-\nu)$ of the expressions can be linearly independent, we must have ν relations of the type
$$F_2 \phi_i(x) = \sum_{j=\nu+1}^{n_1} c_{ij} F_2 \phi_j(x) \quad (i = 1, 2, ..., \nu), \tag{26}$$
where (c_{ij}) are constants. Hence the simultaneous equations $(F_1 y = 0 = F_2 y)$ have ν common solutions
$$y = \phi_i(x) - \sum_{j=\nu+1}^{n_1} c_{ij} \phi_j(x) \quad (i = 1, 2, ..., \nu). \tag{27}$$
If we erase the last $(\nu+1)$ derivatives in each set (21), we have $(n_1+n_2-2\nu)$ expressions linear and homogeneous in $z \equiv F_k y$ and its first $(n_1+n_2-2\nu-1)$ derivatives. If these expressions are not linearly independent, we can construct an identity showing that $(G_1 z = 0 = G_2 z)$ have a common solution other than zero, contrary to our hypothesis that F_k is the H.C.F. of the given operators. We can therefore solve for z from the $(n_1+n_2-2\nu)$ linear expressions, and write
$$z = A_k G_1 z + B_k G_2 z, \tag{28}$$
where the operators (A_k, B_k) are of order not higher than $(n_2-\nu-1, n_1-\nu-1)$. This is equivalent to the identity (15), and indicates the order of the operators concerned.

10. Adjoint Equations[†]

Equations of the First Order. We can readily solve the equation
$$Dy + p(x)y = f(x) \tag{1}$$
by an appropriate use of the identity
$$D(yz) \equiv z[Dy + p(x)y] + y[Dz - p(x)z]. \tag{2}$$

[†] G. Darboux, *Théorie générale des surfaces*, **2**, 112–34; M. Bôcher, *Leçons sur les méthodes de Sturm*, 22–42; F. B. Pidduck, *Proc. Royal Soc.* A, **117** (1927), 201–8.

Let y be indeterminate, but let z be chosen so that
$$Dz = p(x)z, \quad \text{or} \quad z = \exp\left[\int p(x)\, dx\right] = e^{\varpi(x)} \text{ say.}$$
Then we have, on multiplying (1) by z,
$$D[e^{\varpi(x)}y] = e^{\varpi(x)}f(x), \tag{3}$$
and so
$$e^{\varpi(x)}y = \int e^{\varpi(x)}f(x)\, dx + C. \tag{4}$$
But, conversely, we may leave z indeterminate, and choose y, so that $Dy = -p(x)y$, or $y = e^{-\varpi(x)}$. We can then solve any equation of the form
$$Dz - p(x)z = g(x), \tag{5}$$
by writing it as
$$D[e^{-\varpi(x)}z] = e^{-\varpi(x)}g(x), \tag{6}$$
so that
$$e^{-\varpi(x)}z = \int e^{-\varpi(x)}g(x)\, dx + C'. \tag{7}$$

There is complete reciprocity between the pair of equations
$$Dy + p(x)y = 0, \qquad Dz - p(x)z = 0, \tag{8}$$
the solutions of either being integrating factors of the other. Such equations are said to be mutually *adjoint*.

Adjoint Canonical Systems. The more general identity
$$D\left[\sum_{i=1}^n y_i z_i\right] \equiv \sum_{i=1}^n z_i\left[Dy_i - \sum_{j=1}^n a_{ij}(x)y_j\right] + \sum_{i=1}^n y_i\left[Dz_i + \sum_{j=1}^n a_{ji}(x)z_j\right], \tag{9}$$
suggests a similar reciprocal relationship between the two homogeneous systems

(S_n^*) $\qquad\qquad Dy_i = \sum_{j=1}^n a_{ij}(x)y_j,$

(Σ_n^*) $\qquad\qquad Dz_i = -\sum_{j=1}^n a_{ji}(x)z_j.$

If (y_i) are indeterminate, but (z_i) satisfy (Σ_n^*), the first group of terms on the right in (9) becomes an exact derivative; conversely, if (z_i) are indeterminate, but (y_i) satisfy (S_n^*), the second group becomes an exact derivative.

Lagrange's Adjoint Equation. Consider the identity
$$D[p_0 z y_n + z_1 y_{n-1} + \ldots + z_{n-1} y_1]$$
$$\equiv z[p_0 Dy_n + p_1 y_n + p_2 y_{n-1} + \ldots + p_n y_1] +$$
$$+ \sum_{i=1}^{n-1} z_i[Dy_{n-i} - y_{n-i+1}] + [D(p_0 z) - p_1 z + z_1]y_n +$$
$$+ [Dz_1 - p_2 z + z_2]y_{n-1} + \ldots +$$
$$+ [Dz_{n-2} - p_{n-1}z + z_{n-1}]y_2 + [Dz_{n-1} - p_n z]y_1. \tag{10}$$

Let us write $y_i \equiv D^{i-1}y$ $(i = 1, 2, ..., n)$ and also put

$$\left.\begin{aligned}
z_1 &= p_1 z - D(p_0 z), \\
z_2 &= p_2 z - D(p_1 z) + D^2(p_0 z), \\
&\cdot\quad\cdot\quad\cdot\quad\cdot\quad\cdot\quad\cdot\quad\cdot \\
z_{n-1} &= p_{n-1} z - D(p_{n-2} z) + ... + (-)^{n-1} D^{n-1}(p_0 z),
\end{aligned}\right\} \quad (11)$$

so that only y and z remain indeterminate. If we put

$$Fy \equiv p_0 D^n y + p_1 D^{n-1} y + ... + p_n y, \quad (12)$$

$$F^*z \equiv (-)^n D^n(p_0 z) + (-)^{n-1} D^{n-1}(p_1 z) + ... + p_n z, \quad (13)$$

the operators F, F^* are said to be *adjoint* to one another, and we find the identity

$$zFy - yF^*z \equiv D\Psi(y, z), \quad (14)$$

where $\Psi(y, z)$ is the *bilinear concomitant*

$$\Psi(y, z) \equiv p_0 z D^{n-1} y +$$
$$+ [p_1 z - D(p_0 z)] D^{n-2} y +$$
$$\cdot\quad\cdot\quad\cdot\quad\cdot\quad\cdot\quad\cdot\quad\cdot$$
$$+ [p_{n-1} z - D(p_{n-2} z) + ... + (-)^{n-1} D^{n-1}(p_0 z)] y. \quad (15)$$

The same method of interpretation shows that every solution of $F^*z = 0$ is an integrating factor of $Fy = 0$, and vice versa. To show that F^{**}, the operator constructed from F^* by the same rule as F^* was formed from F, is identical with F, we need only write down the identities

$$\left.\begin{aligned}
zFy - yF^*z &\equiv D\Psi(y, z), \\
yF^*z - zF^{**}y &\equiv D\Psi^*(z, y),
\end{aligned}\right\} \quad (16)$$

which show that

$$z(Fy - F^{**}y) \equiv D[\Psi(y, z) + \Psi^*(z, y)] \quad (17)$$

is an exact derivative, identically in y and z. But this is impossible unless both sides vanish identically.

Composite Operators. If (F_1, F_1^*) and (F_2, F_2^*) are pairs of adjoint operators, we have

$$\left.\begin{aligned}
zF_1 y - yF_1^* z &\equiv D\Psi_1(y, z), \\
zF_2 y - yF_2^* z &\equiv D\Psi_2(y, z),
\end{aligned}\right\} \quad (18)$$

hence $\quad z(F_1 + F_2)y - y(F_1^* + F_2^*)z = D[\Psi_1(y, z) + \Psi_2(y, z)]. \quad (19)$

Thus the operators $(F_1 + F_2, F_1^* + F_2^*)$ are also adjoint to one another, and similarly for the sum of any number of operators.

Again, write $F_2 y$ instead of y in the first identity, and $F_1^* z$ instead

of z in the second identity of (18); adding the results, we have the identity
$$zF_1F_2y-yF_2^*F_1^*z \equiv D[\Psi_1(F_2y,z)+\Psi_2(y,F_1^*z)], \tag{20}$$
showing that the products $(F_1F_2, F_2^*F_1^*)$, with the *factors reversed*, are adjoint operators; and similarly for any number of factors.

When two equations $(F_1y = 0, F_2y = 0)$ admit ν linearly independent common integrating factors, their adjoints have ν common solutions; hence they can be written
$$F_1^*z \equiv G_1^*H^*z = 0, \qquad F_2^*z \equiv G_2^*H^*z = 0, \tag{21}$$
where H^* is of order ν. The original equations are therefore $(HG_1y = 0, HG_2y = 0)$. By a process parallel to Brassine's, it would be possible to extract directly the *highest outer common factor* of two operators.

11. Simultaneous Equations with Variable Coefficients

Reduction to a Diagonal System. The procedure of § 8 can be applied to the reduction of systems with variable coefficients, provided that we take account of the presence of non-commutative operators. Consider the system
$$\Phi_i \equiv \sum_{j=1}^{n} F_{ij}y_j = 0 \quad (i = 1, 2, \ldots, n), \tag{1}$$
where (F_{ij}) are operators of the type
$$F \equiv a_0(x)D^m + a_1(x)D^{m-1} + \ldots + a_m(x), \tag{2}$$
whose coefficients are differentiable as often as may be necessary in the course of the reduction. Our object is to reduce the system to be solved to as few equations as possible. If there is any operator $F_{\alpha\beta}$ which does not involve D and is not identically zero, we can solve $\Phi_\alpha = 0$ for y_β and eliminate y_β from the remaining equations. It will be assumed that every opportunity of thus reducing the system is to be taken.

If every operator present involves D, we pick out the one of lowest degree, say F_{11}, and look for any operator in the same column which does not admit F_{11} as an *inner or right-hand factor*. Suppose $F_{21} \equiv QF_{11} + F_{21}^*$, where F_{21}^* is of lower degree than F_{11} but not identically zero. Then on replacing $\Phi_2 = 0$ by $[\Phi_2 - Q\Phi_1] = 0$, the matrix will be *reduced by rows* to one having an operator of lower degree in D. If every operator in the first column is divisible on the right by F_{11}, we look for any operator in the first row which does not admit

F_{11} as an *outer or left-hand factor*. Suppose $F_{12} \equiv F_{11}Q + F_{12}^*$, where F_{12}^* is of lower degree than F_{11} but not identically zero. Then on taking $[y_1 + Qy_2]$ as a new variable instead of y_1, the matrix will be *reduced by columns* to one having an operator of lower degree. If an operator of degree zero appears at any stage, we at once eliminate a variable.

If the lowest operator is an inner factor of every operator in its column and an outer factor of every operator in its row, the system is of the type

$$\left.\begin{aligned} \Phi_1 &\equiv F_{11}\Big[y_1 + \sum_{j=2}^{n} A_j y_j\Big] = 0, \\ \Phi_i &\equiv B_i F_{11} y_1 + \sum_{j=2}^{n} F_{ij} y_j = 0 \quad (i = 2, 3, \ldots, n). \end{aligned}\right\} \quad (3)$$

The first equation may be written as $F_{11} z_1 = 0$; and we can eliminate y_1 from the others by taking the combinations

$$\Phi_i - B_i \Phi_1 = 0 \quad (i = 2, 3, \ldots, n).$$

Proceeding with the reduction of the latter set, we ultimately obtain a diagonal system

$$G_1 z_1 = 0, \qquad G_2 z_2 = 0, \qquad \ldots, \qquad G_n z_n = 0, \quad (4)$$

where the old variables (y_i) are expressible in terms of the new variables (z_i) and vice versa.

Analogues of the Invariant Factors. The reduction (4) is not so complete as that of a system with constant coefficients to the canonical form. But it forms a convenient intermediate stage, after which the equations may be examined two at a time. Any pair $(G_1 z_1 = 0 = G_2 z_2)$ can be further reduced, unless the lower operator G_1 is both an inner and an outer factor of G_2, or unless

$$G_2 \equiv G_2^* G_1 \equiv G_1 G_2^{**};$$

this need not imply that G_2 is of the form $G_1 H G_1$, as is obvious by considering operators with constant coefficients.

If G_1 is not an *outer* factor of G_2, we can write

$$G_1 z_1 + G_2 z_2 = 0, \qquad G_2 z_2 = 0; \quad (5)$$

this system is reducible by *columns*.

If G_1 is not an *inner* factor of G_2, we can write $z_2 \equiv z_1 + z_2^*$, and the system

$$G_1 z_1 = 0, \qquad G_2 z_1 + G_2 z_2^* = 0, \quad (6)$$

will be reducible by *rows*.

After a finite number of steps and eliminations we must arrive at a system where every pair of operators fulfils the conditions. Arranging the operators in order of ascending degree, the system will be of the type

$$H_1 w_1 = 0, \qquad H_2 w_2 = 0, \qquad ..., \qquad H_n w_n = 0, \qquad (7)$$

where $H_{k+1} \equiv H_{k+1}^* H_k \equiv H_k H_{k+1}^{**}$. In other words, the $(k+1)$th equation is satisfied by every solution of the kth, and admits every integrating factor of the kth equation.

Generally speaking, two equations taken at random would have no common solution. We could then construct the H.C.F. identity $[K_1 G_1 + K_2 G_2] \equiv 1$, and the equation satisfied by any solution of either equation, $Lz = 0$, where $L \equiv P_1 G_1 \equiv P_2 G_2$. The general solution of $Lz = 0$ is $z = z_1 + z_2$, where $G_1 z_1 = 0$ and $G_2 z_2 = 0$. But we can also express z_1 and z_2 in terms of z, for we have

$$z_1 = (K_1 G_1 + K_2 G_2) z_1 = K_2 G_2 z_1 = K_2 G_2 (z_1 + z_2) = K_2 G_2 z, \qquad (8)$$

and similarly $z_2 = K_1 G_1 z$.

Conclusion. In certain investigations there is a gain in symmetry and clarity in considering a normal canonical system; and we have seen in § 1 how every equation or system can be reduced to that form. But in studying the permutations of the solutions of an analytic system, it is obviously simpler to consider a set of n analytic functions (and as many derivatives as we please) than an array of n^2 functions, the former being a fundamental system of solutions of a single equation, and the latter that of a normal canonical system, both of the same order n. Knowing how to replace any analytic system by single equations, each involving one unknown, we need only examine directly the properties of the solutions of a single typical equation with analytic coefficients.

EXAMPLES. III

1. **Wronskian Determinants.** If $\phi_i(x) \equiv \phi_i\{\theta(t)\} \equiv \chi_i(t)$, prove that

$$W_x(\phi_1, \phi_2, ..., \phi_n) = \left(\frac{dt}{dx}\right)^{\frac{1}{2}n(n-1)} W_t(\chi_1, \chi_2, ..., \chi_n).$$

2. If λ is any function of x, prove that

$$W_x(\lambda \phi_1, \lambda \phi_2, ..., \lambda \phi_n) = \lambda^n W(\phi_1, \phi_2, ..., \phi_n).$$

Hence show that, if $\phi_1 \neq 0$,

$$W_x(\phi_1, \phi_2, ..., \phi_n) \equiv \phi_1^n W_x\left[D\left(\frac{\phi_2}{\phi_1}\right), D\left(\frac{\phi_3}{\phi_1}\right), ..., D\left(\frac{\phi_n}{\phi_1}\right)\right].$$

3. If $m < n$ and $Gy \equiv D^m y + q_1 D^{m-1} y + \ldots + q_m y$, show that the elements of a column of $W(\phi_1, \phi_2, \ldots, \phi_n)$ may be replaced by

$$\phi_i, D\phi_i, \ldots, D^{m-1}\phi_i, G\phi_i, DG\phi_i, \ldots, D^{n-m-1}G\phi_i.$$

By putting
$$Gy \equiv \frac{W(\phi_1, \phi_2, \ldots, \phi_m, y)}{W(\phi_1, \phi_2, \ldots, \phi_m)},$$
show that
$$W(\phi_1, \phi_2, \ldots, \phi_n, y) = W(\phi_1, \phi_2, \ldots, \phi_m) W(G\phi_{m+1}, G\phi_{m+2}, \ldots, G\phi_n, Gy).$$

4. SYMBOLIC FACTORS OF FROBENIUS. (i) By considering the minors of the determinant $W(\phi_1, \phi_2, \ldots, \phi_n, y)$, show that

$$D\left[\frac{W(\phi_1, \phi_2, \ldots, \phi_{n-1}, y)}{W(\phi_1, \phi_2, \ldots, \phi_{n-1}, \phi_n)}\right] = \frac{W(\phi_1, \phi_2, \ldots, \phi_{n-1}) W(\phi_1, \phi_2, \ldots, \phi_n, y)}{[W(\phi_1, \phi_2, \ldots, \phi_{n-1}, \phi_n)]^2}.$$

(ii) If $\eta_1 \equiv \phi_1(x)$, $\eta_r \equiv \dfrac{W(\phi_1, \phi_2, \ldots, \phi_r)}{W(\phi_1, \phi_2, \ldots, \phi_{r-1})}$, deduce from the above that

$$\frac{W(\phi_1, \phi_2, \ldots, \phi_n, y)}{W(\phi_1, \phi_2, \ldots, \phi_n)} = \eta_n D \frac{\eta_{n-1}}{\eta_n} D \frac{\eta_{n-2}}{\eta_{n-1}} D \ldots \frac{\eta_1}{\eta_2} D \frac{y}{\eta_1}.$$

(iii) Show that the equation

$$\frac{1}{\alpha_{n+1}} D \frac{1}{\alpha_n} D \ldots \frac{1}{\alpha_2} D \frac{y}{\alpha_1} = 0$$

admits the solutions

$$\phi_r(x) = \alpha_1(x) \int^x \alpha_2(x_2)\, dx_2 \int^{x_2} \alpha_3(x_3)\, dx_3 \ldots \int^{x_{r-1}} \alpha_r(x_r)\, dx_r.$$

(iv) Verify that $W(\phi_1, \phi_2, \ldots, \phi_n) = \alpha_1^n \alpha_2^{n-1} \ldots \alpha_n$,

$$\alpha_r = \frac{\eta_r}{\eta_{r-1}} = D\left[\frac{W(\phi_1, \phi_2, \ldots, \phi_{r-2}, \phi_r)}{W(\phi_1, \phi_2, \ldots, \phi_{r-2}, \phi_{r-1})}\right].$$

5. If
$$F_r y \equiv \frac{1}{\alpha_{r+1}} D \frac{1}{\alpha_r} D \ldots \frac{1}{\alpha_2} D \frac{y}{\alpha_1},$$

$$F_r^* z \equiv \frac{(-)^r}{\alpha_{n-r+1}} D \frac{1}{\alpha_{n-r+2}} D \ldots \frac{1}{\alpha_n} D \frac{z}{\alpha_{n+1}},$$

show that $F_n y = 0$ and $F_n^* z = 0$ are adjoint equations, and that their bilinear concomitant is

$$\Psi(y, z) = \sum_{r=0}^{n-1} (F_{n-r-1} y)(F_r^* z). \qquad \text{[DARBOUX.]}$$

6. A self-adjoint equation of *even* order may be written

$$\frac{1}{\phi_1} D \frac{1}{\phi_2} \ldots D \frac{1}{\phi_n} D \frac{1}{\phi_{n+1}} D \frac{1}{\phi_n} \ldots D \frac{1}{\phi_2} D \frac{y}{\phi_1} = 0,$$

and one of *odd* order

$$\frac{1}{\phi_1} D \ldots \frac{1}{\phi_n} D \frac{1}{\phi_{n+1}} D \frac{1}{\phi_{n+1}} D \ldots \frac{1}{\phi_2} D \frac{y}{\phi_1} = 0.$$

[FROBENIUS; DARBOUX.]

7. Verify that the following equations are self-adjoint (except as regards sign):

(i) $\dfrac{1}{\phi} D \dfrac{y}{\phi} = 0,$ (ii) $D^n f D^n y = 0,$

(iii) $D^n f D^{n+1} y + D^{n+1} f D^n y = 0$,   (iv) $Fy \pm F^* y = 0$,

(v) $F f F^* y = 0$,   (vi) $F f D f F^* y = 0$,

where (ϕ, f) are functions of x, and (F, F^*) are adjoint operators.

8. If $\Psi(y, z)$ is the bilinear concomitant of a self-adjoint equation of order n, $Fy = 0$, then $\Psi(y, y) \equiv 0$, if n is even, and $\Psi(y, y) \equiv 2yFy$, if n is odd. By writing $(y + \lambda z)$ for y, show that, if yFy is an exact derivative, n is odd and Fy is self-adjoint. [DARBOUX.]

9. If $Fy \equiv p_0 D^{2n} y + p_1 D^{2n-1} y + \ldots + p_{2n} y = 0$ is self-adjoint, $p_1 \equiv nDp_0$, and $[Fy - D^n p_0 D^n y] = 0$ is also self-adjoint. Hence show that the most general self-adjoint equation of *even* order is of the form
$$D^n \psi_0 D^n y + D^{n-1} \psi_1 D^{n-1} y + \ldots + \psi_n y = 0,$$
and that of *odd* order is of the form
$$\sum_{i=0}^{n} [D^{n-i} \psi_i D^{n-i+1} y + D^{n-i+1} \psi_i D^{n-i} y] = 0.$$
[JACOBI; DARBOUX.]

10. If $F \equiv \sum_{i=0}^{n} \binom{n}{i} a_i D^{n-i}$ and $F^* \equiv \sum_{i=0}^{n} \binom{n}{i} b_i D^{n-i}$ are adjoint operators, show that
$$a_0 = b_0,$$
$a_1 = Db_0 - b_1,$ $b_1 = Da_0 - a_1,$

$a_2 = D^2 b_0 - 2Db_1 + b_2,$ $b_2 = D^2 a_0 - 2Da_1 + a_2,$ etc.

11. If $y_i = \phi_{ij}(x)$ $(i, j = 1, 2, \ldots, n)$ is a fundamental system of solutions of a normal canonical homogeneous system, and $\Delta = |\phi_{ij}(x)|$, then a fundamental system of solutions of the adjoint system is $z_i = \theta_{ij}(x) = \dfrac{1}{\Delta} \dfrac{\partial \Delta}{\partial \phi_{ij}}$, and there is complete reciprocity between the two systems.

12. If $\{\phi_i(x)\}$ is a fundamental system of solutions of a homogeneous equation of order n, and $W \equiv W(\phi_1, \phi_2, \ldots, \phi_n)$, then $\theta_i(x) = \dfrac{1}{W} \dfrac{\partial W}{\partial \phi_i^{(n-1)}}$ are integrating factors. Verify the relations
$$\sum_{i=1}^{n} \theta_i(x) D^j \phi_i(x) = 0 \quad (j = 0, 1, \ldots, n-2), \quad \sum_{i=1}^{n} \theta_i(x) D^{n-1} \phi_i(x) = 1,$$
and by means of these and their derivatives prove that
$$W(\theta_1, \theta_2, \ldots, \theta_n) W(\phi_1, \phi_2, \ldots, \phi_n) = 1,$$
and hence that $\{\theta_i(x)\}$ are linearly independent.

13. Show how the order of a system or of an equation can be depressed by m, when m independent solutions of its adjoint are known.

14. In a self-adjoint normal system, $a_{ij}(x) + a_{ji}(x) \equiv 0$, and $\sum_{i=1}^{n} y_i^2 = $ constant, for all solutions.

15. Solve by means of an integrating factor
$$x(1-x^2) D^2 y + (2 - 5x^2) Dy - 4xy = 0.$$

16. If ρ, λ are constants, show that the equation
$$D^2 y + [\rho^2 - \lambda \psi(x)] y = 0,$$

with the initial conditions $y = A$, $Dy = B$, when $x = 0$, is equivalent to the integral equation
$$y = A\cos\rho x + \frac{B}{\rho}\sin\rho x + \frac{\lambda}{\rho}\int_0^x \sin\rho(x-t)\psi(t)y(t)\,dt.$$

If $\lambda\psi(x)$ is continuous and small compared with ρ in the interval $(0 \leqslant x \leqslant X)$, show that the equation can be satisfied by a series
$$y = u_0(x) + \lambda u_1(x) + \lambda^2 u_2(x) + \dots. \qquad \text{[LIOUVILLE.]}$$

17. **APPELL'S THEOREM.** If $\{\phi_i(x)\}$ is a fundamental system of solutions of a homogeneous equation, any polynomial in $\{D^j\phi_i(x)\}$ which is merely multiplied by a constant, when $\{\phi_i(x)\}$ is replaced by any other fundamental system, is of the form $[W(\phi_1,\phi_2,\dots,\phi_n)]^k P$, where P is a function of the coefficients of the equation and their derivatives.

[E. Picard, *Traité d'analyse*, **3**, 541; L. Schlesinger, *Handbuch*, **1**, 40.]

IV

EQUATIONS WITH UNIFORM ANALYTIC COEFFICIENTS

12. Group of the Equation

Analytical Continuation. Let us now suppose that the coefficients of the equation

$$(E_n^*) \qquad D^n y + p_1(x) D^{n-1} y + \ldots + p_n(x) y = 0$$

are uniform analytic functions, having only isolated singularities in the complex plane. By § 2, every solution $y = \phi(x)$ is analytic, and its only singularities in the finite part of the plane are at singularities of the coefficients; but $\phi(x)$ is *not* in general single-valued.

To examine this question, we draw any closed circuit Γ of finite length, beginning and ending at an ordinary point $x = \xi$, and not passing through any singularity. We can apply the process of analytical continuation simultaneously to the coefficients and to the solution $\phi(x)$, which will continue to satisfy (E_n^*) identically. For if we expand the solution and the coefficients in Taylor series of powers of $(x-\xi)$, convergent in a circle C, and if $x = \xi'$ is any point in C, we can rewrite $\phi(x)$ and $\{p_i(x)\}$ as Taylor series in powers of $(x-\xi')$, convergent in a circle C'. In the region common to the circles (C, C') the two sets of expansions take the same values at every point, so that the second set of power-series satisfy (E_n^*) identically; and they will continue to satisfy the equation in the entire circle of convergence C', which will in general extend beyond C.

After completing the circuit, the coefficients $\{p_i(x)\}$ resume their original forms as power-series in $(x-\xi)$, being by hypothesis uniform functions; but $\phi(x)$ need not resume its original form, though it will be *some* solution, $\Phi(x)$ say, of the equation. To trace these changes, the process of analytical continuation must be simultaneously applied to all the solutions of a fundamental system $\{\phi_i(x)\}$, which assume the new forms

$$\Phi_i(x) = \sum_{j=1}^{n} a_{ij} \phi_j(x) \quad (i = 1, 2, \ldots, n), \tag{1}$$

where (a_{ij}) are constants.

Determinant of the Transformation. The determinant $A \equiv |a_{ij}|$ cannot vanish; for this would imply a linear relation $\sum c_i \Phi_i(x) \equiv 0$ between the new forms; and, on returning along the same path, we should have $\sum c_i \phi_i(x) \equiv 0$, contrary to the hypothesis that $\{\phi_i(x)\}$ are linearly independent. The determinant A was evaluated by Poincaré. We obtain from (1) by differentiation

$$D^k \Phi_i(x) = \sum_{j=1}^{n} a_{ij} D^k \phi_j(x) \quad (i = 1, 2, ..., n), \qquad (2)$$

so that the derivatives of the set $\{\phi_i(x)\}$ undergo *cogredient* transformations. By the multiplication of determinants, we now have

$$W(\Phi_1, \Phi_2, ..., \Phi_n) = A W(\phi_1, \phi_2, ..., \phi_n). \qquad (3)$$

But Liouville's formula gives

$$W(x) \equiv W(\phi_1, \phi_2, ..., \phi_n) = W(\xi) \exp\left[-\int_{\xi}^{x} p_1(t)\, dt\right], \qquad (4)$$

and after a complete circuit $W(x)$ is multiplied by

$$A = \exp\left[-\int p_1(x)\, dx\right], \qquad (5)$$

the contour integral being taken once round Γ. If $p_1(x)$ is holomorphic inside the contour, we have $A = 1$.

Group of the Equation. Let Γ_1 and Γ_2 be two closed circuits starting from the same base-point $x = \xi$; and let (a_{ij}) and (b_{ij}) be the matrices of their respective transformations, for the same initial set of functions $\{\phi_i(x)\}$. If the functions are continued around the combined circuit in the order $\Gamma_1 \Gamma_2$, the matrix of the new transformation is the product (c_{ij}), where

$$c_{ij} = \sum_{k=1}^{n} a_{ik} b_{kj} \quad (i,j = 1, 2, ..., n). \qquad (6)$$

This product of matrices is not commutative, for we must take the *rows* of the matrix belonging to the first circuit with the *columns* of that of the second. All the linear transformations belonging to every possible closed circuit form a group, called the *group of the differential equation*.

If there are m winding points in the finite part of the plane, every closed circuit can be deformed without crossing a singularity into a sequence of standard loops, each starting from the base-point and encircling one singularity. The m transformations belonging to these loops are a set of generating operations of the group, in terms of which every matrix can be expressed.

Riemann's Converse Theorems. A set of n linearly independent analytic functions with isolated singularities, which admit a group of linear transformations with constant coefficients for all closed circuits, must be a fundamental system of solutions of a linear differential equation with uniform coefficients. For by virtue of (1) and (2), every n-rowed determinant of the array

$$\begin{Vmatrix} \phi_1, & D\phi_1, & ..., & D^n\phi_1 \\ \phi_2, & D\phi_2, & ..., & D^n\phi_2 \\ \cdot & \cdot & \cdot & \cdot \\ \phi_n, & D\phi_n, & ..., & D^n\phi_n \end{Vmatrix} \tag{7}$$

is multiplied by the same constant $A \neq 0$, after continuation about a closed circuit. Hence quotients of these determinants are uniform functions of x. But in the equation formally satisfied by the given functions

$$\frac{W(\phi_1, \phi_2, ..., \phi_n, y)}{W(\phi_1, \phi_2, ..., \phi_n)} = 0, \tag{8}$$

every coefficient is a quotient of this kind; so that the coefficients of the equation (8) are in fact uniform.

Every set of functions $\{\psi_i(x)\}$ which undergo cogredient transformations with the set $\{\phi_i(x)\}$ for all closed circuits can be written in the form

$$\psi_i(x) = u_0\phi_i(x) + u_1 D\phi_i(x) + ... + u_{n-1} D^{n-1}\phi_i(x) \quad (i = 1, 2, ..., n), \tag{9}$$

where (u_i) are uniform analytic functions. For since we assume that $\{\phi_i(x)\}$ are linearly independent, we have

$$W(\phi_1, \phi_2, ..., \phi_n) \neq 0, \tag{10}$$

except at isolated points. Hence (u_i) are uniquely determined by the equations (9), each of them being the quotient of two n-rowed determinants of the array

$$\begin{Vmatrix} \phi_1, & D\phi_1, & ..., & D^{n-1}\phi_1, & \psi_1 \\ \phi_2, & D\phi_2, & ..., & D^{n-1}\phi_2, & \psi_2 \\ \cdot & \cdot & \cdot & \cdot & \cdot \\ \phi_n, & D\phi_n, & ..., & D^{n-1}\phi_n, & \psi_n \end{Vmatrix} \tag{11}$$

By hypothesis, the elements of every column of this array are cogredient; and so each n-rowed determinant is multiplied by the same factor $A \neq 0$, corresponding to a given circuit; therefore the quotients (u_i) are uniform analytic functions.

13. Canonical Transformations

Characteristic Determinant. With a view to simplifying the transformation

$$\Phi_i(x) = \sum_{j=1}^{n} a_{ij}\phi_j(x) \quad (i = 1, 2, ..., n) \tag{1}$$

corresponding to any selected circuit Γ, we may look for solutions
$$y = c_1\phi_1(x)+c_2\phi_2(x)+\ldots+c_n\phi_n(x), \tag{2}$$
which are merely multiplied by a constant. Since we are to have
$$\left[\sum_{i=1}^{n} c_i\Phi_i(x)\right] = \lambda\left[\sum_{i=1}^{n} c_i\phi_i(x)\right], \tag{3}$$
we find from (1) and (3) the identity
$$\left[\sum_{i=1}^{n}\sum_{j=1}^{n} c_i a_{ij}\phi_j(x)\right] = \lambda\left[\sum_{j=1}^{n} c_j\phi_j(x)\right]; \tag{4}$$
and, because $\{\phi_j(x)\}$ are linearly independent, this implies that
$$\sum_{i=1}^{n} c_i a_{ij} = \lambda c_j \quad (j = 1, 2, \ldots, n). \tag{5}$$
If this system of equations for (c_j) is compatible, λ must satisfy the equation
$$\Delta_n(\lambda) \equiv \begin{vmatrix} a_{11}-\lambda, & a_{21}, & \ldots, & a_{n1} \\ a_{12}, & a_{22}-\lambda, & \ldots, & a_{n2} \\ \cdot & \cdot & \cdot & \cdot \\ a_{1n}, & a_{2n}, & \ldots, & a_{nn}-\lambda \end{vmatrix} = 0, \tag{6}$$
which is called the *characteristic equation* of the matrix (a_{ij}).

Evidently no root can be zero; for we saw in § 12 that the determinant $A = |a_{ij}| \neq 0$ in the present problem. Each root gives at least one linear form (2), and k such forms $(y_1, y_2, \ldots, y_k)$ belonging respectively to unequal roots $(\lambda_1, \lambda_2, \ldots, \lambda_k)$ are linearly independent. For suppose they were connected by a linear relation
$$c_1 y_1 + c_2 y_2 + \ldots + c_k y_k \equiv 0. \tag{7}$$
After s complete circuits this relation is transformed into
$$c_1 \lambda_1^s y_1 + c_2 \lambda_2^s y_2 + \ldots + c_k \lambda_k^s y_k \equiv 0 \quad (s = 1, 2, \ldots, k-1). \tag{8}$$
But since (λ_i) are unequal, we have
$$\begin{vmatrix} 1, & 1, & \ldots, & 1 \\ \lambda_1, & \lambda_2, & \ldots, & \lambda_k \\ \cdot & \cdot & \cdot & \cdot \\ \lambda_1^{k-1}, & \lambda_2^{k-1}, & \ldots, & \lambda_k^{k-1} \end{vmatrix} \neq 0, \tag{9}$$
so that the relations are incompatible and so the coefficients (c_i) are all zero.

Accordingly, when the characteristic equation has n distinct roots we get n linearly independent combinations of $\{\phi_i(x)\}$, forming a new fundamental system, which undergoes the *canonical transformation*

$$Y_i = \lambda_i y_i \quad (i = 1, 2, \ldots, n). \tag{10}$$

Invariants. The numbers (λ_i) must be independent of the choice of the original fundamental system $\{\phi_i(x)\}$. For a canonical solution y_j has the same multiplier λ_j, whether it is expressed in terms of $\{\phi_i(x)\}$ or of some other fundamental system $\{\psi_i(x)\}$, which undergoes the transformation

$$\Psi_i(x) = \sum_{j=1}^{n} b_{ij} \psi_j(x) \quad (i = 1, 2, \ldots, n). \tag{11}$$

Hence λ_j must be a root of the characteristic equation of the matrix (b_{ij}) also.

To prove this algebraically, let us suppose that the two systems are connected by the relations

$$\psi_i(x) = \sum_{j=1}^{n} c_{ij} \phi_j(x) \quad (i = 1, 2, \ldots, n), \tag{12}$$

whose determinant $C = |c_{ij}| \neq 0$. In terms of the first system, (11) may be written

$$\sum_{j=1}^{n} c_{ij} \Phi_j(x) = \sum_{j=1}^{n} \sum_{k=1}^{n} b_{ij} c_{jk} \phi_k(x), \tag{13}$$

and by (1) we have

$$\sum_{j=1}^{n} \sum_{k=1}^{n} c_{ij} a_{jk} \phi_k(x) = \sum_{j=1}^{n} \sum_{k=1}^{n} b_{ij} c_{jk} \phi_k(x). \tag{14}$$

These n relations between the independent functions $\{\phi_k(x)\}$ must be identically satisfied, and so

$$\sum_{j=1}^{n} c_{ij} a_{jk} = \sum_{j=1}^{n} b_{ij} c_{jk} \quad (i, k = 1, 2, \ldots, n). \tag{15}$$

Thus the matrices (a_{ij}), (b_{ij}), (c_{ij}) are connected by the relations

$$(c_{ij})(a_{ij}) = (b_{ij})(c_{ij}). \tag{16}$$

Let us write the two characteristic determinants in the forms

$$|u_{ij}| \equiv |a_{ij} - \lambda \delta_{ij}|, \tag{17}$$

$$|v_{ij}| \equiv |b_{ij} - \lambda \delta_{ij}|, \tag{18}$$

with the usual notation for the Kronecker deltas. Using (15), we write

$$w_{ik} \equiv \sum_{j=1}^{n} c_{ij} u_{jk},$$

$$= \sum_{j=1}^{n} c_{ij} a_{jk} - \lambda c_{ik},$$

$$= \sum_{j=1}^{n} b_{ij} c_{jk} - \lambda c_{ik},$$

$$= \sum_{j=1}^{n} v_{ij} c_{jk}. \tag{19}$$

We thus have $\quad (w_{ij}) = (c_{ij})(u_{ij}) = (v_{ij})(c_{ij}). \tag{20}$

Now the determinant of the product of two matrices is equal to the product of their determinants; so we have

$$|w_{ij}| = |c_{ij}|.|u_{ij}| = |v_{ij}|.|c_{ij}|, \quad \text{where } |c_{ij}| \neq 0; \tag{21}$$

hence $\quad |a_{ij} - \lambda \delta_{ij}| = \Delta_n(\lambda) = |b_{ij} - \lambda \delta_{ij}|. \tag{22}$

Again, each s-rowed minor of $|w_{ij}|$ is a linear homogeneous function of the s-rowed minors of $|u_{ij}|$, or of $|v_{ij}|$, and vice versa. For instance,

$$\begin{vmatrix} w_{11}, & w_{12} \\ w_{21}, & w_{22} \end{vmatrix} = \begin{vmatrix} \sum_{i=1}^{n} c_{1i} u_{i1}, & \sum_{j=1}^{n} c_{1j} u_{j2} \\ \sum_{i=1}^{n} c_{2i} u_{i1}, & \sum_{j=1}^{n} c_{2j} u_{j2} \end{vmatrix},$$

$$= \sum_{i=1}^{n} \sum_{j=1}^{n} \begin{vmatrix} c_{1i}, & c_{1j} \\ c_{2i}, & c_{2j} \end{vmatrix} u_{i1} u_{j2},$$

$$= \frac{1}{2!} \sum_{i=1}^{n} \sum_{j=1}^{n} \begin{vmatrix} c_{1i}, & c_{1j} \\ c_{2i}, & c_{2j} \end{vmatrix} . \begin{vmatrix} u_{i1}, & u_{i2} \\ u_{j1}, & u_{j2} \end{vmatrix}, \tag{23}$$

and the method of proof is general. The converse follows from the relations

$$(u_{ij}) = (c_{ij})^{-1}(w_{ij}) = (c'_{ij})(w_{ij}), \quad (v_{ij}) = (w_{ij})(c_{ij})^{-1} = (w_{ij})(c'_{ij}), \tag{24}$$

where (c'_{ij}) is the inverse matrix. The elements of a *row* in (c'_{ij}) are proportional to the first minors of those of a *column* in (c_{ij}); and we have the relations

$$\left. \begin{array}{c} c'_{ij} = \dfrac{1}{C} \dfrac{\partial C}{\partial c_{ji}}, \quad \text{where } C \equiv |c_{ij}| \neq 0, \\[1em] \displaystyle\sum_{j=1}^{n} c_{ij} c'_{jk} = \delta_{ik} = \sum_{j=1}^{n} c'_{ij} c_{jk}. \end{array} \right\} \tag{25}$$

From the relations (23), the H.C.F. of all s-rowed minors of $\Delta_n(\lambda) \equiv |u_{ij}|$ divides every s-rowed minor of $|w_{ij}|$; hence the H.C.F.

of all s-rowed minors of $|w_{ij}|$ is also divisible by that of all s-rowed minors of $|u_{ij}|$. But, from the converse relations (24), we see that the H.C.F. of all s-rowed minors of $|u_{ij}|$ is divisible by the H.C.F. of all s-rowed minors of $|w_{ij}|$. Hence the two expressions are identical. But a similar connexion holds between $|v_{ij}|$ and $|w_{ij}|$. If therefore $\Delta_s(\lambda)$ is the H.C.F. of all s-rowed minors of the characteristic determinant $\tilde{\Delta}_n(\lambda)$, it has the same form, whether it is calculated from $|a_{ij}-\lambda\delta_{ij}|$ or from $|b_{ij}-\lambda\delta_{ij}|$.

It is obvious that $\Delta_s(\lambda)$ must be divisible by $\Delta_{s-1}(\lambda)$, the corresponding H.C.F. of the $(s-1)$-rowed minors. We have already met with the rational *invariant factors*

$$E_s(\lambda) = \Delta_s(\lambda)/\Delta_{s-1}(\lambda) \quad (s = 1, 2, ..., n), \qquad (26)$$

in connexion with systems of simultaneous equations with constant coefficients. These can be found by elementary algebraic operations without solving the characteristic equation. If, however, the roots of $\Delta_n(\lambda)$ are known, and we write

$$E_s(\lambda) \equiv (\lambda-\lambda_s)^{e_s}(\lambda-\lambda_s')^{e_s'}..., \qquad (27)$$

where $(\lambda_s, \lambda_s', ...)$ are unequal, the irrational invariants $\{(\lambda-\lambda_s)^{e_s}\}$ are called *elementary divisors* of $\Delta_n(\lambda)$.

Following a less complete reduction by Jacobi, Jordan reduced the most general linear transformation to a canonical form, which is the analogue of Weierstrass's reduction of a pair of quadratic forms. The canonical variables are divided into sets; each set is transformed independently of the others, and corresponds to one elementary divisor of $\Delta_n(\lambda)$.

Jordan's Canonical Form. Every linear transformation (whose determinant is not zero) is equivalent to one or more mutually independent ones of the type

$$U_1 = \lambda_1 u_1, \ U_2 = \lambda_1(u_1+u_2), \ U_3 = \lambda_1(u_2+u_3), \ ..., \ U_\alpha = \lambda_1(u_{\alpha-1}+u_\alpha), \qquad (28)$$

where λ_1 is a root of the characteristic equation and $(\lambda-\lambda_1)^\alpha$ an elementary divisor.

If there is only one variable, we must have $Y_1 = \lambda_1 y_1$ as the only possible type of transformation, and the theorem is true. Assuming its truth for n variables, we will prove it for $(n+1)$. Consider the transformation

$$Y_i = \sum_{j=1}^{n+1} a_{ij}y_j \quad (i = 1, 2, ..., n+1), \qquad (29)$$

and let λ' be any root of $\Delta_n(\lambda) = 0$. We can construct a linear form $z \equiv \sum c_i y_i$, such that $Z = \lambda' z$. Without loss of generality, let us suppose that $c_{n+1} \neq 0$. We may then eliminate y_{n+1} and obtain an equivalent transformation in the independent variables $(z, y_1, y_2, ..., y_n)$

$$\left. \begin{array}{l} Z = \lambda' z, \\ Y_i = \sum_{j=1}^{n} b_{ij} y_j + p_i z \quad (i = 1, 2, ..., n). \end{array} \right\} \quad (30)$$

If we ignore z and apply the theorem, assumed true for n variables, to the (y_i), we obtain a number of sets of the type

$$\left. \begin{array}{l} U_1 = \lambda_1 u_1 + a_1 z, \\ U_2 = \lambda_1 (u_1 + u_2) + a_2 z, \\ \cdots \cdots \cdots \cdots \cdots \\ U_\alpha = \lambda_1 (u_{\alpha-1} + u_\alpha) + a_\alpha z. \end{array} \right\} \quad (31)$$

Now let
$$u'_i \equiv u_i + A_i z \quad (i = 1, 2, ..., \alpha), \quad (32)$$
so that

$$\left. \begin{array}{l} U'_1 = \lambda_1 u'_1 + [a_1 + A_1(\lambda' - \lambda_1)] z, \\ U'_2 = \lambda_1 (u'_1 + u'_2) + [a_2 - A_1 \lambda_1 + A_2(\lambda' - \lambda_1)] z, \\ \cdots \cdots \cdots \cdots \cdots \cdots \cdots \cdots \cdots \\ U'_\alpha = \lambda_1 (u'_{\alpha-1} + u'_\alpha) + [a_\alpha - A_{\alpha-1} \lambda_1 + A_\alpha(\lambda' - \lambda_1)] z. \end{array} \right\} \quad (33)$$

(i) If $\lambda' \neq \lambda_1$, we can choose $(A_1, A_2, ..., A_\alpha)$ in turn so that z shall disappear from these formulae (33). The set (U'_i) is then of the canonical type.

(ii) If $\lambda' = \lambda_1$, A_α does not appear in (33); but since, of course, $\lambda' \neq 0$, we can choose $(A_1, A_2, ..., A_{\alpha-1})$ so that z appears *only once*, in the first equation of the set, whose form is now

$$\left. \begin{array}{l} U'_1 = \lambda' u'_1 + a_1 z, \\ U'_2 = \lambda' (u'_1 + u'_2), \\ \cdots \cdots \cdots \cdots \\ U'_\alpha = \lambda' (u'_{\alpha-1} + u'_{\alpha-2}). \end{array} \right\} \quad (34)$$

(iii) If, by chance, $a_1 = 0$, the set (34) is canonical and no further reduction is needed.

(iv) If there is just one set with the multiplier λ' and $a_1 \neq 0$, we put $a_1 z \equiv \lambda' u'_0$, and adjoin u'_0 to the set, which is now canonical in $(\alpha + 1)$ variables

$$U'_0 = \lambda' u'_0, \quad U'_1 = \lambda'(u'_0 + u'_1), \quad ..., \quad U'_\alpha = \lambda'(u'_{\alpha-1} + u'_\alpha). \quad (35)$$

(v) If there are several sets with the multiplier λ', where the coefficient corresponding to a_1 is not zero, we select the one with the *greatest* number of variables and reduce it to the canonical form (35). If $\beta \leqslant \alpha$ and another such set is (after eliminating z) of the form

$$V'_1 = \lambda'(v'_1 + b_1 u'_0), \quad V'_2 = \lambda'(v'_1 + v'_2), \quad ..., \quad V'_\beta = \lambda'(v'_{\beta-1} + v'_\beta), \quad (36)$$

we have only to write

$$v''_i \equiv v'_i - b_1 u'_i \quad (i = 1, 2, ..., \beta) \quad (37)$$

to obtain the canonical set of β variables

$$V''_1 = \lambda' v''_1, \quad V''_2 = \lambda'(v''_1 + v''_2), \quad ..., \quad V''_\beta = \lambda'(v''_{\beta-1} + v''_\beta). \quad (38)$$

We have thus completed the reduction for $(n+1)$ variables, and the theorem follows by induction.

14. Hamburger Sets of Solutions

Euler's Homogeneous Equation. The interchange of solutions around a circuit is well illustrated by Euler's equation

$$F(\delta)y \equiv \left[\prod_{r=1}^{n}(\delta - \rho_r)\right]y = 0 \quad (\delta \equiv xD), \quad (1)$$

which has an isolated singularity at $x = 0$, and which is soluble by elementary methods. If all the (ρ_r) are unequal, a fundamental system of solutions is given by $y_r \equiv x^{\rho_r}$ $(r = 1, 2, ..., n)$. After a simple positive circuit about the origin, these undergo the canonical transformation

$$Y_r = \lambda_r y_r \quad \{\lambda_r = \exp(2i\pi\rho_r)\}. \quad (2)$$

If $F(\delta)$ has a multiple linear factor, $(\delta - \rho_1)^{n_1}$ say, the corresponding solutions are known to be

$$x^{\rho_1}, x^{\rho_1}\log x, ..., x^{\rho_1}(\log x)^{n_1 - 1}. \quad (3)$$

These solutions are linearly transformed among themselves, but the transformation is not of Jordan's canonical type. But we can easily write down an equivalent system which does undergo a canonical transformation. For this purpose let us introduce the notations

$$\left. \begin{array}{c} L \equiv \dfrac{\log x}{2i\pi}, \\ L_0 \equiv 1, \; L_1 \equiv \dfrac{L}{1!}, \; L_2 \equiv \dfrac{L(L-1)}{2!}, \; ..., \; L_s \equiv \dfrac{L(L-1)...(L-s+1)}{s!}. \end{array} \right\} \quad (4)$$

After the circuit, L becomes $(L+1)$ and the polynomials L_s, whose form is suggested by the calculus of finite differences, become

$$(L+1)_s = L_{s-1} + L_s, \quad (5)$$

as is immediately verified. Since $(L_0, L_1, ..., L_s)$ are respectively of

degree $(0, 1,...,s)$ in L, they are linearly independent; and the set of solutions (3) may be replaced by the equivalent set
$$y_s \equiv x^{\rho_1} L_{s-1} \quad (s = 1, 2,..., n_1). \tag{6}$$
These undergo the canonical Jordan transformation
$$Y_1 = \lambda_1 y_1, \quad Y_2 = \lambda_1(y_1+y_2), \quad ..., \quad Y_{n_1} = \lambda_1(y_{n_1-1}+y_{n_1}). \tag{7}$$
Such a set of solutions is called a *Hamburger set*.

Solutions at an Isolated Singularity. Let $x = 0$ be a typical isolated singular point of a linear differential equation, whose coefficients are uniform analytic functions; and let Γ be a simple positive circuit enclosing only this singular point. We shall suppose that the corresponding linear transformation of the solutions has been reduced to the canonical form, and that one of the canonical sets undergoes the transformation (7).

Let ρ_1 be any finite root of the equation $\exp(2i\pi\rho_1) = \lambda_1$, and let us put $y_r = x^{\rho_1} z_r$, $(r = 1, 2,..., n_1)$. Then the corresponding transformation of the (z_r) takes the form
$$Z_1 = z_1, \quad Z_2 = z_1+z_2, \quad ..., \quad Z_{n_1} = z_{n_1-1}+z_{n_1}. \tag{8}$$

Now we can write down n_1 sets of functions which are transformed in this manner, namely
$$\left.\begin{array}{ccccc} L_0, & L_1, & L_2, & ..., & L_{n_1-1} \\ 0, & L_0, & L_1, & ..., & L_{n_1-2} \\ \cdot & \cdot & \cdot & \cdot & \cdot \\ 0, & 0, & 0, & ..., & L_0. \end{array}\right\} \tag{9}$$

Since the determinant of this system is $L_0^{n_1} = 1$, we can find n_1 functions (w_r) satisfying the equations
$$\left.\begin{array}{l} z_1 = L_0 w_1, \\ z_2 = L_1 w_1 + L_0 w_2, \\ \cdot \cdot \cdot \cdot \cdot \cdot \cdot \cdot \\ z_{n_1} = L_{n_1-1} w_1 + ... + L_0 w_{n_1}. \end{array}\right\} \tag{10}$$

If we substitute these forms in (8) and use (5), we find the relations
$$\left.\begin{array}{l} L_0 W_1 = L_0 w_1, \\ L_1 W_1 + L_0 W_2 = L_1 w_1 + L_0 w_2, \\ \cdot \cdot \cdot \cdot \cdot \cdot \cdot \cdot \cdot \cdot \cdot \cdot \\ L_{n_1-1} W_1 + ... + L_0 W_{n_1} = L_{n_1-1} w_1 + ... + L_0 w_{n_1}, \end{array}\right\} \tag{11}$$
which are equivalent to
$$W_r = w_r \quad (r = 1, 2,..., n_1), \tag{12}$$

ANALYTIC COEFFICIENTS

and imply that (w_r) are single-valued in the neighbourhood of $x = 0$. They can be expanded as Laurent series

$$w_r \equiv \Pi_r(x) = \sum_{s=-\infty}^{\infty} c_{rs} x^s \quad (r = 1, 2, ..., n_1) \tag{13}$$

converging in a ring $(0 < \epsilon < |x| < R)$, whose inner radius is as small as we please.† The solutions of a canonical Hamburger set can therefore always be expressed in the form

$$y_r = x^{\rho_1} \cdot \sum_{s=1}^{n_1} L_{n_1-s} \Pi_s(x) \quad (r = 1, 2, ..., n_1), \tag{14}$$

where $\{\Pi_s(x)\}$ are locally uniform near the isolated singular point $x = 0$.

Regular Solutions. It may happen that each of the Laurent series has only a finite number of negative powers of x, which can be removed by an adjustment of the value chosen for ρ_1; the auxiliary functions $\{\Pi_r(x)\}$ will then be holomorphic. A solution of this type is said to be *regular*, and a singular point where *all* the solutions are of this type is called a *regular singularity*.

15. Fuchs's Conditions for a Regular Singularity

THEOREM. *The necessary and sufficient conditions that an isolated singularity, say $x = 0$, of the homogeneous equation (E_n^*) should be regular are that it should be at most a pole of order r of the coefficient $p_r(x)$ of $D^{n-r}y$, in the canonical form where the highest coefficient is $p_0(x) \equiv 1$.*

Following Thomé, we proceed by induction. If the equation of the first order

$$Dy + p_1(x)y = 0 \tag{1}$$

has the regular solution

$$y = \phi_1(x) = A x^{\rho_1}[1 + c_1 x + c_2 x^2 + ...], \tag{2}$$

where $\sum c_n x^n$ converges when $|x| \leqslant R_1$, say, then we have

$$p_1(x) = -\phi_1'(x)/\phi_1(x)$$

$$= -\frac{\rho_1}{x} - \left[\frac{c_1 + 2c_2 x + 3c_3 x^2 + ...}{1 + c_1 x + c_2 x^2 + ...}\right]$$

$$= -\frac{\rho_1}{x} - b_0 - b_1 x - b_2 x^2 ..., \tag{3}$$

---

† The same method can be applied if the coefficients of the differential equation are single-valued and holomorphic in a ring $(r < |x| < R)$ whose inner boundary is finite and encloses several singularities, and if Γ is a circuit enclosing the inner boundary.

where $\sum b_n x^n$ converges in some circle $|x| \leqslant R_2$, where $0 < R_2 < R_1$. Hence $p_1(x)$ has at most a pole of the first order.

Conversely, if the coefficient $p_1(x)$ of (1) is of the form (3) the solution $\exp[-\int p_1(x)\,dx]$ is regular and has an expansion of the form (2). Thus Fuchs's theorem holds for equations of the first order.

Now the general equation with uniform coefficients (E_n^*) has at least one solution, which is merely multiplied by a constant after a circuit about the origin. Let this be chosen as the first solution of a certain fundamental system, which undergoes the transformation

$$\left.\begin{aligned}\Phi_1(x) &= \lambda_1 \phi_1(x), \\ \Phi_i(x) &= \sum_{j=1}^{n} a_{ij}\phi_j(x) \quad (i=2,3,...,n),\end{aligned}\right\} \quad (4)$$

whose determinant is not zero. Then the expressions

$$\psi_i(x) \equiv D\left[\frac{\phi_i(x)}{\phi_1(x)}\right] \quad (i=2,3,...,n) \quad (5)$$

undergo a corresponding transformation with a determinant not equal to zero:

$$\Psi_i(x) = \sum_{j=2}^{n} \frac{a_{ij}}{\lambda_1}\psi_j(x) \quad (i=2,3,...,n). \quad (6)$$

They therefore satisfy an equation of order $(n-1)$, whose coefficients are uniform at $x = 0$,

$$\frac{W(\psi_2,\psi_3,...,\psi_n,z)}{W(\psi_2,\psi_3,...,\psi_n)} = 0, \quad (7)$$

or $\qquad D^{n-1}z + q_1(x)D^{n-2}z + ... + q_{n-1}(x)z = 0. \quad (8)$

We can pass from the given equation (E_n^*) to (8) by putting

$$y = \phi_1(x)\int z\,dx, \quad (9)$$

and the coefficients are connected by the relations

$$\left.\begin{aligned}q_1(x) &= \binom{n}{1}\frac{D\phi_1}{\phi_1} + p_1(x), \\ q_2(x) &= \binom{n}{2}\frac{D^2\phi_1}{\phi_1} + \binom{n-1}{1}p_1(x)\frac{D\phi_1}{\phi_1} + p_2(x), \\ &\cdot\cdot\cdot\cdot\cdot\cdot\cdot\cdot\cdot\cdot\cdot\cdot\cdot\cdot\cdot\cdot\cdot\cdot\cdot\cdot\cdot\cdot\cdot \\ q_{n-1}(x) &= \binom{n}{n-1}\frac{D^{n-1}\phi_1}{\phi_1} + \binom{n-1}{n-2}p_1(x)\frac{D^{n-2}\phi_1}{\phi_1} + ... + p_{n-1}(x),\end{aligned}\right\} \quad (10)$$

to which we may add the identity

$$0 = \frac{D^n\phi_1}{\phi_1} + p_1(x)\frac{D^{n-1}\phi_1}{\phi_1} + \ldots + p_n(x), \tag{11}$$

expressing that (E_n^*) is satisfied by $\phi_1(x)$.

First, suppose that the solutions $\{\phi_r(x)\}$ are all regular. Then the expansion of the leading solution $\phi_1(x)$ is of the form (2); its reciprocal has a regular expansion

$$1/\phi_1(x) = x^{-\rho_1}[c_0' + c_1'x + c_2'x^2 + \ldots], \tag{12}$$

and so $(D^r\phi_1)/\phi_1$ has at most a pole of order r at the origin. Moreover, the $(n-1)$ functions $\{\psi_r(x)\}$ will also be regular; if the conditions are assumed to be *necessary* for equations of order $(n-1)$, the coefficients $\{q_r(x)\}$ of (8) will have respectively poles of order r at most at $x = 0$. The relations (10) and (11) taken in turn then show that $\{p_r(x)\}$ will have respectively poles of order r at most at $x = 0$. Thus the conditions will also be *necessary* for equations of order n. But we know them to be necessary for equations of the first order, and so it follows by induction that they are *necessary* in general.

Conversely, suppose that $\{p_r(x)\}$ have respectively poles of order r at most at $x = 0$. We shall see in the next chapter how to construct a fundamental system of n regular solutions; but let us assume provisionally that there is *one* regular solution $\phi_1(x)$ of the form (2). Then the relations (10) show that $\{q_r(x)\}$ have respectively poles of order r at most at $x = 0$. If the conditions are assumed to be *sufficient* for equations of order $(n-1)$, all the solutions $\{\psi_r(x)\}$ of (8) will be regular. By (9), we then see that all the solutions $\{\phi_r(x)\}$ of (E_n^*) will be regular; hence the conditions will also be *sufficient* for equations of order n. But we know them to be sufficient for equations of the first order, and so it follows by induction that they are *sufficient* in general.

Alternative Method.† A direct proof of the sufficiency of the conditions has been given by Birkhoff, on the lines of an investigation by Liapounoff. If the conditions are satisfied, we may write the equation in the *normal form*

$$x^n D^n y + x^{n-1} P_1(x) D^{n-1} y + \ldots + P_n(x) y = 0, \tag{13}$$

where $\{P_r(x)\}$ are holomorphic at $x = 0$. If we put

$$y_i \equiv x^{i-1} D^{i-1} y \quad (i = 1, 2, \ldots, n), \tag{14}$$

---

† G. D. Birkhoff, *Trans. American Math. Soc.* **11** (1910), 199–202.

this equation is equivalent to the system
$$xDy_i = (i-1)y_i + y_{i+1} \quad (i = 1, 2, ..., n-1),$$
$$xDy_n = -P_n y_1 - P_{n-1} y_2 - ... - (P_1 + 1 - n)y_n. \quad \biggr\} \quad (15)$$
This is a system of the type
$$xDy_i = \sum_{j=1}^{n} A_{ij}(x) y_j \quad (i = 1, 2, ..., n), \quad (16)$$
where $\{A_{ij}(x)\}$ are holomorphic and admit an upper bound $|A_{ij}(x)| < M$ in the circle $|x| \leqslant R$ (say).

Now the equation (13) has at least one solution, which is merely multiplied by a constant after a circuit about the origin; and this can be written in the form $x^{\rho_1} \Pi(x)$, where $\Pi(x) = \sum_{m=-\infty}^{\infty} c_m x^m$ is a Laurent series convergent in a ring ($\epsilon \leqslant |x| \leqslant R$), whose inner radius may be as small as we please. The system (15) will have a corresponding solution $y_i = x^{\rho_1} \Pi_i(x)$ $(i = 1, 2, ..., n)$. On the circumference $|x| = r_0$ of a circle lying within the ring, $\Pi(x)$ or $\{\Pi_i(x)\}$ are single-valued and bounded. If we can find a real number κ such that $|\Pi_i(x)| = O(|x|^{-\kappa})$ as $x \to 0$ along a fixed radius vector, we can show that $c_m = 0$, if $m < -\kappa$. For if Γ is the inner boundary of the ring, we have by Cauchy's integral
$$|c_m| = \left| \frac{1}{2i\pi} \int_\Gamma \Pi(x) x^{-m-1} \, dx \right| = O(\epsilon^{-\kappa-m}); \quad (17)$$
and, if $(\kappa+m)$ is negative, this tends to zero as ϵ is made arbitrarily small.

Since (y_i) are analytic functions of x, or of $\log x$, we have $x \dfrac{dy_i}{dx} \equiv r \dfrac{\partial y_i}{\partial r}$, where $r = |x|$. Hence (16) gives
$$\left| r \frac{\partial y_i}{\partial r} \right| = \left| \sum_{j=1}^{n} A_{ij}(x) y_j \right| \leqslant M \sum_{j=1}^{n} |y_j|. \quad (18)$$
If $(\bar{y}_i)$ are complex numbers conjugate to (y_i), let
$$S \equiv \sum_{i=1}^{n} |y_i|^2 = \sum_{i=1}^{n} y_i \bar{y}_i;$$
then we have from (18)
$$\left| r \frac{\partial S}{\partial r} \right| = \left| \sum_{i=1}^{n} \left(r \frac{\partial y_i}{\partial r} \bar{y}_i + y_i \, r \frac{\partial \bar{y}_i}{\partial r} \right) \right|,$$
$$\leqslant 2M \sum_{i=1}^{n} \sum_{j=1}^{n} |y_i| \cdot |y_j|,$$

$$\leqslant M \sum_{i=1}^{n} \sum_{j=1}^{n} \{|y_i|^2 + |y_j|^2\}$$
$$\leqslant 2nMS. \tag{19}$$

Along a fixed radius vector ($r_0 > r > 0$), we have by integration

$$-2nM \leqslant \frac{\log S(r) - \log S(r_0)}{\log r - \log r_0} \leqslant 2nM, \tag{20}$$

and so, as $r \to 0$,
$$S(r) = O(r^{-2nM}), \tag{21}$$

provided that $S(r_0)$ is bounded. Now, if ρ_1 is complex, the factor x^{ρ_1} can become infinitely large as am(x) increases or decreases indefinitely. But, if am(x) is bounded, $S(r_0)$ remains bounded; we then have, *a fortiori*, $|y_1| = O(r^{-nM})$ as $r \to 0$. After removing the factor x^{ρ_1}, we can choose κ so that we have uniformly

$$\Pi(x), \Pi_i(x) = O(r^{-\kappa}) \quad (|x| = r \to 0). \tag{22}$$

It then follows from (17) that the number of negative powers of x in these Laurent series is limited. The particular solution $\phi_1(x)$ is thus regular and the proof may be completed by induction as before.

EXAMPLES. IV

1. JACOBI'S FORM. Show that every linear transformation is equivalent to one of the form

$$Y_1 = \lambda_1 y_1,$$
$$Y_2 = b_{21} y_1 + \lambda_2 y_2,$$
$$\cdots \cdots \cdots \cdots \cdots$$
$$Y_n = b_{n1} y_1 + b_{n2} y_2 + \ldots + b_{n,n-1} y_{n-1} + \lambda_n y_n.$$

2. ELEMENTARY DIVISORS. (i) If $\Delta_s(\lambda)$ is the H.C.F. of all s-rowed minors of $\Delta_n(\lambda)$, show by differentiation that every root of $\Delta_{n-1}(\lambda) = 0$ is a root of higher multiplicity of $\Delta_n(\lambda) = 0$.

(ii) If $\Delta_s(\lambda_1) = 0$, $\Delta_{s-1}(\lambda_1) \neq 0$, then $\lambda = \lambda_1$ is a root of multiplicity $(n-s+1)$ at least of $\Delta_n(\lambda) = 0$; and the system $\sum_{i=1}^{n}(a_{ij} - \lambda_1 \delta_{ij})c_i = 0$ $(j = 1, 2, \ldots, n)$ has exactly $(n-s+1)$ linearly independent solutions $(c_i^{(k)})$.

3. Show that the elementary divisors of the canonical characteristic determinant

$$\Delta_\alpha(\lambda) \equiv \begin{vmatrix} \lambda_1 - \lambda, & \lambda_1, & 0, & 0, & \ldots, & 0 \\ 0, & \lambda_1 - \lambda, & \lambda_1, & 0, & \ldots, & \\ \cdot & \cdot & \cdot & \cdot & \cdot & \cdot \\ 0, & 0, & 0, & 0, & \ldots, & \lambda_1 \\ 0, & 0, & 0, & 0, & \ldots, & \lambda_1 - \lambda \end{vmatrix}$$

are $[(\lambda - \lambda_1)^\alpha, 1, 1, 1, \ldots, 1]$.

4. If the characteristic determinant is of the form

$$\Delta_n(\lambda) \equiv \begin{vmatrix} A & * & * & . & . & . \\ * & B & * & . & . & . \\ * & * & C & . & . & . \\ . & . & . & . & . & . \end{vmatrix},$$

where $(A, B, C,...)$ are canonical subdeterminants belonging to the elementary divisors $\{(\lambda-\lambda_1)^\alpha, (\lambda-\lambda_2)^\beta, (\lambda-\lambda_3)^\gamma,...\}$ and where the asterisks represent blocks of zeros, show that:

(i) There is a first minor $\pm\lambda_1^{\alpha-1}\Delta_n(\lambda)/(\lambda-\lambda_1)^\alpha$,

a second minor $\pm\lambda_1^{\alpha-1}\lambda_2^{\beta-1}\Delta_n(\lambda)/(\lambda-\lambda_1)^\alpha(\lambda-\lambda_2)^\beta$, etc.

(ii) If $(\lambda-\lambda_1)^\alpha$ is the only elementary divisor belonging to the root $\lambda = \lambda_1$, it is a factor in $E_n(\lambda)$.

(iii) If $(\lambda-\lambda_1)^\alpha$, $(\lambda-\lambda_1)^{\alpha'}$, $(\lambda-\lambda_1)^{\alpha''},...$, all belong to the same root and $\alpha \geqslant \alpha' \geqslant \alpha'' \geqslant ...$, then $(\lambda-\lambda_1)^\alpha$ is a factor in $E_n(\lambda)$, $(\lambda-\lambda_1)^{\alpha'}$ is a factor in $E_{n-1}(\lambda)$, $(\lambda-\lambda_1)^{\alpha''}$ is a factor in $E_{n-2}(\lambda)$, etc.

(iv) Show that $E_n(\lambda)$ is divisible by $E_{n-1}(\lambda)$, $E_{n-1}(\lambda)$ by $E_{n-2}(\lambda)$, etc.

[Cf. M. Bôcher, *Higher Algebra*, 262–78; H. Hilton, *Linear Substitutions*, 1–33, 50–9; H. W. Turnbull and A. C. Aitken, *Canonical Matrices*, 64–73.]

5. TWO LIMITING CASES. (i) An analytic function has singular points at $x = a_1, a_2,..., a_m, \infty$, and it is merely multiplied by a constant after any closed circuit. Show that its logarithmic derivative is a uniform function, and that it is of the form

$$f(x) \equiv (x-a_1)^{\rho_1}(x-a_2)^{\rho_2}...(x-a_m)^{\rho_m}\Pi(x),$$

where $\Pi(x)$ is uniform.

(ii) An analytic function has singular points at $x = 0$, ∞ only, and has n linearly independent branches undergoing an arbitrarily assigned linear transformation after a circuit about $x = 0$. Show that its branches are cogredient with those of an equation of Euler's type $F(\delta)y = 0$ ($\delta \equiv xD$).

6. By repeated use of Fuchs's substitution, show that a fundamental system $\{\phi_r(x)\}$ at a regular singularity can be written in the form

$$\phi_r(x) = \alpha_1(x)\int^x \alpha_2(x_1)\,dx_1\int^{x_1}...\int^{x_{r-2}}\alpha_r(x_{r-1})\,dx_{r-1},$$

where $\{\alpha_i(x)\}$ are regular and free from logarithms. Discuss how logarithms make their appearance in the solutions.

7. LINEARLY INDEPENDENT SOLUTIONS. If (λ_p) are unequal, if A_{pq} are constants, and if the expression

$$\phi(x) \equiv \sum_{p=1}^{m}\sum_{q=0}^{n} A_{pq}\lambda_p^x x^q$$

vanishes when $x = 1, 2, 3, ...$, show that

$$A_{pq} = 0 \quad (p = 1, 2, ..., m;\ q = 0, 1, ..., n).$$

8. If (ρ_p) do not differ by integers, and if $\{\Pi_{pq}(x)\}$ are single-valued near the origin, show that if the expression
$$F(x) \equiv \sum_{p=1}^{m} \sum_{q=0}^{n} x^{\rho_p}(\log x)^q \Pi_{pq}(x)$$
vanishes identically, then $\{\Pi_{pq}(x)\}$ all vanish identically.

[Consider the form of $F(x)$ after s complete circuits about the origin.]

9. If $\{\Pi_i(x)\}$ are single-valued near $x = 0$, and if $x^\rho \sum_{i=0}^{p} (\log x)^i \Pi_i(x)$ satisfies a linear equation with uniform coefficients, then so does the coefficient of each power of A in the expression $x^\rho \sum_{i=0}^{p} (A + \log x)^i \Pi_i(x)$.

10. If $\{\Pi_{ij}(x)\}$ are single-valued near $x = 0$, and (ρ_i) do not differ by integers, show that if an equation with uniform coefficients is satisfied by
$$\sum_{i=1}^{k} \sum_{j=0}^{l} x^{\rho_i}(\log x)^j \Pi_{ij}(x),$$
it is satisfied by each group of terms
$$x^{\rho_i} \sum_{j=0} (\log x)^j \Pi_{ij}(x).$$

V
REGULAR SINGULARITIES

16. Formal Solutions in Power-Series

Indicial Equation and Recurrence Formulae. LET $x = 0$ be a typical singularity where Fuchs's conditions are satisfied. To expand the solutions we use the *normal form* of Fuchs and Frobenius

$$x^n D^n y + P_1(x) x^{n-1} D^{n-1} y + \ldots + P_n(x) y = 0, \tag{1}$$

where $\{P_i(x)\}$ are holomorphic at $x = 0$. Let $\delta \equiv xD$, so that $x^m D^m \equiv \delta(\delta-1)\ldots(\delta-m+1)$, and let (1) be written in the form

$$F(x, \delta) y \equiv \delta^n y + Q_1(x) \delta^{n-1} y + \ldots + Q_n(x) y = 0. \tag{2}$$

The coefficients (P_i) and (Q_i) are mutually expressible as linear functions of one another with constant coefficients, so that both sets are holomorphic together. We now expand the coefficients in Taylor series

$$Q_i(x) = \sum_{j=0}^{\infty} Q_{ij} x^j \quad (i = 1, 2, \ldots, n), \tag{3}$$

and write

$$\left.\begin{aligned}
F_0(\delta) &= \delta^n + Q_{10} \delta^{n-1} + Q_{20} \delta^{n-2} + \ldots + Q_{n0}, \\
F_j(\delta) &= Q_{1j} \delta^{n-1} + Q_{2j} \delta^{n-2} + \ldots + Q_{nj}.
\end{aligned}\right\} \tag{4}$$

The equation may now be written

$$F(x, \delta) y \equiv \sum_{j=0}^{\infty} x^j F_j(\delta) y = 0, \tag{5}$$

and on substituting as a trial regular solution

$$y = x^\rho [c_0 + c_1 x + c_2 x^2 + \ldots], \tag{6}$$

we have

$$F(x, \delta) y = \sum_{i=0}^{\infty} \sum_{j=0}^{\infty} x^j F_j(\delta) c_i x^{\rho+i}$$

$$= \sum_{i=0}^{\infty} \sum_{j=0}^{\infty} x^{\rho+i+j} F_j(\rho+i) c_i$$

$$= \sum_{k=0}^{\infty} x^{\rho+k} \left[\sum_{i=0}^{k} F_{k-i}(\rho+i) c_i \right] = 0. \tag{7}$$

This vanishes identically if each coefficient is zero, or if (c_i) satisfy the relations

$$\sum_{i=0}^{k} F_{k-i}(\rho+i) c_i = 0 \quad (k = 0, 1, 2, \ldots). \tag{8}$$

In particular, if $c_0 \neq 0$, ρ must be a root of the equation

$$F_0(\rho) \equiv \rho^n + Q_{10}\rho^{n-1} + \dots + Q_{n0} = 0. \tag{9}$$

This is called the *indicial equation* and its roots are called the *exponents* of the given singularity.

Sets of Exponents. The number of linearly independent regular solutions (6) cannot exceed the number of distinct roots of the indicial equation and may fall short of it. For if two independent solutions belong to the same exponent ρ, we can form by subtraction a solution belonging to a higher exponent ρ', which must itself satisfy the indicial equation. Thus one distinct power-series (6) at the most is associated with each distinct exponent.

If $F_0(\rho) = 0$, $F_0(\rho+k) \neq 0$ $(k = 1, 2, \dots)$, we may choose $c_0 \neq 0$, and every subsequent coefficient (c_k) is then uniquely determined. But if several exponents differ by integers, the relations (8) may or may not be compatible, and the number of distinct series may be smaller than the number of distinct exponents.

Whenever there is a shortage of regular solutions of the type (6), on account of multiple roots of the indicial equation or roots differing by integers, the deficiency is supplied by solutions involving logarithms. In practice these may be constructed either by a direct method, which has been fully worked out by Heffter,[†] or by an artifice such as the method of Frobenius.[‡] We shall give modified versions of both methods in this chapter.

Let all the roots of the indicial equation which differ by integers be collected in sets, and let each set be arranged in order of *ascending* real parts. We shall show how to construct all the solutions corresponding to a typical set of h distinct exponents (ρ_i) of multiplicity (κ_i). All these solutions are associated with the same multiplier $\lambda = \exp(2i\pi\rho)$, which is a root of multiplicity $\sum \kappa_i$ of the characteristic equation belonging to a circuit about the origin.

Heffter's procedure consists in arranging the solution as a polynomial in $\log x$ of degree less than n, with coefficients which are regular series of the type (6); he constructs first all solutions not involving $\log x$, then those of the first degree, and so on. Instead of this we assume at once a trial solution of the form

$$y = x^\rho[u_0 + u_1 x + u_2 x^2 + \dots], \tag{10}$$

[†] L. Heffter, *Einleitung in die Theorie der linearen Differentialgleichungen* (Leipzig, 1894), 20–34, 104–32.

[‡] G. Frobenius, *Crelle's J. für Math.* **76** (1873), 214–35.

where (u_i) are polynomials in $\log x$ of degree less than n. We then have
$$F(x,\delta)y = \sum_{i=0}^{\infty}\sum_{j=0}^{\infty} x^j F_j(\delta) x^{\rho+i} u_i,$$
$$= \sum_{i=0}^{\infty}\sum_{j=0}^{\infty} x^{\rho+i+j} F_j(\delta+\rho+i) u_i,$$
$$= \sum_{k=0}^{\infty} x^{\rho+k}\left[\sum_{i=0}^{k} F_{k-i}(\delta+\rho+i) u_i\right], \qquad (11)$$
which vanishes identically if (u_i) satisfy the relations
$$\sum_{i=0}^{k} F_{k-i}(\delta+\rho+i) u_i = 0 \quad (k=0,1,2,\ldots). \qquad (12)$$

This may be regarded as a system of linear differential equations with constant coefficients in the independent variable $t \equiv \log x$. We do not require the complete primitive, but only the most general solution in *polynomials*. We can accordingly simplify the system as follows.

The Auxiliary Systems. The first equation may be written
$$F_0(\delta+\rho)u_0 \equiv F_0(\rho)u_0 + \frac{1}{1!}F_0'(\rho)\,\delta u_0 + \frac{1}{2!}F_0''(\rho)\,\delta^2 u_0 + \ldots = 0, \qquad (13)$$
and is the *generalized indicial equation*. If u_0 is a polynomial in t not identically zero, this expression is a polynomial of the same degree unless $F_0(\rho) = 0$. For an effective solution, ρ must satisfy the indicial equation; and if we identify it with the lowest root of the set $\rho = \rho_1$, the series belonging to the higher roots will appear in due course. Since $F_0(\delta+\rho_1)u_0 \equiv G_1(\delta)\delta^{\kappa_1} u_0$, where $G_1(0) \neq 0$, the polynomial u_0 must satisfy the reduced equation
$$\delta^{\kappa_1} u_0 = 0. \qquad (14)$$

Again, suppose that the polynomials $(u_0, u_1, \ldots, u_{k-1})$ have been found. If $F_0(\rho_1+k) \neq 0$, u_k is uniquely given as a polynomial, whose degree does not exceed the highest degree of any earlier coefficient, by the symbolic formula
$$u_k = -\frac{1}{F_0(\delta+\rho_1+k)}\sum_{i=0}^{k-1} F_{k-i}(\delta+\rho_1+i) u_i,$$
$$= -(A_0 + A_1\delta + A_2\delta^2 + \ldots)\sum_{i=0}^{k-1} F_{k-i}(\delta+\rho_1+i) u_i,$$
$$= L_k(u_0, u_1, \ldots, u_{k-1}). \qquad (15)$$
But if $k = \rho_i - \rho_1$, we have
$$F_0(\delta+\rho_1+k) = F_0(\delta+\rho_i) = G_i(\delta)\delta^{\kappa_i} \quad \{G_i(0) \neq 0\}, \qquad (16)$$

and then we have instead of (15) the relation
$$\delta^{\kappa_i} u_{\rho_i - \rho_1} = L_k(u_0, u_1, ..., u_{k-1}) \quad (k = \rho_i - \rho_1). \tag{17}$$

The structure of the solution is completely determined by the first $(N+1)$ relations, where N is the difference between the highest and lowest exponents of the set. For every subsequent relation can be reduced to the standard form (15). We may distinguish the h critical polynomials $U_i \equiv u_{\rho_i - \rho_1}$, and express all the others explicitly in the form
$$u_k = \Lambda_k(U_1, U_2, ..., U_h), \tag{18}$$
where (U_i) satisfy a system of equations of the form
$$\left. \begin{array}{l} \delta^{\kappa_1} U_1 = 0, \\ f_{21}(\delta) U_1 + \delta^{\kappa_2} U_2 = 0, \\ f_{31}(\delta) U_1 + f_{32}(\delta) U_2 + \delta^{\kappa_3} U_3 = 0, \\ \cdots \cdots \cdots \cdots \cdots \end{array} \right\} \tag{19}$$

Invariant Factors. Hamburger Sets. To each exponent ρ' of multiplicity κ' correspond solutions beginning with
$$x^{\rho'} (\log x)^s \quad (s = 0, 1, ..., \kappa' - 1),$$
which may involve higher powers of $\log x$ in the later terms. To resolve the aggregate of solutions into Hamburger sets, we have only to eliminate as many (U_i) as we can express explicitly in terms of the rest, and to replace the reduced system of equations by an equivalent diagonal system exhibiting the invariant factors of the matrix of operators (19),
$$\delta^{e_i} V_i = 0 \quad (i = 1, 2, ..., h' \leqslant h; \ \sum e_i = \sum \kappa_i). \tag{20}$$
The (V_i) are explicit linear functions of (U_i) and their derivatives, and vice versa; neither set can contain a polynomial of higher degree than all the other set. Hence each invariant factor $\delta^{e'}$ yields a solution whose degree in $\log x$ is $(e'-1)$. If we take this as the highest solution of a Hamburger set, the other solutions of the set are constructed by taking successive differences. Two different invariant factors yield distinct solutions. For we cannot make (u_i) all identically zero unless (U_i) and (V_i) are also zero, and so no combination of (V_i) yields a nul solution except $(V_i \equiv 0)$.

17. Convergence

Rearrangement of the Series. When the critical terms at the beginning of the series have been found and the degree s of the solution in $\log x$ has been determined, we can calculate as many more coefficients as we require by means of recurrence formulae. We shall suppose that the lowest exponent of the set has been made zero, by putting $y = x^{\rho_1} y^*$, to simplify the notation. We introduce binomial coefficients and write

$$y = \sum_{i=0}^{\infty} u_i x^i = \sum_{j=0}^{s} \binom{s}{j} Y_j (\log x)^{s-j}, \tag{1}$$

where $\qquad u_i = \sum_{j=0}^{s} \binom{s}{j} c_{ji} (\log x)^{s-j}, \qquad Y_j = \sum_{i=0}^{\infty} c_{ji} x^i. \tag{2}$

By Leibnitz's theorem, we have

$$\delta^m[uv] = \sum \frac{m!}{i!(m-i)!} \delta^{m-i} u . \delta^i v; \tag{3}$$

and hence, for any linear operator $F \equiv F(x, \delta)$ of order n,

$$F[uv] = \sum_{i=0}^{n} \frac{1}{i!} \{F^{(i)} u\} \delta^i v, \tag{4}$$

where $F^{(i)}$ is an operator of order $(n-i)$, formally defined by differentiation as though δ were a variable

$$F^{(i)} \equiv F^{(i)}(x, \delta) \equiv \frac{\partial^i}{\partial \delta^i} F(x, \delta). \tag{5}$$

Now $\qquad Fy = F\left[\sum_{j=0}^{s} \frac{s!}{j!(s-j)!} Y_j (\log x)^{s-j} \right]$

$$= \sum_{i=0}^{n} \sum_{j=0}^{s} \frac{s!}{i! j! (s-j)!} \{F^{(i)} Y_j\} \delta^i (\log x)^{s-j}$$

$$= \sum_{i=0}^{n} \sum_{j=0}^{s-i} \frac{s!}{i! j! (s-i-j)!} \{F^{(i)} Y_j\} (\log x)^{s-i-j}$$

$$= \sum_{k=0}^{s} \binom{s}{k} (\log x)^{s-k} \left[\sum_{j=0}^{k} \binom{k}{j} F^{(k-j)} Y_j \right]. \tag{6}$$

We must therefore have

$$\sum_{j=0}^{k} \binom{k}{j} F^{(k-j)} Y_j = 0 \quad (k = 0, 1, \ldots, s); \tag{7}$$

and, by re-combining these relations, where s does not appear explicitly, we find that the expressions

$$Y_0, \quad [Y_0 \log x + Y_1], \quad [Y_0 (\log x)^2 + 2 Y_1 (\log x) + Y_2], \text{ etc.}, \tag{8}$$

are each a solution of $Fy = 0$.

REGULAR SINGULARITIES

We now arrange (7) in powers of x, and find

$$\sum_{j=0}^{k}\sum_{p=0}^{\infty}\sum_{i=0}^{\infty}\binom{k}{j}x^p F_p^{(k-j)}(\delta)c_{ji}x^i$$

$$=\sum_{j=0}^{k}\sum_{p=0}^{\infty}\sum_{i=0}^{\infty}\binom{k}{j}x^{p+i}F_p^{(k-j)}(i)c_{ji}$$

$$=\sum_{q=0}^{\infty}x^q\bigg[\sum_{j=0}^{k}\sum_{i=0}^{q}\binom{k}{j}F_{q-i}^{(k-j)}(i)c_{ji}\bigg]=0. \quad (9)$$

Since each coefficient vanishes, we have

$$\sum_{j=0}^{k}\sum_{i=0}^{q}\binom{k}{j}F_{q-i}^{(k-j)}(i)c_{ji}=0 \quad (k=0,1,...,s;\ q=0,1,...,\infty), \quad (10)$$

and in particular, for $q=0$,

$$\sum_{j=0}^{k}\binom{k}{j}F_0^{(k-j)}(0)c_{j0}=0 \quad (k=0,1,...,s); \quad (11)$$

these relations are compatible because $[F_0(0)]^{s+1}=0$, since zero is a root of the indicial equation.

If $\rho = N$ is the highest exponent of the set, we assume that coefficients of all powers of x up to x^N have been determined by the methods of § 16. If $q > N$, $F_0(q) \neq 0$, and the coefficients of x^q are uniquely given by (10) in terms of the earlier coefficients.

Convergence. Let d be the shortest distance between $x = 0$ and any other singularity. Then the series $\{Q_i(x)\}$ are all convergent when $|x| < d$; the convergence of (Y_j) may then be proved by means of the following lemma. Let $\sum b_k x^k$ be any series convergent when $|x| < d$, and such that the equation

$$F(x,\delta)Y = \sum_{k=0}^{\infty} b_k x^k \quad (12)$$

is formally satisfied by a power-series $Y = \sum c_k x^k$; then that power-series is convergent when $|x| < d$.

Assuming for a moment the truth of this lemma, we first establish the convergence of Y_0, by putting zero on the right-hand side of (12). Then, if we have proved the convergence of $(Y_0, Y_1, ..., Y_{k-1})$, we can reduce the equation (7) giving Y_k to the above form (12); and another application of the lemma establishes the convergence of Y_k. The convergence of the complete set of power-series follows by induction with respect to k.

Now (12) gives the recurrence formulae

$$\sum_{j=0}^{k} F_{k-j}(j)c_j = b_k \quad (k = 0, 1, 2,...), \tag{13}$$

which are (by hypothesis) algebraically compatible. Let x be any number such that $|x| = r < d$; and let $R = \frac{1}{2}(r+d) < d$. Then because $\sum Q_{ik} x^k$ and $\sum b_k x^k$ are absolutely convergent when $|x| = R$, their terms have upper bounds

$$|Q_{ik}|R^k, \quad |b_k|R^k < M. \tag{14}$$

Hence, for positive integers j, we have

$$|F_{k-j}(j)| = \Big| \sum_{m=1}^{n} Q_{m(k-j)} j^{n-m} \Big|$$
$$\leqslant MR^{j-k}(1+j+j^2+...+j^{n-1})$$
$$\leqslant MR^{j-k}(j+1)^{n-1}. \tag{15}$$

Again, since $F_0(k) \neq 0$ $(k > N)$, and since $\lim_{k \to \infty}[F_0(k)/k^n] = 1$, we can find a positive θ such that

$$F_0(k) \geqslant \theta k^n \quad (k > N). \tag{16}$$

From (13), (14), (16) we get the inequality

$$\theta k^n |c_k| \leqslant \sum_{j=0}^{k-1} MR^{j-k}(j+1)^{n-1}|c_j| + MR^{-k} \quad (k > N). \tag{17}$$

We now construct a dominant series $\sum C_k x^k \gg \sum c_k x^k$ by putting

$$C_k \geqslant |c_k| \quad (k = 0, 1,..., N), \tag{18}$$

$$\theta k^n R^k C_k = M\Big[1 + \sum_{j=0}^{k-1}(j+1)^{n-1}R^j C_j\Big]$$
$$= \theta(k-1)^n R^{k-1} C_{k-1} + Mk^{n-1}R^{k-1}C_{k-1} \quad (k > N). \tag{19}$$

Since $\lim_{k \to \infty}(C_k/C_{k-1}) = 1/R < 1/r$, the dominant series $\sum C_k x^k$ is absolutely convergent for the value of x in question, and *a fortiori* the series $\sum c_k x^k$. The lemma has accordingly been proved; and the convergence of all the series (Y_j) follows.

18. Apparent Singularities

Example. Consider the equation

$$xD^2y - (1+x)Dy + y = 0, \tag{1}$$

which has a regular singularity at $x = 0$. If we substitute $y = \sum c_n x^{\rho+n}$, we have the indicial equation and recurrence formulae

$$\left. \begin{array}{l} \rho(\rho-2)c_0 = 0, \\ (\rho+n-2)[(\rho+n)c_n - c_{n-1}] = 0 \quad (n = 1, 2,...). \end{array} \right\} \tag{2}$$

As the exponents (0, 2) differ by an integer, we must examine whether the solution belonging to the lower exponent is free from logarithms. In fact, on putting $\rho = 0$, we find that c_0 and c_2 are both arbitrary, and that
$$\left. \begin{array}{l} c_1 = c_0, \\ nc_n = c_{n-1} \quad (n = 3, 4, \ldots). \end{array} \right\} \quad (3)$$
The solution is $y = A(1+x)+Be^x$, as may easily be seen by elementary methods. Now this is always holomorphic at $x = 0$, so that we are led to ask why this should be a singular point. The reason is that, when $x = 0$, we have $(y = A+B, Dy = A+B)$, so that arbitrary values cannot be assigned to y and Dy at the origin.

Conditions for an Apparent Singularity. A singular point where the complete primitive of the differential equation is holomorphic is called an *apparent singularity*. If $x = 0$ is an apparent singularity of an equation of order n, there are n linearly independent ordinary power-series (y_i) satisfying the equation, and we can arrange by subtraction that no two of them shall belong to the same exponent. Accordingly the exponents of the singularity are n unequal non-negative integers (ρ_i). Now the Wronskian determinant
$$W(y_1, y_2, \ldots, y_n)$$
is also regular, and belongs to the exponent
$$\sigma = \rho_1 + \rho_2 + \ldots + \rho_n - \tfrac{1}{2}n(n-1). \quad (4)$$
This exponent σ is a positive integer for every admissible set of exponents except $(0, 1, 2, \ldots, n-1)$, when we have $\sigma = 0$. We then have, by Liouville's formula,
$$\begin{aligned} p_1(x) &= -D \log W(y_1, y_2, \ldots, y_n) \\ &= -\frac{\sigma}{x} + b_0 + b_1 x + b_2 x^2 + \ldots. \end{aligned} \quad (5)$$
If $\sigma \neq 0$, $p_1(x)$ has a pole at $x = 0$ and the point cannot be an ordinary point. But if $\sigma = 0$ and if, as we are assuming, every solution is holomorphic, every coefficient of the equation
$$\frac{W(y_1, y_2, \ldots, y_n, y)}{W(y_1, y_2, \ldots, y_n)} = 0 \quad (6)$$
is holomorphic at $x = 0$, and the point is an *ordinary point* of the equation.

We can of course have a singular point where the exponents are $(0, 1, 2, \ldots, n-1)$, but in that case the solution involves logarithms.

As an example, the reader may examine the equation
$$xD^2y - y = 0. \tag{7}$$
If the roots of the indicial equation at $x = 0$ are unequal non-negative integers, we must examine the most general holomorphic solution $y = \sum c_n x^n$. If $N = (\rho_n - \rho_1)$ is the difference between the highest and lowest exponents, we must examine the compatibility of the $(N+1)$ recurrence formulae satisfied by the coefficients of $(x^{\rho_1}, x^{\rho_1+1},..., x^{\rho_n})$. The necessary and sufficient condition that the point should be an apparent singularity is that the n critical coefficients of the powers (x^{ρ_i}) should be arbitrary, or that the rank of the system should be $(N-n+1)$. We can always write the coefficients of the non-critical powers of x explicitly in terms of these n critical ones, and obtain a reduced system of relations between the latter only:

$$\left.\begin{aligned} A_{11}c_{\rho_1} &= 0, \\ A_{21}c_{\rho_1} + A_{22}c_{\rho_2} &= 0, \\ &\cdots\cdots\cdots \\ A_{n1}c_{\rho_1} + ... + A_{nn}c_{\rho_n} &= 0. \end{aligned}\right\} \tag{8}$$

The necessary and sufficient condition for an apparent singularity is that all the coefficients (A_{ij}) should vanish. Further details of the procedure will be found in Heffter's treatise.†

It may happen that all the exponents at a regular singularity are unequal, but differ only by integers, and that the solution is found to be free from logarithms. In that case the solutions are rendered holomorphic by a transformation of the type $y = (x-\xi)^{\rho_1} y^*$, and the point is said to be *reducible to an apparent singularity*.

19. The Method of Frobenius

D'Alembert's Method. Suppose we know that Euler's equation $F(\delta)y = 0$ is satisfied by $y = \sum C_r x^{\rho_r}$, where (ρ_r) are the roots of the indicial equation $F(\rho) = 0$. If the roots are unequal, this solution is the complete primitive; but if there are multiple roots, fresh solutions of a different type must be found. The latter can be obtained by evaluating the limits of such expressions as

$$\frac{x^{\rho_1} - x^{\rho_2}}{\rho_1 - \rho_2}, \quad \begin{vmatrix} x^{\rho_1} & x^{\rho_2} & x^{\rho_3} \\ \rho_1 & \rho_2 & \rho_3 \\ 1 & 1 & 1 \end{vmatrix} \div \begin{vmatrix} \rho_1^2 & \rho_2^2 & \rho_3^2 \\ \rho_1 & \rho_2 & \rho_3 \\ 1 & 1 & 1 \end{vmatrix}, \tag{1}$$

† Loc. cit. 20–34.

when $\rho_1 \to \rho_2 \to \rho_3$. The required solutions are, as we know, of the type $x^{\rho_1} \log x$, $x^{\rho_1}(\log x)^2$, etc., when ρ_1 is a double or triple root. D'Alembert's method consists in introducing a parameter into the equation, and studying the behaviour of the complete primitive as that parameter approaches some critical value. Frobenius has elaborated a method of this kind for obtaining all the solutions of a linear equation at a regular singularity, when the exponents become equal or differ by integers. We retain the same notation as before.

The Auxiliary Equation. Frobenius's plan is to consider an equation of the form
$$F(x, \delta)y = f(\alpha)x^\alpha, \qquad (2)$$
where α is a parameter at our disposal. We shall choose $f(\alpha)$ slightly differently from Frobenius, so that the coefficients of the solution shall be integral functions of α. If the indicial polynomial at $x = 0$ is
$$F_0(\rho) \equiv (\rho-\rho_1)(\rho-\rho_2)\ldots(\rho-\rho_n), \qquad (3)$$
we write
$$\Psi(z) \equiv \Gamma(z-\rho_1)\Gamma(z-\rho_2)\ldots\Gamma(z-\rho_n), \qquad (4)$$
where
$$\frac{1}{\Gamma(z)} = e^{\gamma z} z \prod_{r=1}^{\infty} \left[\left(1+\frac{z}{r}\right)e^{-z/r}\right], \qquad (5)$$
and γ is Euler's constant. We now consider the equation
$$F(x, \delta)y = x^\alpha/\Psi(\alpha), \qquad (6)$$
whose right-hand side vanishes when $\alpha = \rho_i - m$, where m is zero or any positive integer.

We substitute formally
$$y = \phi(x, \alpha) = x^\alpha \sum_{k=0}^{\infty} c_k(\alpha) x^k, \qquad (7)$$
and obtain the relations
$$\left. \begin{array}{l} F_0(\alpha)c_0(\alpha) = 1/\Psi(\alpha), \\ \sum_{i=0}^{k} F_{k-i}(\alpha+i)c_i(\alpha) = 0 \quad (k = 1, 2, 3, \ldots). \end{array} \right\} \qquad (8)$$
Since we know that
$$\Gamma(z+1) \equiv z\Gamma(z), \quad \Psi(z+1) \equiv F_0(z)\Psi(z), \qquad (9)$$
we find that
$$c_0(\alpha) = 1/\Psi(\alpha+1), \qquad (10)$$
and we observe that $c_0(\rho_h) \neq 0$, if $\alpha = \rho_h$ is the highest of a set of roots differing by integers of the indicial equation $F_0(\alpha) = 0$. We have further
$$c_k(\alpha) = \frac{(-)^k h_k(\alpha) c_0(\alpha)}{F_0(\alpha+1)F_0(\alpha+2)\ldots F_0(\alpha+k)} = \frac{(-)^k h_k(\alpha)}{\Psi(\alpha+k+1)}, \qquad (11)$$

by (9) and (10), where $h_k(\alpha)$ stands for the determinant

$$h_k(\alpha) = \begin{vmatrix} F_1(\alpha) & F_0(\alpha+1) & 0 & \ldots & 0 \\ F_2(\alpha) & F_1(\alpha+1) & F_0(a+2) & \ldots & 0 \\ \cdot & \cdot & \cdot & \cdot & \cdot \\ F_{k-1}(\alpha) & F_{k-2}(\alpha+1) & F_{k-3}(\alpha+2) & \ldots & F_0(\alpha+k-1) \\ F_k(\alpha) & F_{k-1}(\alpha+1) & F_{k-2}(\alpha+2) & \ldots & F_1(\alpha+k-1) \end{vmatrix}. \quad (12)$$

Since this is a polynomial in α and $1/\Psi(\alpha+k+1)$ is an integral function, $c_k(\alpha)$ is an integral function, for which we require a uniform upper bound in a circle $|\alpha| < K$. To obtain a dominant polynomial for $h_k(\alpha)$, we use the relations § 17, (15),

$$\left. \begin{aligned} F_0(\alpha) &\ll \alpha^n + M(\alpha+1)^{n-1} = \alpha^n + \phi(\alpha) \quad \text{say,} \\ F_q(\alpha) &\ll R^{-q} M(\alpha+1)^{n-1} = R^{-q} \phi(\alpha) \quad (q > 0), \end{aligned} \right\} \quad (13)$$

where R is any fixed number smaller than the distance from $x = 0$ to any other singularity, and M depends only on R. We now observe that in the expansion of the determinant

$$\begin{vmatrix} a_{11}, & -b_1, & 0, & \ldots, & 0 \\ a_{21}, & a_{22}, & -b_2, & \ldots, & 0 \\ \cdot & \cdot & \cdot & \cdot & \cdot \\ a_{(k-1)1}, & a_{(k-1)2}, & a_{(k-1)3}, & \ldots, & -b_{(k-1)} \\ a_{k1}, & a_{k2}, & a_{k3}, & \ldots, & a_{kk} \end{vmatrix} \quad (14)$$

every term is positive. Hence

$$h_k(\alpha) \ll \begin{vmatrix} R^{-1}\phi(\alpha), & -(\alpha+1)^n - \phi(\alpha+1), & \ldots, & 0 \\ R^{-2}\phi(\alpha), & R^{-1}\phi(\alpha+1), & \ldots, & 0 \\ \cdot & \cdot & \cdot & \cdot \\ R^{1-k}\phi(\alpha), & R^{2-k}\phi(\alpha+1), & \ldots, & -(\alpha+k-1)^n - \phi(\alpha+k-1) \\ R^{-k}\phi(\alpha), & R^{1-k}\phi(\alpha+1), & \ldots, & R^{-1}\phi(\alpha+k-1) \end{vmatrix}$$

$$= R^{-k}\phi(\alpha) \prod_{r=1}^{k-1} [(\alpha+r)^n + 2\phi(\alpha+r)]. \quad (15)$$

Now we can choose a positive number A such that

$$\phi(\alpha), \quad \alpha^n + 2\phi(\alpha) \ll (\alpha+A)^n; \quad (16)$$

hence we have, uniformly in the circle $|\alpha| \leqslant K$,

$$h_k(\alpha) \leqslant R^{-k}[\Gamma(K+A+k)/\Gamma(K+A)]^n. \quad (17)$$

Again, when $\sum \alpha_s = \sum \beta_s$, we have, from the definition (5) of $\Gamma(z)$ as a product,

$$\lim_{k\to\infty} \prod_{s=1}^{m} \left[\frac{\Gamma(\alpha_s+k)}{\Gamma(\beta_s+k)}\right] = \lim_{k\to\infty} \prod_{r=0}^{\infty} \left[\prod_{s=1}^{m} \left(\frac{\beta_s+k+r}{\alpha_s+k+r}\right)\right]$$

$$= \lim_{k\to\infty} \prod_{r=k}^{\infty} \left[1 + O\left(\frac{1}{r^2}\right)\right] = 1. \qquad (18)$$

Hence, as $k \to \infty$, we have

$$\frac{[\Gamma(K+A+k)]^n}{\Psi(\alpha+k+1)} \sim \frac{\Gamma(nK+nA-n\alpha-n+\sum \rho_i+k)}{\Gamma(k)}. \qquad (19)$$

From (11), (17), and (19) we have now, when $|\alpha| \leqslant K$,

$$|c_k(\alpha)| \leqslant R^{-k} H(K) \Gamma(2nK+B+k)/\Gamma(k), \qquad (20)$$

where $H(K)$ does not depend on k.

The integral functions $\{c_k(\alpha)\}$ may be differentiated as often as we please; and upper bounds for the derivatives when $|\alpha| < K' < K$ are given by Cauchy's integral

$$c_k^{(s)}(\alpha) = \frac{s!}{2i\pi} \int \frac{c_k(z)\,dz}{(z-\alpha)^{s+1}}, \qquad (21)$$

taken around the circle $|z| = K$; we have accordingly

$$|c_k^{(s)}(\alpha)| \leqslant \frac{s!\, R^{-k} H(K) \Gamma(2nK+B+k)}{(K-K')^{s+1} \Gamma(k)}. \qquad (22)$$

The expressions (20) and (22) suggest dominant series of the type $G(K)\left(1-\dfrac{x}{R}\right)^{-2nK-B}$ for $\sum c_k(\alpha)x^k$ and its derivatives with respect to α. Since R is any number we please smaller than d and K is any finite constant, the series $\sum c_k^{(s)}(\alpha)x^k$ converge uniformly with respect to α in any finite domain, provided that $|x| \leqslant d-\epsilon < d$. The solution $\phi(x,\alpha)$ is given by multiplying this series by x^α, which is an integral function of α for any value of $x \neq 0$.

Solutions given by a Set of Exponents. In a circle $|x-x_0| < \eta$, which lies entirely within $|x| < d$ but excludes the point $x = 0$, the relation

$$F(x,\delta)\phi(x,\alpha) = x^\alpha/\Psi(\alpha) \qquad (23)$$

is an identity between integral functions of α, which may be differentiated as often as we please. If ρ_h is the exponent with the highest real part in our typical set of exponents differing by integers, we see from the definition (4) that $\alpha = \rho_h$ is a zero of multiplicity κ_h of the right-hand side; but from (10) we know that $c_0(\rho_h) \neq 0$. We can thus obtain κ_h solutions of $F(x,\delta)y = 0$ from the relations

$$F(x,\delta)\left[\frac{\partial^s}{\partial \alpha^s}\phi(x,\alpha)\right]_{\alpha=\rho_h} = 0 \quad (s = 0, 1,...,\kappa_h-1). \quad (24)$$

These solutions are necessarily independent, since they are respectively of degree $(0, 1,...,\kappa_h-1)$ in $\log x$, which appears when we differentiate x^α.

The second highest exponent of the set ρ_{h-1} is a zero of multiplicity $(\kappa_{h-1}+\kappa_h)$ of the right-hand side of (23), and yields $(\kappa_{h-1}+\kappa_h)$ solutions

$$F(x,\delta)\left[\frac{\partial^s}{\partial \alpha^s}\phi(x,\alpha)\right]_{\alpha=\rho_{h-1}} = 0 \quad (s = 0, 1,...,\kappa_{h-1}+\kappa_h-1). \quad (25)$$

But $\alpha = \rho_{h-1}$ is also a zero of multiplicity κ_h of $c_0(\alpha)$, so that the first κ_h of the solutions (25) turn out to belong to the exponent ρ_h and to be combinations of the solutions already given by (24). The last κ_{h-1} solutions are new and linearly independent.

In the same way, the third exponent yields $(\kappa_{h-2}+\kappa_{h-1}+\kappa_h)$ solutions

$$F(x,\delta)\left[\frac{\partial^s}{\partial \alpha^s}\phi(x,\alpha)\right]_{\alpha=\rho_{h-2}} = 0 \quad (s = 0, 1,...,\kappa_{h-2}+\kappa_{h-1}+\kappa_h-1), \quad (26)$$

of which the first $(\kappa_{h-1}+\kappa_h)$ are combinations of the solutions (24) and (25) and the last κ_{h-2} are new, and so on. Every solution can thus be deduced from the expression $\phi(x,\alpha)$ by suitable operations.

20. The Point at Infinity. Equations of the Fuchsian Class

Change of the Independent Variable. If the equation (E_n^*) has m isolated singularities $(x = a_1, a_2,..., a_m)$ in the finite part of the plane, they can all be enclosed in a circle $|x| = R$, outside of which the coefficients are holomorphic, except at $x = \infty$, which is an isolated singular point or an ordinary point. To examine its character, we put $x = 1/z$, and we say that $x = \infty$ is an ordinary, regular, or irregular point (as the case may be) of the given equation, according as $z = 0$ is an ordinary, regular, or irregular point of the transformed equation. Thus $x = \infty$ is placed on exactly the same footing as any other point; this is graphically illustrated when we make a stereographic projection of the plane of the complex variable x (in the usual Argand diagram) upon the surface of Neumann's sphere.

Chap. V, § 20   REGULAR SINGULARITIES   75

A convergent series of the form

$$\phi(x) \equiv \frac{c_0}{x^\rho} + \frac{c_1}{x^{\rho+1}} + \frac{c_2}{x^{\rho+2}} + \ldots \quad (c_0 \neq 0) \tag{1}$$

is said to belong to the exponent ρ at infinity.

Conditions for a Regular Singularity. If we use the form

$$\delta^n y + Q_1(x)\delta^{n-1} y + \ldots + Q_n(x) y = 0, \tag{2}$$

and put $\quad x = \dfrac{1}{z}, \qquad \delta' = z\dfrac{d}{dz} = -x\dfrac{d}{dx} = -\delta, \tag{3}$

we have at once

$$\delta'^n y - Q_1\!\left(\frac{1}{z}\right)\delta'^{n-1} y + \ldots + (-)^n Q_n\!\left(\frac{1}{z}\right) y = 0. \tag{4}$$

The necessary and sufficient conditions that $z = 0$ (or $x = \infty$) should be regular are that $\{Q_i(\infty)\}$ should be finite. In the canonical form, this means that $\{x^i p_i(x)\}$ must remain finite as $x \to \infty$, or that $p_i(x) = O(x^{-i})$.

Conditions for an Ordinary Point. To transform the canonical form (E_n^*), we put

$$\begin{aligned}
D^k y &\equiv x^{-k}\delta(\delta-1)\ldots(\delta-k+1)y, \\
&\equiv x^{-1}(\delta+k-1)(\delta+k-2)\ldots\delta(x^{1-k}y), \\
&\equiv (-)^k z(\delta'-k+1)(\delta'-k+2)\ldots\delta'(z^{k-1}y), \\
&\equiv (-)^k z^{k+1} D'^k(z^{k-1} y);
\end{aligned} \tag{5}$$

$$\begin{aligned}
Fy &\equiv \sum_{r=0}^{n} p_r(x) D^{n-r} y \quad \{p_0(x) \equiv 1\}, \\
&\equiv \sum_{r=0}^{n} (-)^{n-r} p_r\!\left(\frac{1}{z}\right) z^{n-r+1} D'^{n-r}(z^{n-r-1} y), \\
&\equiv \sum_{r=0}^{n} \sum_{k=0}^{n-r} (-)^{n-r} p_r\!\left(\frac{1}{z}\right) z^{n-r+1} \binom{n-r}{k} (D'^{n-r-k} z^{n-r-1})(D'^k y), \\
&\equiv \sum_{r=0}^{n} \sum_{k=0}^{n-r} (-)^{n-r} p_r\!\left(\frac{1}{z}\right) \binom{n-r}{k} \frac{(n-r-1)!}{(k-1)!} z^{n-r+k} D'^k y.
\end{aligned} \tag{6}$$

We now put $k = (n-s)$ and change the order of summation,

$$Fy \equiv \sum_{r=0}^{n} \sum_{s=r}^{n} (-)^{n-r} p_r\!\left(\frac{1}{z}\right) \binom{n-r}{n-s} \frac{(n-r-1)!}{(n-s-1)!} z^{2n-r-s} D'^{n-s} y,$$

$$= \sum_{s=0}^{n} \left[\sum_{r=0}^{s} (-)^{n-r} \binom{n-r}{n-s} \frac{(n-r-1)!}{(n-s-1)!} z^{2n-r-s} p_r\!\left(\frac{1}{z}\right) \right] D'^{n-s} y. \tag{7}$$

The leading term here is $(-)^n z^{2n} D'^n y$, so that the canonical form corresponding to the independent variable z is

$$F^*y \equiv D'^n y + p_1^*(z) D'^{n-1} y + \ldots + p_n^*(z) y, \tag{8}$$

where

$$\left. \begin{aligned} p_s^*(z) &\equiv \sum_{r=0}^{s} (-)^r \binom{n-r}{n-s} \frac{(n-r-1)!}{(n-s-1)!} z^{-r-s} p_r\!\left(\frac{1}{z}\right) \quad (s = 1, 2, \ldots, n-1), \\ p_n^*(z) &\equiv (-)^n z^{-2n} p_n\!\left(\frac{1}{z}\right). \end{aligned} \right\} \tag{9}$$

Now $z = 0$ is an ordinary point of the new equation, if all these expressions are holomorphic; hence the necessary and sufficient conditions that $x = \infty$ should be an ordinary point of the original equation (E_n^*) are that the analytic functions

$$\left. \begin{aligned} \sum_{r=0}^{s} (-)^r \binom{n-r}{n-s} \frac{(n-r-1)!}{(n-s-1)!} x^{r+s} p_r(x) \quad (s = 1, 2, \ldots, n-1), \\ (-)^n x^{2n} p_n(x) \quad \{p_0(x) \equiv 1\}, \end{aligned} \right\} \tag{10}$$

and

should remain finite there.

The Indicial Equation. If infinity is a regular singularity, we find, on introducing an expansion of the form (1), that the coefficient of the dominant term gives us the indicial equation in either of the forms

$$\rho^n - Q_1(\infty) \rho^{n-1} + Q_2(\infty) \rho^{n-2} - \ldots + (-)^n Q_n(\infty) = 0,$$

or

$$\rho(\rho+1)\ldots(\rho+n-1) - P_1(\infty) \rho(\rho+1)\ldots(\rho+n-2) + \ldots + (-)^n P_n(\infty) = 0, \tag{11}$$

where $P_r(x) \equiv x^r p_r(x)$ as before.

Equations of the Fuchsian Class. If the only singularities are m isolated regular singularities in the finite part of the plane and a regular singularity at infinity, the equation is said to be of the *Fuchsian class*. If we put

$$\psi(x) \equiv (x-a_1)(x-a_2)\ldots(x-a_m), \tag{12}$$

where (a_i) are the affixes of the singularities at a finite distance from the origin, the necessary and sufficient conditions that these should be regular are that $\{\psi^r p_r(x)\}$ should be holomorphic for all finite values of x. These expressions are accordingly integral functions of x; but $\{x^r p_r(x)\}$ are finite at infinity, so that we have

$$[\psi(x)]^r p_r(x) \equiv \Pi_{(m-1)r}(x) \quad (r = 1, 2, \ldots, n), \tag{13}$$

where $\Pi_\alpha(x)$ means a polynomial of degree α.

We can now write the equation in the form
$$\psi^n D^n y + \Pi_{m-1}\psi^{n-1}D^{n-1}y + \ldots + \Pi_{mn-n}y = 0, \qquad (14)$$
with *polynomial coefficients*; the advantage of working with this form in practice, in preference to the canonical one, is that the recurrence formulae have only a *finite number of terms*.

Fuchs's Relation between the Exponents. If we put
$$p_1(x) \equiv \frac{\Pi_{m-1}(x)}{\psi(x)} = \sum_{i=1}^n \frac{A_i}{(x-a_i)}, \qquad (15)$$
the indicial equation at $x = a_i$ takes the form
$$\rho(\rho-1)\ldots(\rho-n+1) + A_i\rho(\rho-1)\ldots(\rho-n+2) + \ldots = 0, \qquad (16)$$
so that the sum of the exponents at $x = a_i$ is $[\tfrac{1}{2}n(n-1) - A_i]$. The indicial equation at infinity is
$$\rho(\rho+1)\ldots(\rho+n-1) - (\sum A_i)\rho(\rho+1)\ldots(\rho+n-2) + \ldots = 0; \qquad (17)$$
so that the sum of the exponents at infinity is $[\sum A_i - \tfrac{1}{2}n(n-1)]$. Hence we have the result that the sum of *all* the $(m+1)n$ exponents is
$$\sum_{i=1}^{m+1}\sum_{j=1}^n \rho_{ij} = \tfrac{1}{2}(m-1)n(n-1). \qquad (18)$$

Equations of the Second Order. The most general equation of the Fuchsian class and of the second order is
$$D^2 y + \frac{\Pi_{m-1}}{\psi}Dy + \frac{\Pi_{2m-2}}{\psi^2}y = 0, \qquad (19)$$
or
$$D^2 y + \left\{\sum_{i=1}^m \frac{A_i}{(x-a_i)}\right\}Dy + \left\{\sum_{i=1}^m \frac{B_i}{(x-a_i)^2} + \sum_{i=1}^m \frac{C_i}{(x-a_i)}\right\}y = 0, \qquad (20)$$
where $\sum C_i = 0$, as we see by expanding the last coefficient in descending powers of x. Now, at $x = a_i$, the indicial equation is $\rho(\rho-1) + A_i\rho + B_i = 0$. If the roots of this are $\rho = \rho_i$ and $\rho = \rho_i'$, we have
$$A_i = 1 - \rho_i - \rho_i', \quad B_i = \rho_i\rho_i' \quad (i=1,2,\ldots,m), \qquad (21)$$
which gives the canonical form
$$D^2 y + \left\{\sum_{i=1}^m \left(\frac{1-\rho_i-\rho_i'}{x-a_i}\right)\right\}Dy + \left\{\sum_{i=1}^m \frac{\rho_i\rho_i'}{(x-a_i)^2} + \frac{\Pi_{m-2}}{\psi}\right\}y = 0. \qquad (22)$$
The indicial equation at infinity is
$$\rho(\rho+1) - \left\{\sum_{i=1}^m A_i\right\}\rho + \sum_{i=1}^m (B_i + C_i a_i) = 0, \qquad (23)$$

and Fuchs's relation (18) may be written in the easily remembered form

$$\sum_{i=1}^{m}(1-\rho_i-\rho_i')+(1-\rho_\infty-\rho_\infty') = 2. \qquad (24)$$

If infinity is an ordinary point of the equation of the second order, we find from (10) that

$$p_1(x) = \frac{2}{x}+O\!\left(\frac{1}{x^2}\right), \qquad p_2(x) = O\!\left(\frac{1}{x^4}\right), \qquad (25)$$

or, with the form (20) of the coefficients,

$$\sum_{i=1}^{m}A_i = 2;\ \sum_{i=1}^{m}C_i = 0;\ \sum_{i=1}^{m}(B_i+C_i a_i) = 0;\ \sum_{i=1}^{m}(2B_i a_i+C_i a_i^2) = 0. \qquad (26)$$

Reduced Forms. If we put $y = \phi(x)u$, we have

$$D^2 u+\left\{p_1(x)+\frac{2\phi'(x)}{\phi(x)}\right\}Du+\left\{p_2(x)+p_1(x)\frac{\phi'(x)}{\phi(x)}+\frac{\phi''(x)}{\phi(x)}\right\}u = 0. \qquad (27)$$

We can choose $\phi(x)$ in 2^m ways, so as to reduce to zero one exponent at every singularity in the finite part of the plane; one choice is

$$\phi(x) \equiv \prod_{i=1}^{m}(x-a_i)^{\rho_i}, \qquad (28)$$

and the others are obtained by interchanging (ρ_i, ρ_i'). If we introduce the exponent-differences $\delta_i = (\rho_i'-\rho_i)$, we find that (27) takes the form

$$D^2 u+\left\{\sum_{i=1}^{m}\left(\frac{1-\delta_i}{x-a_i}\right)\right\}Du+\frac{\Pi^*_{m-2}}{\psi}u = 0, \qquad (29)$$

exhibiting the exponent-differences (δ_i). We may call this the *first reduced form*.

We can also choose $\phi(x)$ so as to make the middle term of (27) disappear; this requires that

$$\phi(x) = \exp[-\tfrac{1}{2}\!\int p_1(x)\,dx], \qquad (30)$$

and leads to the *second reduced form*

$$D^2 u+Iu = 0, \qquad (31)$$

where I is the *invariant*

$$I \equiv p_2(x)-\tfrac{1}{2}Dp_1(x)-\tfrac{1}{4}\{p_1(x)\}^2, \qquad (32)$$

or

$$I \equiv \frac{1}{4}\sum_{i=1}^{m}\frac{(1-\delta_i^2)}{(x-a_i)^2}+\frac{\Pi^{**}_{m-2}}{\psi}. \qquad (33)$$

The necessary and sufficient condition that the equations
$$\left.\begin{array}{l} D^2y+p_1(x)Dy+p_2(x)y = 0, \\ D^2z+q_1(x)Dz+q_2(x)z = 0, \end{array}\right\} \quad (34)$$
should be transformable into one another by putting $y = z\chi(x)$ is that they should have the same invariant, or that
$$p_2-\tfrac{1}{2}Dp_1-\tfrac{1}{4}p_1^2 \equiv I \equiv q_2-\tfrac{1}{2}Dq_1-\tfrac{1}{4}q_1^2. \quad (35)$$

EXAMPLES. V

1. BESSEL'S EQUATION. If ν is not an integer, the solution of
$$x^2D^2y+xDy+(x^2-\nu^2)y = 0$$
is $y = AJ_\nu(x)+BJ_{-\nu}(x)$, where
$$J_\nu(x) = \sum_{m=0}^{\infty} \frac{(-)^m(\tfrac{1}{2}x)^{2m+\nu}}{m!\,\Gamma(m+\nu+1)}.$$

2. Bessel's equation of order zero
$$xD^2y+Dy+xy = 0$$
can be satisfied by $y = \sum_{m=0}^{\infty} u_{2m}x^{2m}$, where (u_{2m}) are polynomials of the first degree in $\log x$, determined by the relations
$$\delta^2 u_0 = 0, \quad (\delta+2m)^2 u_{2m}+u_{2m-2} = 0.$$
The reduced equations for polynomials are of the form
$$\begin{aligned}
u_{2m} &= -\frac{1}{(2m+\delta)^2}u_{2m-2}, \\
&= -\frac{1}{4m^2}\left(1-\frac{\delta}{m}\right)u_{2m-2}, \\
&= \frac{(-)^m}{4^m(m!)^2}\left(1-\frac{\delta}{m}\right)\left(1-\frac{\delta}{m-1}\right)\cdots\left(1-\frac{\delta}{1}\right)u_0, \\
&= \frac{(-)^m}{4^m(m!)^2}\left[1-\left(1+\frac{1}{2}+\frac{1}{3}+\ldots+\frac{1}{m}\right)\delta\right]u_0.
\end{aligned}$$
The solution $J_0(x)$ corresponds to $u_0 \equiv 1$. If we put $u_0 \equiv 2\log(\tfrac{1}{2}x)+2\gamma$,
$$\psi(m+1) \equiv \frac{\Gamma'(m+1)}{\Gamma(m+1)} = 1+\frac{1}{2}+\frac{1}{3}+\ldots+\frac{1}{m}-\gamma,$$
where γ is Euler's constant, we have Hankel's function
$$Y_0(x) = \sum_{m=0}^{\infty} \frac{(-)^m(\tfrac{1}{2}x)^{2m}}{(m!)^2}[2\log(\tfrac{1}{2}x)-2\psi(m+1)].$$
Since $\psi(m+1) \sim \log m$, the series converges for all finite values of x, except $x = 0$.

3. If n is a positive integer, $J_n(x) = (-)^n J_{-n}(x)$. The second solution $y = x^{-n}\left[\sum_{m=0}^{\infty} u_{2m}x^{2m}\right]$, where (u_{2m}) are polynomials of the first degree in $\log x$,

is determined by the relations
$$\delta(\delta-2n)u_0 = 0, \qquad (\delta+2n)\delta u_{2n}+u_{2n-2} = 0,$$
$$(\delta+2k)(\delta+2k-2n)u_{2k}+u_{2k-2} = 0 \quad (k \neq 0, n).$$
Show that the equation is satisfied by
$$Y_n(x) = -\sum_{m=0}^{n-1} \frac{(n-m-1)!(\tfrac{1}{2}x)^{2m-n}}{m!} +$$
$$+ \sum_{m=0}^{\infty} \frac{(-)^m(\tfrac{1}{2}x)^{2m+n}}{m!(m+n)!}[2\log(\tfrac{1}{2}x)-\psi(m+1)-\psi(m+n+1)].$$

4. KUMMER'S EQUATIONS. The doubly confluent hypergeometric equation
$$xD^2y+cDy-y = 0$$
is satisfied, when c is not an integer, by
$$y = A \,_0F_1(c;x)+Bx^{1-c}\,_0F_1(2-c;x),$$
where
$$_0F_1(c;x) \equiv 1+\frac{x}{1.c}+\frac{x^2}{1.2.c(c+1)}+\dots.$$
Show that the equation is connected with Bessel's by the relation
$$J_\nu(x) = \frac{(\tfrac{1}{2}x)^\nu}{\Gamma(\nu+1)} \cdot {}_0F_1\!\left(\nu+1; -\frac{x^2}{4}\right).$$

5. The simply confluent hypergeometric equation
$$xD^2y+(c-x)Dy-ay = 0$$
is satisfied, when c is not an integer, by
$$y = A \,_1F_1(a;c;x)+Bx^{1-c}\,_1F_1(a-c+1;2-c;x),$$
where
$$_1F_1(a;c;x) = 1+\frac{a}{1.c}x+\frac{a(a+1)}{1.2.c(c+1)}x^2+\dots.$$

If c is an integer, the condition that the solution should be free from logarithms is

(i) $c > 1$, $\quad (a-c+1)(a-c+2)\dots(a-1) = 0,$

or (ii) $c < 1$, $\quad a(a+1)(a+2)\dots(a-c) = 0.$

6. If $c = 1$, the logarithmic solution is in general
$$y = \sum_{m=0}^{\infty} \frac{\Gamma(a+m)}{(m!)^2}x^m[\log x+\psi(a+m)-2\psi(m+1)].$$
Examine the form when a is a negative integer or zero.

7. If c is an integer greater than unity, the solution involving logarithms is in general
$$y = \sum_{m=0}^{c-2} \frac{(-)^{c-m}\Gamma(a-c+1+m)(c-2-m)!}{m!}x^{m-c+1}+$$
$$+ \sum_{m=0}^{\infty} \frac{\Gamma(a+m)x^m}{m!(m+c-1)!}[\log x+\psi(a+m)-\psi(c+m)-\psi(m+1)].$$
By putting $y = x^{1-c}y'$ or otherwise, obtain the corresponding form when c is an integer less than unity.

8. Hypergeometric Equation. If c is not an integer, the equation
$$x(1-x)D^2y + [c-(a+b+1)x]Dy - aby = 0$$
is satisfied by
$$y = AF(a,b;c;x) + Bx^{1-c}F(a-c+1, b-c+1; 2-c; x),$$
where
$$F(a,b;c;x) \equiv 1 + \frac{ab}{1.c}x + \frac{a(a+1)b(b+1)}{1.2c(c+1)}x^2 + \ldots.$$

9. If c is a positive integer, the equation is in general satisfied by
$$y = \sum_{m=0}^{c-2} (-)^{c-m} \frac{\Gamma(a-c+1+m)\Gamma(b-c+1+m)(c-2-m)! \, x^{m-c+1}}{m!} +$$
$$+ \sum_{m=0}^{\infty} \frac{\Gamma(a+m)\Gamma(b+m)x^m}{m!(m+c-1)!} [\log x + \psi(a+m) + \psi(b+m) - \psi(c+m) - \psi(m+1)].$$

10. If c is an integer less than unity, the equation is in general satisfied by
$$y = \sum_{m=0}^{-c} (-)^{c+m} \frac{\Gamma(a+m)\Gamma(b+m)(-c-m)!}{m!} x^m +$$
$$+ \sum_{m=1-c}^{\infty} \frac{\Gamma(a+m)\Gamma(b+m)}{m!(m+c-1)!} x^m [\log x + \psi(a+m) + \psi(b+m) - \psi(c+m) - \psi(m+1)].$$

11. (i) The exponents of the hypergeometric equation at $x=0$ are $(0, 1-c)$. If c is an integer other than unity, the singularity will be free from logarithms if
$$c > 1 \quad \text{and} \quad \prod_{r=1}^{c-1} [(a-r)(b-r)] = 0,$$
or
$$c < 1 \quad \text{and} \quad \prod_{r=0}^{-c} [(a+r)(b+r)] = 0.$$

(ii) The exponents at $x=1$ are $(0, c-a-b)$. If $(c-a-b)$ is an integer other than zero, the singularity will be free from logarithms if
$$(a+b-c) > 0 \quad \text{and} \quad \prod_{r=1}^{a+b-c} [(a-r)(b-r)] = 0,$$
or
$$(a+b-c) < 0 \quad \text{and} \quad \prod_{r=0}^{c-a-b-1} [(a+r)(b+r)] = 0.$$

(iii) The exponents at $x = \infty$ are (a, b). If $(a-b)$ is an integer other than zero, the singularity will be free from logarithms if
$$a > b \quad \text{and} \quad \prod_{r=0}^{a-b-1} [(b+r)(b-c+r)] = 0,$$
or
$$a < b \quad \text{and} \quad \prod_{r=0}^{b-a-1} [(a+r)(a-c+r)] = 0.$$

12. Examine the singularities of the associated Legendre equation
$$(1-x^2)D^2y - 2xDy + \left[n(n+1) - \frac{m^2}{(1-x^2)}\right]y = 0.$$

13. Show that Laplace's tidal equation

$$\frac{d}{dx}\left[\frac{(1-x^2)}{(f^2-x^2)}\frac{dy}{dx}\right]+\beta y = 0$$

has regular singularities at $x = \pm 1$, apparent singularities at $x = \pm f$, and an irregular singularity at $x = \infty$.

[It is simplest to use Prof. A. E. H. Love's transformation

$$z = \frac{(1-x^2)}{(f^2-x^2)}\frac{dy}{dx}, \qquad \beta y = -\frac{dz}{dx},$$

which gives $(1-x^2)D^2z + \beta(f^2-x^2)z = 0.$

See H. Lamb, *Hydrodynamics* (5th ed.), 313.]

14. Construct a linear differential equation of the second order satisfied by $y = (x-1)^p$ and $y = (x+1)^q$. If $p \neq q$, show that there are regular singularities at $x = \pm 1, \infty$, and an apparent singularity at $x = (p+q)/(q-p)$.

15. Show that the exponents of an equation of the Fuchsian class are unchanged by a linear transformation $x' = (Ax+B)/(Cx+D)$. Hence show that the most general equation with only *one* regular singularity can be transformed into $D^n y = 0$, and that with *two* into Euler's homogeneous equation $F(\delta)y = 0$.

16. Verify the expansions in the preceding examples by the method of Frobenius, where it is applicable.

VI
THE HYPERGEOMETRIC EQUATION
21. Riemann's P-Function†

Definition. The most celebrated equation of the Fuchsian class is the hypergeometric equation, and it is instructive to begin by showing how it is determined by certain quite general properties of the solution. Following Riemann, we denote by

$$P\begin{Bmatrix} a & b & c & \\ \alpha & \beta & \gamma & x \\ \alpha' & \beta' & \gamma' & \end{Bmatrix}$$

any branch of a certain many-valued analytic function of x with the following properties.

(i) Every branch is finite and holomorphic, except at the three singular points $x = a, b, c$.

(ii) Any three branches are linearly connected.

(iii) At $x = a$ there are two principal branches $(P^{(\alpha)}, P^{(\alpha')})$ which are 'regular' and belong to the exponents (α, α'). Similarly there are two regular branches $(P^{(\beta)}, P^{(\beta')})$ belonging to the exponents (β, β') at $x = b$, and $(P^{(\gamma)}, P^{(\gamma')})$ belonging to the exponents (γ, γ') at $x = c$.

(iv) The exponent-differences $(\alpha'-\alpha)$, $(\beta'-\beta)$, $(\gamma'-\gamma)$ are not integers; and the six exponents are always connected by the relation

$$\alpha+\alpha'+\beta+\beta'+\gamma+\gamma' = 1. \tag{1}$$

It is evident that the meaning of the P-symbol is unaltered if we permute the first three columns, or if we exchange the two exponents (α, α') in the same column, and similarly (β, β') or (γ, γ').

Linear Transformation. If we put

$$x' = \frac{Ax+B}{Cx+D} \quad (AD-BC \neq 0), \tag{2}$$

we obtain a P-function of the new independent variable x', with singularities at the points $x' = a', b', c'$, corresponding to $x = a, b, c$, and with the same exponents, so that

$$P\begin{Bmatrix} a & b & c & \\ \alpha & \beta & \gamma & x \\ \alpha' & \beta' & \gamma' & \end{Bmatrix} = P\begin{Bmatrix} a' & b' & c' & \\ \alpha & \beta & \gamma & x' \\ \alpha' & \beta' & \gamma' & \end{Bmatrix}. \tag{3}$$

† B. Riemann, *Mathematische Werke* (1892), 66–83.

In particular, we can make the singularities coincide with $x' = 0, \infty, 1$ by putting
$$x' = \frac{(x-a)(b-c)}{(x-b)(a-c)}, \qquad (4)$$
and we observe that this is one of the anharmonic ratios of the set of numbers (a, b, c, x). For the P-function with the singularities in the standard position we may write more simply
$$P\begin{Bmatrix} 0 & \infty & 1 \\ \alpha & \beta & \gamma & x \\ \alpha' & \beta' & \gamma' \end{Bmatrix} \equiv P\begin{pmatrix} \alpha & \beta & \gamma \\ \alpha' & \beta' & \gamma' \end{pmatrix} x \end{pmatrix}. \qquad (5)$$

By permuting (a, b, c) the transformation (4) may be effected in six different ways, the new variables being connected by the relations
$$x' = x, \quad 1-x, \quad \frac{1}{x}, \quad \frac{1}{1-x}, \quad \frac{x}{x-1}, \quad \frac{x-1}{x}. \qquad (6)$$

The same function is represented by six schemes with different independent variables

$$\left. \begin{aligned} &P\begin{pmatrix} \alpha & \beta & \gamma \\ \alpha' & \beta' & \gamma' \end{pmatrix} x\end{pmatrix}, \quad P\begin{pmatrix} \beta & \gamma & \alpha \\ \beta' & \gamma' & \alpha' \end{pmatrix} \frac{1}{1-x}\end{pmatrix}, \quad P\begin{pmatrix} \gamma & \alpha & \beta \\ \gamma' & \alpha' & \beta' \end{pmatrix} \frac{x-1}{x}\end{pmatrix}, \\ &P\begin{pmatrix} \alpha & \gamma & \beta \\ \alpha' & \gamma' & \beta' \end{pmatrix} \frac{x}{x-1}\end{pmatrix}, \quad P\begin{pmatrix} \beta & \alpha & \gamma \\ \beta' & \alpha' & \gamma' \end{pmatrix} \frac{1}{x}\end{pmatrix}, \quad P\begin{pmatrix} \gamma & \beta & \alpha \\ \gamma' & \beta' & \alpha' \end{pmatrix} 1-x\end{pmatrix}. \end{aligned} \right\} \qquad (7)$$

Change of Exponents. It also follows from the definition that

$$\left(\frac{x-a}{x-b}\right)^\delta \left(\frac{x-c}{x-b}\right)^\epsilon P\begin{Bmatrix} a & b & c \\ \alpha & \beta & \gamma & x \\ \alpha' & \beta' & \gamma' \end{Bmatrix} = P\begin{Bmatrix} a & b & c \\ \alpha+\delta & \beta-\delta-\epsilon & \gamma+\epsilon & x \\ \alpha'+\delta & \beta'-\delta-\epsilon & \gamma'+\epsilon \end{Bmatrix}, \qquad (8)$$

and similarly

$$x^\delta(1-x)^\epsilon P\begin{pmatrix} \alpha & \beta & \gamma \\ \alpha' & \beta' & \gamma' \end{pmatrix} x\end{pmatrix} = P\begin{pmatrix} \alpha+\delta & \beta-\delta-\epsilon & \gamma+\epsilon \\ \alpha'+\delta & \beta'-\delta-\epsilon & \gamma'+\epsilon \end{pmatrix} x\end{pmatrix}. \qquad (9)$$

We can thus assign arbitrary values to two exponents at two distinct singularities, without introducing a new singularity or disturbing the relation (1). These transformations leave invariant the exponent-differences $(\alpha'-\alpha)$, $(\beta'-\beta)$, $(\gamma'-\gamma)$; we may therefore write $P(\alpha'-\alpha, \beta'-\beta, \gamma'-\gamma, x)$ for the family of functions
$$x^\delta(1-x)^\epsilon P\begin{pmatrix} \alpha & \beta & \gamma \\ \alpha' & \beta' & \gamma' \end{pmatrix} x\end{pmatrix}.$$

THE HYPERGEOMETRIC EQUATION

The Differential Equation. After describing any closed circuit, two linearly independent branches (y_1, y_2) of the P-function are transformed into branches of the form $(a_{11}y_1 + a_{12}y_2, a_{21}y_1 + a_{22}y_2)$, where $(a_{11}a_{22} - a_{12}a_{21}) \neq 0$. Hence the determinants

$$\begin{Vmatrix} y_1 & Dy_1 & D^2y_1 \\ y_2 & Dy_2 & D^2y_2 \end{Vmatrix} \tag{10}$$

are each multiplied by $(a_{11}a_{22} - a_{12}a_{21})$, and so any other branch satisfies an equation of the second order with uniform coefficients

$$\frac{W(y_1, y_2, y)}{W(y_1, y_2)} = 0. \tag{11}$$

This has regular singularities at $x = 0, \infty, 1$; any other singularities must be apparent, since the solutions are holomorphic. But apparent singularities are excluded by the condition (1). For consider the Wronskian of any two independent solutions $W(y_1, y_2)$; in the neighbourhood of $x = 0$ this may be written as a numerical multiple of the Wronskian of the principal branches $W(P^{(\alpha)}, P^{(\alpha')})$, which is regular and belongs to the exponent $(\alpha + \alpha' - 1)$. Hence and by similar reasoning

$$\left. \begin{aligned} W(y_1, y_2) &= O(x^{\alpha+\alpha'-1}) & \text{as } x \to 0, \\ &= O(x^{-\beta-\beta'-1}) & \text{as } x \to \infty, \\ &= O\{(1-x)^{\gamma+\gamma'-1}\} & \text{as } x \to 1. \end{aligned} \right\} \tag{12}$$

Accordingly the expression

$$\phi(x) \equiv x^{1-\alpha-\alpha'}(1-x)^{1-\gamma-\gamma'} W(y_1, y_2) \tag{13}$$

is holomorphic for all finite values of x; and, as $x \to \infty$, we find from (12) that

$$\phi(x) = O(x^{(1-\alpha-\alpha')+(1-\gamma-\gamma')-(1+\beta+\beta')}) = O(1), \tag{14}$$

on account of the relation (1). By Liouville's theorem $\phi(x)$ is a constant; and so $W(y_1, y_2)$ cannot vanish for any finite value of x other than zero or unity, and hence there are *no apparent singularities*, for the reasons explained in § 18.

The differential equation is now found to be uniquely determined by its singularities and exponents, by the method of § 20, (20)–(23), in the form

$$D^2y + \left[\frac{1-\alpha-\alpha'}{x} + \frac{1-\gamma-\gamma'}{x-1}\right]Dy + \\ + \left[\frac{\alpha\alpha'}{x^2} + \frac{\gamma\gamma'}{(x-1)^2} + \frac{\beta\beta'-\alpha\alpha'-\gamma\gamma'}{x(x-1)}\right]y = 0. \tag{15}$$

For the general scheme, Papperitz obtained the elegant canonical form

$$D^2y + \left[\sum \frac{1-\alpha-\alpha'}{x-a}\right]Dy +$$
$$+ \left[\sum \frac{\alpha\alpha'(a-b)(a-c)}{(x-a)}\right]\frac{y}{(x-a)(x-b)(x-c)} = 0. \quad (16)$$

Reduced Forms. We can reduce one exponent to zero at each of the singularities $x = 0, 1$, in accordance with (9), by putting

$$P\begin{pmatrix} \alpha & \beta & \gamma \\ \alpha' & \beta' & \gamma' \end{pmatrix} x\bigg) = x^\alpha(1-x)^\gamma P\begin{pmatrix} 0 & \alpha+\beta+\gamma & 0 \\ \alpha'-\alpha & \alpha+\beta'+\gamma & \gamma'-\gamma \end{pmatrix} x\bigg). \quad (17)$$

By interchanging (α, α') or (γ, γ') we can effect the reduction in *four* ways; and since the method is applicable to each of the *six* schemes (7), we obtain altogether *twenty-four* reduced forms in six different independent variables.

If we introduce the exponent-differences

$$\lambda = \alpha'-\alpha, \qquad \mu = \beta'-\beta, \qquad \nu = \gamma'-\gamma, \quad (18)$$

we find from (1) that

$$\alpha+\beta+\gamma = \tfrac{1}{2}(1-\lambda-\mu-\nu); \quad (19)$$

and so the reduced scheme (17) may be written

$$P\begin{pmatrix} 0 & \tfrac{1}{2}(1-\lambda-\mu-\nu) & 0 \\ \lambda & \tfrac{1}{2}(1-\lambda+\mu-\nu) & \nu \end{pmatrix} x\bigg), \quad (20)$$

and on inserting these values in (15) we have the reduced equation

$$D^2y + \left[\frac{1-\lambda}{x} + \frac{1-\nu}{x-1}\right]Dy + \frac{(1-\lambda-\mu-\nu)(1-\lambda+\mu-\nu)}{4x(x-1)}y = 0. \quad (21)$$

The other reduced forms are found by changing the signs of λ or ν.

The second reduced form, or *invariant form*, is found by removing the middle term of the equation. We must therefore make the sum of the exponents unity at $x = 0$ and $x = 1$, and this can only be done in one way for a given scheme. The new scheme is

$$P\begin{pmatrix} \tfrac{1}{2}(1-\lambda) & -\tfrac{1}{2}(1+\mu) & \tfrac{1}{2}(1-\nu) \\ \tfrac{1}{2}(1+\lambda) & -\tfrac{1}{2}(1-\mu) & \tfrac{1}{2}(1+\nu) \end{pmatrix} x\bigg), \quad (22)$$

and corresponds to the differential equation

$$D^2y + \left[\frac{1-\lambda^2}{4x^2} + \frac{1-\nu^2}{4(x-1)^2} + \frac{\lambda^2-\mu^2+\nu^2-1}{4x(x-1)}\right]y = 0, \quad (23)$$

which involves only the squares of the exponent-differences.

The Hypergeometric Equation. While Riemann's equation clearly

exhibits the exponents, it is not the most convenient form for numerical calculations. We identify (21) with the standard hypergeometric equation
$$x(1-x)D^2y+[c-(a+b+1)x]Dy-aby = 0, \qquad (24)$$
by writing
$$\left.\begin{array}{l}\lambda = 1-c, \quad \mu = \pm(a-b), \quad \nu = c-a-b, \\ a,b = \tfrac{1}{2}(1-\lambda\pm\mu-\nu), \quad c = 1-\lambda.\end{array}\right\} \qquad (25)$$
The scheme of the hypergeometric equation (24) is therefore
$$P\begin{pmatrix} 0 & a & 0 \\ 1-c & b & c-a-b \end{pmatrix} x, \qquad (26)$$
and we observe that the condition (1) is automatically satisfied.

22. Kummer's Twenty-four Series

Solutions at the Origin. To construct the regular solutions of the type $y = x^\rho \sum c_n x^n$, we use the form
$$\delta(\delta+c-1)y - x(\delta+a)(\delta+b)y = 0 \quad (\delta \equiv xD), \qquad (1)$$
and obtain the indicial equation
$$\rho(\rho+c-1) = 0, \qquad (2)$$
and the recurrence formulae
$$(n+\rho+1)(n+\rho+c)c_{n+1} = (n+\rho+a)(n+\rho+b)c_n. \qquad (3)$$
If c is not an integer, the principal branches are
$$P^{(\alpha)} = F(a,b;c;x), \qquad P^{(\alpha')} = x^{1-c}F(a-c+1, b-c+1; 2-c; x), \qquad (4)$$
with the usual notation for the hypergeometric series
$$F(a,b;c;x) \equiv 1 + \frac{ab}{1!\,c}x + \frac{a(a+1)b(b+1)}{2!\,c(c+1)}x^2 + \dots. \qquad (5)$$
From (3) we have $\lim_{n\to\infty}(c_{n+1}/c_n) = 1$, so that the solutions are convergent when $|x| < 1$. The first series reduces to a polynomial if a or b, the second if $(a-c+1)$ or $(b-c+1)$, is zero or a negative integer. From the four equivalent Riemann schemes
$$\left.\begin{array}{c} P\begin{pmatrix} 0 & a & 0 \\ 1-c & b & c-a-b \end{pmatrix} x, \quad (1-x)^{c-a-b}P\begin{pmatrix} 0 & c-a & 0 \\ 1-c & c-b & a+b-c \end{pmatrix} x, \\ (1-x)^{-a}P\begin{pmatrix} 0 & a & 0 & \frac{x}{x-1} \\ 1-c & c-b & b-a & \end{pmatrix}, \\ (1-x)^{-b}P\begin{pmatrix} 0 & b & 0 & \frac{x}{x-1} \\ 1-c & c-a & a-b & \end{pmatrix}, \end{array}\right\} \qquad (6)$$

we can write down four equivalent expansions of the branch $P^{(\alpha)}$, which is holomorphic at $x = 0$, namely

$$\left.\begin{array}{c} F(a,b;c;x), \qquad (1-x)^{c-a-b}F(c-a,c-b;c;x), \\ (1-x)^{-a}F\left(a,c-b;c;\dfrac{x}{(x-1)}\right), \quad (1-x)^{-b}F\left(b,c-a;c;\dfrac{x}{(x-1)}\right). \end{array}\right\} \quad (7)$$

The expansions in powers of x converge when $|x| < 1$; those in powers of $x/(x-1)$ in the half-plane $|x| < |x-1|$. In the same way, by a linear transformation of the Riemann scheme, each of the six principal branches can be expanded in four ways, in ascending or descending powers of x, $(1-x)$, or $x/(x-1)$. The domains of convergence of the various series are the interior or exterior of the circles $|x| = 1$ or $|1-x| = 1$, or the half-planes bounded by the line $|x| = |x-1|$. These series, which were obtained by Kummer, are given in the following table.

Table of Kummer's Series

$P^{(\alpha)}$	$F(a,b;c;x)$ $(1-x)^{c-a-b}F(c-a,c-b;c;x)$ $(1-x)^{-a}F(a,c-b;c;x/(x-1))$ $(1-x)^{-b}F(b,c-a;c;x/(x-1))$
$P^{(\alpha')}$	$x^{1-c}F(a-c+1,b-c+1;2-c;x)$ $x^{1-c}(1-x)^{c-a-b}F(1-a,1-b;2-c;x)$ $x^{1-c}(1-x)^{c-a-1}F(a-c+1,1-b;2-c;x/(x-1))$ $x^{1-c}(1-x)^{c-b-1}F(b-c+1,1-a;2-c;x/(x-1))$
$P^{(\beta)}$	$x^{-a}F(a,a-c+1;a-b+1;1/x)$ $x^{-a}(1-1/x)^{c-a-b}F(1-b,c-b;a-b+1;1/x)$ $x^{-a}(1-1/x)^{-a}F(a,c-b;a-b+1;1/(1-x))$ $x^{-a}(1-1/x)^{c-a-1}F(1-b,a-c+1;a-b+1;1/(1-x))$
$P^{(\beta')}$	$x^{-b}F(b,b-c+1;b-a+1;1/x)$ $x^{-b}(1-1/x)^{c-a-b}F(1-a,c-a;b-a+1;1/x)$ $x^{-b}(1-1/x)^{-b}F(b,c-a;b-a+1;1/(1-x))$ $x^{-b}(1-1/x)^{c-b-1}F(1-a,b-c+1;b-a+1;1/(1-x))$
$P^{(\gamma)}$	$F(a,b;a+b-c+1;1-x)$ $x^{1-c}F(a-c+1,b-c+1;a+b-c+1;1-x)$ $x^{-a}F(a,a-c+1;a+b-c+1;(x-1)/x)$ $x^{-b}F(b,b-c+1;a+b-c+1;(x-1)/x)$
$P^{(\gamma')}$	$(1-x)^{c-a-b}F(c-a,c-b;c-a-b+1;1-x)$ $x^{1-c}(1-x)^{c-a-b}F(1-a,1-b;c-a-b+1;1-x)$ $x^{a-c}(1-x)^{c-a-b}F(1-a,c-a;c-a-b+1;(x-1)/x)$ $x^{b-c}(1-x)^{c-a-b}F(1-b,c-b;c-a-b+1;(x-1)/x)$.

23. Group of Riemann's Equation

Invariants. The principal branches of a Riemann function (whose exponent-differences are not integers) are single-valued in the upper half-plane, and are connected by the relations

$$P^{(\alpha)} = \alpha_\beta P^{(\beta)} + \alpha_{\beta'} P^{(\beta')} = \alpha_\gamma P^{(\gamma)} + \alpha_{\gamma'} P^{(\gamma')},$$
$$P^{(\alpha')} = \alpha'_\beta P^{(\beta)} + \alpha'_{\beta'} P^{(\beta')} = \alpha'_\gamma P^{(\gamma)} + \alpha'_{\gamma'} P^{(\gamma')}.$$
(1)

The mutual ratios
$$\frac{\alpha_\beta}{\alpha'_\beta} : \frac{\alpha_{\beta'}}{\alpha'_{\beta'}} : \frac{\alpha_\gamma}{\alpha'_\gamma} : \frac{\alpha_{\gamma'}}{\alpha'_{\gamma'}}$$
(2)

are independent of the choice of the constant multiplier belonging to each branch, and were determined by Riemann as follows.

Let us consider the effect upon the two solutions (1) of a simple positive circuit, beginning and ending at a point of the upper half-plane, and enclosing $x = 0$ and $x = 1$. We may regard this as a sequence of two positive loops, first about $x = 0$ and then about $x = 1$, or else as a negative loop about $x = \infty$. The new branches obtained from (1) can therefore be expressed in the two alternative forms

$$\alpha_\beta e^{-2i\pi\beta} P^{(\beta)} + \alpha_{\beta'} e^{-2i\pi\beta'} P^{(\beta')} = e^{2i\pi\alpha}[\alpha_\gamma e^{2i\pi\gamma} P^{(\gamma)} + \alpha_{\gamma'} e^{2i\pi\gamma'} P^{(\gamma')}],$$
$$\alpha'_\beta e^{-2i\pi\beta} P^{(\beta)} + \alpha'_{\beta'} e^{-2i\pi\beta'} P^{(\beta')} = e^{2i\pi\alpha'}[\alpha'_\gamma e^{2i\pi\gamma} P^{(\gamma)} + \alpha'_{\gamma'} e^{2i\pi\gamma'} P^{(\gamma')}].$$
(3)

We may take the first equation of each pair (1) and (3) and solve for $(P^{(\beta)}, P^{(\beta')})$, and we may do the same with the second equation of each pair. On eliminating $(P^{(\beta)}, P^{(\beta')})$, we have two linear relations between $(P^{(\gamma)}, P^{(\gamma')})$ which are identically satisfied for all values of the latter; this will give four relations between the coefficients. Thus the two expressions for $P^{(\beta)}$ are

$$\alpha_\beta \sin \pi(\beta'-\beta) P^{(\beta)}$$
$$= \alpha_\gamma e^{i\pi(\alpha+\beta+\gamma)} \sin \pi(\alpha+\beta'+\gamma) P^{(\gamma)} +$$
$$+ \alpha_{\gamma'} e^{i\pi(\alpha+\beta+\gamma')} \sin \pi(\alpha+\beta'+\gamma') P^{(\gamma')},$$
$$\alpha'_\beta \sin \pi(\beta'-\beta) P^{(\beta)}$$
$$= \alpha'_\gamma e^{i\pi(\alpha'+\beta+\gamma)} \sin \pi(\alpha'+\beta'+\gamma) P^{(\gamma)} +$$
$$+ \alpha'_{\gamma'} e^{i\pi(\alpha'+\beta+\gamma')} \sin \pi(\alpha'+\beta'+\gamma') P^{(\gamma')},$$
(4)

and from these and the corresponding forms for $P^{(\beta')}$ we get the relations

$$\frac{\alpha_\beta}{\alpha'_\beta} = \frac{\alpha_\gamma e^{i\pi\alpha} \sin \pi(\alpha+\beta'+\gamma)}{\alpha'_\gamma e^{i\pi\alpha'} \sin \pi(\alpha'+\beta'+\gamma)} = \frac{\alpha_{\gamma'} e^{i\pi\alpha} \sin \pi(\alpha+\beta'+\gamma')}{\alpha'_{\gamma'} e^{i\pi\alpha'} \sin \pi(\alpha'+\beta'+\gamma')},$$
$$\frac{\alpha_{\beta'}}{\alpha'_{\beta'}} = \frac{\alpha_\gamma e^{i\pi\alpha} \sin \pi(\alpha+\beta+\gamma)}{\alpha'_\gamma e^{i\pi\alpha'} \sin \pi(\alpha'+\beta+\gamma)} = \frac{\alpha_{\gamma'} e^{i\pi\alpha} \sin \pi(\alpha+\beta+\gamma')}{\alpha'_{\gamma'} e^{i\pi\alpha'} \sin \pi(\alpha'+\beta+\gamma')}.$$
(5)

These relations are compatible, because $\sum \alpha = 1$. If we write

(λ,μ,ν) for the exponent-differences $(\alpha'-\alpha,\beta'-\beta,\gamma'-\gamma)$, we have $\sin\pi(\alpha+\beta+\gamma) = \cos\tfrac{1}{2}\pi(\lambda+\mu+\nu)$, etc. The relations then become

$$e^{i\pi\lambda}\frac{\alpha_\beta}{\alpha'_\beta}\cos\tfrac{1}{2}\pi(\lambda+\mu+\nu)\cos\tfrac{1}{2}\pi(\lambda+\mu-\nu)$$
$$= e^{i\pi\lambda}\frac{\alpha_{\beta'}}{\alpha'_{\beta'}}\cos\tfrac{1}{2}\pi(\lambda-\mu+\nu)\cos\tfrac{1}{2}\pi(\lambda-\mu-\nu)$$
$$= \frac{\alpha_\gamma}{\alpha'_\gamma}\cos\tfrac{1}{2}\pi(\lambda+\mu+\nu)\cos\tfrac{1}{2}\pi(\lambda-\mu+\nu)$$
$$= \frac{\alpha_{\gamma'}}{\alpha'_{\gamma'}}\cos\tfrac{1}{2}\pi(\lambda+\mu-\nu)\cos\tfrac{1}{2}\pi(\lambda-\mu-\nu). \qquad (6)$$

The coefficients will be evaluated below in § 26.

Exceptional Cases. If we have $\cos\tfrac{1}{2}\pi(\lambda+\mu+\nu) = 0$, we must also have $(\alpha_{\beta'} = 0 = \alpha_{\gamma'})$ or else $(\alpha'_\beta = 0 = \alpha'_\gamma)$. We can then arrange the notation and the constant multipliers of the principal branches so that the relations (1) become

$$P^{(\alpha)} = P^{(\beta)} = P^{(\gamma)},$$
$$P^{(\alpha')} = \alpha'_\beta P^{(\beta)} + \alpha'_{\beta'} P^{(\beta')} = \alpha'_\gamma P^{(\gamma)} + \alpha'_{\gamma'} P^{(\gamma')}. \qquad (7)$$

The first solution is merely multiplied by a constant after any closed circuit; and, since it is regular at infinity, its form can only be

$$P_1 = x^\alpha(1-x)^\gamma \text{ [a polynomial].} \qquad (8)$$

Similarly, there is a solution expressible in finite terms whenever $\cos\tfrac{1}{2}\pi(\lambda\pm\mu\pm\nu) = 0$. In the hypergeometric form, we find that

$$\lambda\pm\mu\pm\nu = (1-c)\pm(b-a)\pm(a+b-c) \qquad (9)$$

must be an odd integer. Hence one of the numbers $(a, b, a-c, b-c)$ must be an integer. We shall see that these cases are soluble by elementary methods. When the hypergeometric equation is satisfied by a polynomial, its group is generated by the transformations

$$\begin{array}{ll} Y_1 = y_1, & Y_1 = y_1, \\ Y_2 = e^{2i\pi\lambda}y_2, & Y_2 = Ay_1+e^{2i\pi\nu}y_2, \end{array} \qquad (10)$$

corresponding to circuits about $x = 0$ and $x = 1$ respectively, where y_1 is the polynomial and y_2 the other principal branch at $x = 0$.

General Case. If $\cos\tfrac{1}{2}\pi(\lambda\pm\mu\pm\nu) \neq 0$, we can reduce the relations between the solutions at $x = 0$ and $x = 1$ to the form

$$P^{(\alpha)} = P^{(\gamma)} + P^{(\gamma')}, \quad P^{(\alpha')} = P^{(\gamma)} + \kappa P^{(\gamma')}, \qquad (11)$$

where $\kappa = \dfrac{\alpha_\gamma \alpha'_{\gamma'}}{\alpha'_\gamma \alpha_{\gamma'}} = \dfrac{\cos\tfrac{1}{2}\pi(\lambda+\mu-\nu)\cos\tfrac{1}{2}\pi(\lambda-\mu-\nu)}{\cos\tfrac{1}{2}\pi(\lambda+\mu+\nu)\cos\tfrac{1}{2}\pi(\lambda-\mu+\nu)}. \qquad (12)$

Chap. VI, § 23 THE HYPERGEOMETRIC EQUATION

The generating transformations expressed in terms of the principal branches at the origin are

$$Y_1 = e^{2i\pi\alpha}y_1, \quad \Biggr\} \qquad Y_1 = \frac{(\kappa e^{2i\pi\gamma}-e^{2i\pi\gamma'})y_1+(e^{2i\pi\gamma}-e^{2i\pi\gamma'})y_2}{(\kappa-1)},$$
$$Y_2 = e^{2i\pi\alpha'}y_2, \qquad Y_2 = \frac{\kappa(e^{2i\pi\gamma}-e^{2i\pi\gamma'})y_1+(e^{2i\pi\gamma}-\kappa e^{2i\pi\gamma'})y_2}{(\kappa-1)}. \Biggr\} \quad (13)$$

The latter is obtained by expressing the transformation in terms of $(P^{(\gamma)}, P^{(\gamma')})$ and eliminating $(P^{(\gamma)}, P^{(\gamma')})$ by means of (11). It is evident that $\kappa \neq 0, \infty$; and it may be verified that $\kappa \neq 1$, because $\sin \pi\lambda \sin \pi\nu \neq 0$. Thus the relations (13) are well determined.

Associated P-Functions. In the relations (5) the exponents appear only through the multipliers $(e^{2i\pi\alpha})$, so that P-functions with the same multipliers have the same group. Consider three functions

$$P_i\begin{pmatrix} \alpha_i & \beta_i & \gamma_i \\ \alpha_i' & \beta_i' & \gamma_i' \end{pmatrix} x \Biggr) \qquad (\sum \alpha_i = 1) \quad (i=1,2,3), \tag{14}$$

whose respective exponents are congruent *modulo* 1. If $(\lambda_i, \mu_i, \nu_i)$ are the exponent-differences $(\alpha_i'-\alpha_i, \beta_i'-\beta_i, \gamma_i'-\gamma_i)$, we have, on account of the relations $\sum \alpha_i = 1$,

$$\lambda_i+\mu_i+\nu_i = 1-2(\alpha_i+\beta_i+\gamma_i), \tag{15}$$

and so $\qquad \lambda_i-\lambda_j+\mu_i-\mu_j+\nu_i-\nu_j =$ an even integer. $\tag{16}$

We may restrict ourselves to the reduced forms

$$P_i\begin{pmatrix} 0 & \frac{1}{2}(1-\lambda_i-\mu_i-\nu_i) & 0 \\ \lambda_i & \frac{1}{2}(1-\lambda_i+\mu_i-\nu_i) & \nu_i \end{pmatrix} x \Biggr) \quad (i=1,2,3), \tag{17}$$

where the necessity of the condition (16) is apparent, if the multipliers at infinity are equal. We suppose the branches so chosen that the coefficients in (1) are the same for $i = 1, 2, 3$; and we consider the expression

$$S_{ij} = [P_i^{(\alpha_i)}P_j^{(\alpha_j)}-P_i^{(\alpha_i')}P_j^{(\alpha_j)}]$$
$$= (\alpha_\beta\alpha_{\beta'}'-\alpha_\beta'\alpha_{\beta'})[P_i^{(\beta_i)}P_j^{(\beta_j)}-P_i^{(\beta_i')}P_j^{(\beta_j)}]$$
$$= (\alpha_\gamma\alpha_{\gamma'}'-\alpha_\gamma'\alpha_{\gamma'})[P_i^{(\gamma_i)}P_j^{(\gamma_j)}-P_i^{(\gamma_i')}P_j^{(\gamma_j)}]. \tag{18}$$

This belongs to the lower of the exponents $(\alpha_i+\alpha_j', \alpha_i'+\alpha_j)$ at $x = 0$, $(\beta_i+\beta_j', \beta_i'+\beta_j)$ at $x = \infty$, $(\gamma_i+\gamma_j', \gamma_i'+\gamma_j)$ at $x = 1$, and these may be written

$$\alpha_{ij} = \tfrac{1}{2}[\alpha_i+\alpha_j+\alpha_i'+\alpha_j'-|\lambda_i-\lambda_j|],$$
$$\beta_{ij} = \tfrac{1}{2}[\beta_i+\beta_j+\beta_i'+\beta_j'-|\mu_i-\mu_j|], \Biggr\} \tag{19}$$
$$\gamma_{ij} = \tfrac{1}{2}[\gamma_i+\gamma_j+\gamma_i'+\gamma_j'-|\nu_i-\nu_j|].$$

Thus $x^{-\alpha_{ij}}(x-1)^{-\gamma_{ij}} S_{ij}$ belongs at infinity to the exponent
$$(\alpha_{ij}+\beta_{ij}+\gamma_{ij})$$
and is holomorphic for all finite values of x, so that it is a polynomial of degree
$$N_{ij} = \tfrac{1}{2}[|\lambda_i-\lambda_j|+|\mu_i-\mu_j|+|\nu_i-\nu_j|-2], \tag{20}$$
which is an integer because of (16).

Three corresponding branches of the three functions are of the form
$$P_i = A P_i^{(\alpha_i)} + A' P_i^{(\alpha'_i)} \quad (i = 1, 2, 3), \tag{21}$$
with the same coefficients (A, A'). These are connected by the relation
$$P_1 S_{23} + P_2 S_{31} + P_3 S_{12} = 0. \tag{22}$$

If $(\bar{\alpha}, \bar{\gamma})$ are the lowest exponents of the triads (α_{ij}) and (γ_{ij}) differing among themselves by integers, we can remove a factor $x^{\bar{\alpha}}(1-x)^{\bar{\gamma}}$ and find a relation
$$P_1 \chi_1 + P_2 \chi_2 + P_3 \chi_3 = 0, \tag{23}$$
where χ_1 is a polynomial of degree $(\alpha_{23}-\bar{\alpha}+\gamma_{23}-\bar{\gamma}+N_{23})$, and so on.

24. Recurrence Formulae. Hypergeometric Polynomials

Contiguous Series. We can illustrate Riemann's theorems on associated hypergeometric functions by means of Gauss's relations between contiguous functions. The six functions
$$F(a\pm 1, b; c; x), \quad F(a, b\pm 1; c; x), \quad F(a, b; c\pm 1; x)$$
are said to be *contiguous* to $F \equiv F(a, b; c; x)$, and are denoted by F_{a+}, etc. Each of them can be simply expressed in terms of F and DF, and on eliminating the latter we get fifteen relations between F and any two contiguous series, which were given by Gauss. To pass from F to F_{a+}, we must multiply the coefficient of x^n by $(a+n)$ and remove the factor a; this is readily effected by the operator $(\delta+a)$, where $\delta \equiv xD$. We thus have at once three of the required formulae
$$aF_{a+} = (\delta+a)F, \tag{1}$$
$$bF_{b+} = (\delta+b)F, \tag{2}$$
$$(c-1)F_{c-} = (\delta+c-1)F. \tag{3}$$

We next write the equation for F_{a-} in the form
$$[\delta(\delta+c-1)-x(\delta+a-1)(\delta+b)]F_{a-} = 0,$$
or $\quad [(\delta+c-a)-x(\delta+b)](\delta+a-1)F_{a-} = (c-a)(a-1)F_{a-};$

on using (1), with $(a-1)$ written for a, we now get
$$(c-a)F_{a-} = [(\delta+c-a)-x(\delta+b)]F,$$
i.e.
$$(c-a)F_{a-} = (1-x)\delta F+(c-a-bx)F, \qquad (4)$$
and
$$(c-b)F_{b-} = (1-x)\delta F+(c-b-ax)F. \qquad (5)$$

Finally we have the equation
$$[\delta(\delta+c)-x(\delta+a)(\delta+b)]F_{c+} = 0,$$
i.e. $\quad [\delta-x(\delta+a+b-c)](\delta+c)F_{c+} = (c-a)(c-b)xF_{c+},$
which gives by means of (3)
$$(c-a)(c-b)F_{c+} = c[(1-x)DF+(c-a-b)F]. \qquad (6)$$

Our six formulae may now be arranged as follows:
$$xDF = a(F_{a+}-F), \qquad (1')$$
$$xDF = b(F_{b+}-F), \qquad (2')$$
$$xDF = (c-1)(F_{c-}-F), \qquad (3')$$
$$x(1-x)DF = (c-a)F_{a-}+(a-c+bx)F, \qquad (4')$$
$$x(1-x)DF = (c-b)F_{b-}+(b-c+ax)F, \qquad (5')$$
$$c(1-x)DF = (c-a)(c-b)F_{c+}+c(a+b-c)F. \qquad (6')$$

Gauss's fifteen relations now follow by equating two values of DF.

Associated Series. Gauss showed, by constructing chains of contiguous functions, that any three series $F(a+l, b+m; c+n; x)$, where l, m, n are integers, are connected by a linear homogeneous relation with polynomial coefficients. By differentiating the hypergeometric equation a certain number of times and eliminating intermediate derivatives, we can show that this is true for any three derivatives of $F(a, b; c; x)$. Any other associated series can be expressed in terms of F and
$$DF \equiv (ab/c)F(a+1, b+1; c+1; x)$$
by repeated use of the operations (1)–(6), the results being very similar to a celebrated formula of Jacobi given below.

If k is a positive integer, we have from (1)
$$a_k F(a+k, b; c; x) = (\delta+a)(\delta+a+1)...(\delta+a+k-1)F$$
$$= x^{1-a}D^k[x^{a+k-1}F(a, b; c; x)], \qquad (7)$$
where $a_k \equiv a(a+1)...(a+k-1)$.

Similarly from (3),
$$(c-k)_k F(a, b; c-k; x) = x^{1-c+k}D^k[x^{c-1}F(a, b; c; x)]. \qquad (8)$$

Again, we may write (4) in the form
$$(c-a)x^{c-a+1}(1-x)^{a+b-c-1}F(a-1,b;c;x)$$
$$= x^2 D[x^{c-a}(1-x)^{a+b-c}F(a,b;c;x)],$$
and hence we have
$$(c-a)_k x^{c-a+k}(1-x)^{a+b-c-k}F(a-k,b;c;x)$$
$$= (x^2 D)^k[x^{c-a}(1-x)^{a+b-c}F(a,b;c;x)].$$
But
$$(x^2 D)^k = (x\delta)^k$$
$$= x^k(\delta+k-1)(\delta+k-2)\ldots\delta$$
$$= x\delta(\delta-1)\ldots(\delta-k+1)x^{k-1}$$
$$= x^{k+1}D^k x^{k-1}, \tag{9}$$
and so we find that
$$(c-a)_k F(a-k,b;c;x)$$
$$= x^{a-c+1}(1-x)^{c-a-b+k}D^k[x^{c-a+k-1}(1-x)^{a+b-c}F(a,b;c;x)]. \tag{10}$$
In the same way we have
$$(c-a)_k(c-b)_k F(a,b;c+k;x)$$
$$= c_k(1-x)^{c-a-b+k}D^k[(1-x)^{a+b-c}F(a,b;c;x)]. \tag{11}$$

Jacobi's Formulae.† If $M \equiv x^{c-1}(1-x)^{a+b-c}$, the hypergeometric equation may be written in the form
$$D[x(1-x)MDy] = ab\, My. \tag{12}$$
Similarly, we may differentiate the hypergeometric equation $(k-1)$ times and write
$$D[x^k(1-x)^k M D^k y] = (a+k-1)(b+k-1)x^{k-1}(1-x)^{k-1}MD^{k-1}y, \tag{13}$$
whence we have the recurrence formula
$$D^k[x^k(1-x)^k M D^k y] = (a+k-1)(b+k-1)D^{k-1}[x^{k-1}(1-x)^{k-1}MD^{k-1}y]$$
$$= a_k b_k My. \tag{14}$$
This is true for every solution of the hypergeometric equation. If we put, in particular, $y = F(a,b;c;x)$, and so
$$D^k y = (a_k b_k/c_k)F(a+k,b+k;c+k;x),$$
we may remove the factors $a_k b_k$, provided that F is not a polynomial of degree less than k, and we then have
$$D^k[x^k(1-x)^k MF(a+k,b+k;c+k;x)] = c_k MF(a,b;c;x). \tag{15}$$
If $b = -n$ is a negative integer, so that F is a polynomial of degree n,

† C. G. J. Jacobi, *Werke*, 6, 184–202.

$F(a+n, b+n; c+n; x)$ is unity, and then we have a finite expression for the hypergeometric polynomial

$$c_n F(a, -n; c; x) = x^{1-c}(1-x)^{n+c-a} D^n[x^{c+n-1}(1-x)^{a-c}]. \quad (16)$$

Again, if $b = -n$, we have on differentiating n times the hypergeometric equation the relation

$$x(1-x)D^{n+2}y + [(c+n) - (a+n+1)x]D^{n+1}y = 0. \quad (17)$$

For the polynomial (16) we have $D^{n+1}y = 0$; and for the general solution

$$x^{c+n}(1-x)^{a-c+1} D^{n+1} y = \text{constant}, \quad (18)$$

$$D^n y = C \int x^{-c-n}(1-x)^{c-a-1} dx. \quad (19)$$

In the general formula (14) we may put $k = n$ and introduce this expression for $D^n y$; the second solution is then expressed in a finite form involving a single quadrature

$$My = AD^n\left[x^n(1-x)^n M \int x^{-c-n}(1-x)^{c-a-1} dx\right]. \quad (20)$$

By expanding in descending powers of x, we see that this is the solution belonging to the exponent a at $x = \infty$. We give among the exercises formulae showing that the hypergeometric equation is soluble in finite terms, if any of the numbers $(a, b, c-a, c-b)$ is an integer.

Another Notation.† To exhibit the polynomials of Legendre and Tschebyscheff as particular cases of Jacobi's, it is convenient to place the singularities of the hypergeometric function at $x = \pm 1, \infty$, and to modify the exponents. We write

$$P_n^{(\alpha,\beta)}(x) \equiv \frac{(x-1)^{-\alpha}(x+1)^{-\beta}}{2^n \cdot n!} D^n[(x-1)^{\alpha+n}(x+1)^{\beta+n}], \quad (21)$$

and find from (16) after a little manipulation

$$P_n^{(\alpha,\beta)}(x) \equiv \frac{(\alpha+1)_n}{n!} F\left(\alpha+\beta+n+1, -n; \alpha+1; \frac{1-x}{2}\right), \quad (22)$$

the scheme of the differential equation being

$$P\begin{Bmatrix} -1 & \infty & 1 & \\ 0 & -n & 0 & x \\ -\beta & (\alpha+\beta+n+1) & -\alpha & \end{Bmatrix}. \quad (23)$$

Thus $P_n^{(\alpha,\beta)}(x)$ is a solution of

$$(1-x^2)D^2 y + [(\beta-\alpha) - (\alpha+\beta+2)x]Dy + n(\alpha+\beta+n+1)y = 0, \quad (24)$$

† G. Pólya and G. Szegö, *Aufgaben und Lehrsätze ...*, **1**, 127; **2**, 93. R. Courant and D. Hilbert, *Methoden der mathematischen Physik*, **1**, 66–9, 72–5.

and a second solution is
$$y = A(x-1)^{-\alpha}(x+1)^{-\beta} \times$$
$$\times D^n\Big[(x-1)^{\alpha+n}(x+1)^{\beta+n} \int (x-1)^{-\alpha-n-1}(x+1)^{-\beta-n-1}\,dx\Big]. \quad (25)$$

Generating Function. If $\phi(z)$ is holomorphic at $z = x$ and $\phi(x) \neq 0$, and if
$$z = x + w\phi(z), \quad (26)$$
then an analytic function which is holomorphic at $z = x$ can be expanded by Lagrange's formula
$$f(z) = f(x) + \sum_{n=1}^{\infty} \frac{w^n}{n!} D^{n-1}\{[\phi(x)]^n f'(x)\}. \quad (27)$$

If we differentiate with respect to w and afterwards write $f(z)$ instead of $\phi(z)f'(z)$, we find the expansion
$$\frac{f(z)}{1-w\phi'(z)} = \sum_{n=0}^{\infty} \frac{w^n}{n!} D^n\{[\phi(x)]^n f(x)\}. \quad (28)$$

Let us here write
$$\phi(x) \equiv \tfrac{1}{2}(x^2-1), \qquad f(x) \equiv (x-1)^{\alpha}(x+1)^{\beta}, \atop z = x + \tfrac{1}{2}w(z^2-1). \quad (29)$$

The root which tends to x as w tends to zero is
$$z = w^{-1}[1-(1-2xw+w^2)^{\frac{1}{2}}], \quad (30)$$
and (21) and (28) together give
$$\sum_{n=0}^{\infty} w^n (x-1)^{\alpha}(x+1)^{\beta} P_n^{(\alpha,\beta)}(x) = \frac{(z-1)^{\alpha}(z+1)^{\beta}}{1-wz}$$
$$= \frac{[1-w-(1-2xw+w^2)^{\frac{1}{2}}]^{\alpha}[1+w-(1-2xw+w^2)^{\frac{1}{2}}]^{\beta}}{w^{\alpha+\beta}(1-2xw+w^2)^{\frac{1}{2}}}$$
$$= \frac{(x-1)^{\alpha}(x+1)^{\beta}[1-w+(1-2xw+w^2)^{\frac{1}{2}}]^{-\alpha}[1+w+(1-2xw+w^2)^{\frac{1}{2}}]^{-\beta}}{(1-2xw+w^2)^{\frac{1}{2}}}. \quad (31)$$

Hence $P_n^{(\alpha,\beta)}(x)$ is the coefficient of w^n in the expansion of
$$\sum_{n=0}^{\infty} w^n P_n^{(\alpha,\beta)}(x) = \frac{[1-w+(1-2xw+w^2)^{\frac{1}{2}}]^{-\alpha}[1+w+(1-2xw+w^2)^{\frac{1}{2}}]^{-\beta}}{(1-2xw+w^2)^{\frac{1}{2}}}. \quad (32)$$

If $\alpha = \beta = 0$, we have the well-known formulae for the Legendre polynomials
$$(1-2xw+w^2)^{-\frac{1}{2}} = \sum_{n=0}^{\infty} w^n P_n(x) = \sum_{n=0}^{\infty} \frac{w^n}{n!\,2^n} D^n(x^2-1)^n. \quad (33)$$

Orthogonal Relations. If $u \equiv P_m^{(\alpha,\beta)}(x)$ and $v \equiv P_n^{(\alpha,\beta)}(x)$, we have by cross-multiplying the differential equations the relation

$$D[(1-x)^{\alpha+1}(1+x)^{\beta+1}(vDu-uDv)]$$
$$= (n-m)(n+m+\alpha+\beta+1)(1-x)^{\alpha}(1+x)^{\beta}uv. \quad (34)$$

If the real parts of α, β are greater than -1 and if $m \neq n$, we may integrate between the limits ± 1, and we have

$$I_{m,n} = \int_{-1}^{1} (1-x)^{\alpha}(1+x)^{\beta} P_m^{(\alpha,\beta)}(x) P_n^{(\alpha,\beta)}(x) \, dx = 0 \quad (m \neq n). \quad (35)$$

To evaluate the integral when $m = n$, we write

$$I_{n,n} = \int_{-1}^{1} \frac{(-)^n}{2^n . n!} [D^n\{(1-x)^{\alpha+n}(1+x)^{\beta+n}\}] P_n^{(\alpha,\beta)}(x) \, dx$$

$$= \int_{-1}^{1} \frac{(1-x)^{\alpha+n}(1+x)^{\beta+n}}{2^n . n!} D^n P_n^{(\alpha,\beta)}(x) \, dx$$

$$= \frac{(\alpha+\beta+n+1)_n}{2^{2n} . n!} \int_{-1}^{1} (1-x)^{\alpha+n}(1+x)^{\beta+n} \, dx$$

$$= \frac{2^{\alpha+\beta+1} \Gamma(\alpha+n+1) \Gamma(\beta+n+1)}{n!(\alpha+\beta+2n+1)\Gamma(\alpha+\beta+n+1)}. \quad (36)$$

25. Quadratic and Cubic Transformations

Quadratic Transformation. There are many known cases of the transformation of one hypergeometric equation into another by relations of the type
$$x' = \phi(x), \qquad y' = wy, \quad (1)$$

and their systematic enumeration was begun by Kummer[†] and completed by Goursat.[‡] The only possible transformations of the equation with *three* arbitrary exponent-differences are *linear*; and those of the equation with *two* unrestricted exponent-differences are at most quadratic. The higher transformations are connected with the theory of the polyhedral functions and apply to equations having only one free parameter or none.

[†] E. E. Kummer, *J. für Math.* **15** (1836), 39–83, 127–72.
[‡] See particularly:—E. Goursat, *Annales de l'École Normale*, (2) **10** (1881), Suppl. 3–142; *Acta Soc. Sc. Fennicae*, **15** (1884–8), 45–127.

The quadratic transformation is most clearly expressed by Riemann's scheme

$$P\begin{Bmatrix} 0 & \infty & 1 \\ 0 & \beta & \gamma & x \\ \tfrac{1}{2} & \beta' & \gamma' \end{Bmatrix} = P\begin{Bmatrix} -1 & \infty & 1 \\ \gamma & 2\beta & \gamma & \sqrt{x} \\ \gamma' & 2\beta' & \gamma' \end{Bmatrix}$$

$$= P\begin{Bmatrix} -1 & \infty & 1 \\ \beta & 2\gamma & \beta & \left(\dfrac{x}{x-1}\right)^{\tfrac{1}{2}} \\ \beta' & 2\gamma' & \beta' \end{Bmatrix}, \qquad (2)$$

where $(\beta+\beta'+\gamma+\gamma') = \tfrac{1}{2}$. It is applicable whenever one exponent-difference is $\tfrac{1}{2}$, or when two are equal. The canonical form of the equation with two equal exponent-differences is that of the associated Legendre equation

$$(1-x^2)D^2y - 2xDy + \left[n(n+1) - \dfrac{m^2}{1-x^2}\right]y = 0, \qquad (3)$$

whose scheme is

$$P\begin{Bmatrix} -1 & \infty & 1 \\ \tfrac{1}{2}m & n+1 & \tfrac{1}{2}m & x \\ -\tfrac{1}{2}m & -n & -\tfrac{1}{2}m \end{Bmatrix}. \qquad (4)$$

If we put $x = \cos\theta$ in (4), we have the relations

$$P\begin{pmatrix} \tfrac{1}{2}m & n+1 & \tfrac{1}{2}m & \sin^2\tfrac{1}{2}\theta \\ -\tfrac{1}{2}m & -n & -\tfrac{1}{2}m \end{pmatrix}$$

$$= P\begin{pmatrix} 0 & \tfrac{1}{2}(n+1) & \tfrac{1}{2}m & \cos^2\theta \\ \tfrac{1}{2} & -\tfrac{1}{2}n & -\tfrac{1}{2}m \end{pmatrix}$$

$$= P\begin{pmatrix} \tfrac{1}{2}(n+1) & m & \tfrac{1}{2}(n+1) & \tfrac{1}{2}(1+i\cot\theta) \\ -\tfrac{1}{2}n & -m & -\tfrac{1}{2}n \end{pmatrix}. \qquad (5)$$

We thus obtain Olbricht's 72 hypergeometric series[†] in ascending or descending powers of

$$\left.\begin{matrix} \cos^2\tfrac{1}{2}\theta, & \sin^2\tfrac{1}{2}\theta, & -\cot^2\tfrac{1}{2}\theta, \\ \cos^2\theta, & \sin^2\theta, & -\cot^2\theta, \\ \tfrac{1}{2}(1\pm i\cot\theta), & e^{2i\theta}. & \end{matrix}\right\} \qquad (6)$$

Whipple's Formula.[‡] We can also write (5) in the equivalent form

---

[†] E. W. Hobson, *Spherical and Ellipsoidal Harmonics* (1931), 284–8.
[‡] F. J. W. Whipple, *Proc. London Math. Soc.* (2), **16** (1917), 301–14; Hobson, loc. cit. 245.

Chap. VI, § 25 THE HYPERGEOMETRIC EQUATION

$$P\begin{Bmatrix} -1 & \infty & 1 & \\ \tfrac{1}{2}m & n+1 & \tfrac{1}{2}m & \cosh\alpha \\ -\tfrac{1}{2}m & -n & -\tfrac{1}{2}m & \end{Bmatrix}$$

$$= \sinh^{-\tfrac{1}{2}}\alpha \, P\begin{Bmatrix} -1 & \infty & 1 & \\ \tfrac{1}{2}n+\tfrac{1}{4} & \tfrac{1}{2}+m & \tfrac{1}{2}n+\tfrac{1}{4} & \coth\alpha \\ -\tfrac{1}{2}n-\tfrac{1}{4} & \tfrac{1}{2}-m & -\tfrac{1}{2}n-\tfrac{1}{4} & \end{Bmatrix}. \quad (7)$$

Since the formula is symmetrical in $\pm(n+\tfrac{1}{2})$, we may assume that the real part of $(n+1)$ is positive. We then obtain two alternative expressions for the branch belonging to the exponent $(n+1)$ at $\cosh\alpha = \infty$ or $\coth\alpha = 1$. We have by definition, at $x = \infty$,

$$Q_n^m(x) = \frac{e^{i\pi m}\Gamma(m+n+1)\Gamma(\tfrac{1}{2})}{2^{n+1}\Gamma(n+\tfrac{3}{2})} \times$$
$$\times (x^2-1)^{\tfrac{1}{2}m} x^{-m-n-1} F\left(\frac{m+n+1}{2}, \frac{m+n+2}{2}; n+\tfrac{3}{2}; \frac{1}{x^2}\right), \quad (8)$$

and so

$$Q_n^m(\cosh\alpha) = \frac{e^{i\pi m}\Gamma(m+n+1)\Gamma(\tfrac{1}{2})}{2^{n+1}\Gamma(n+\tfrac{3}{2})} \times$$
$$\times \sinh^m\alpha \cosh^{-m-n-1}\alpha \, F\left(\frac{m+n+1}{2}, \frac{m+n+2}{2}; n+\tfrac{3}{2}; \operatorname{sech}^2\alpha\right)$$
$$\sim \frac{e^{i\pi m}\Gamma(m+n+1)\Gamma(\tfrac{1}{2})}{\Gamma(n+\tfrac{3}{2})} e^{-(n+1)\alpha}, \quad \text{as } \alpha \to +\infty. \quad (9)$$

Again, we have by definition, at $x = 1$,

$$P_n^m(x) = \frac{1}{\Gamma(1-m)}\left(\frac{x+1}{x-1}\right)^{\tfrac{1}{2}m} F\left(-n, n+1; 1-m; \frac{1-x}{2}\right), \quad (10)$$

and so

$$P_{-m-\tfrac{1}{2}}^{-n-\tfrac{1}{2}}(\coth\alpha) = \frac{1}{\Gamma(n+\tfrac{3}{2})} e^{-(n+\tfrac{1}{2})\alpha} F\left(\tfrac{1}{2}+m, \tfrac{1}{2}-m; n+\tfrac{3}{2}; \frac{1-\coth\alpha}{2}\right)$$
$$\sim \frac{e^{-(n+\tfrac{1}{2})\alpha}}{\Gamma(n+\tfrac{3}{2})}, \quad \text{as } \alpha \to +\infty. \quad (11)$$

On introducing the expressions (9) and (11) on the left and right of the relation (7), we obtain Whipple's formula, which was originally established by transforming contour integrals, namely

$$Q_n^m(\cosh\alpha) = \frac{e^{i\pi m}\pi^{\tfrac{1}{2}}\Gamma(m+n+1)}{2^{\tfrac{1}{2}}\sinh^{\tfrac{1}{2}}\alpha} P_{-m-\tfrac{1}{2}}^{-n-\tfrac{1}{2}}(\coth\alpha). \quad (12)$$

Cubic Transformation. When two exponent-differences are $\tfrac{1}{3}$, or when all three are equal to one another, we can apply the

transformation
$$P\begin{Bmatrix} 0 & \infty & 1 & \\ 0 & 0 & \gamma & x \\ \tfrac{1}{3} & \tfrac{1}{3} & \gamma' & \end{Bmatrix} = P\begin{Bmatrix} 1 & \omega & \omega^2 & \\ \gamma & \gamma & \gamma & \sqrt[3]{x} \\ \gamma' & \gamma' & \gamma' & \end{Bmatrix}, \tag{13}$$

where $(\gamma+\gamma') = \tfrac{1}{3}$ and $\omega = \exp(2i\pi/3)$. We can also apply the quadratic transformation to either of the two expressions (13), and Riemann thus obtains six equivalent P-functions

$$\left.\begin{aligned} &P(\nu,\nu,\nu,x_3), \quad P(\nu,\tfrac{1}{2}\nu,\tfrac{1}{2},x_2), \quad P(\tfrac{1}{2}\nu,2\nu,\tfrac{1}{2}\nu,x_1), \\ &P(\tfrac{1}{3},\nu,\tfrac{1}{3},x_4), \quad P(\tfrac{1}{3},\tfrac{1}{2}\nu,\tfrac{1}{2},x_5), \quad P(\tfrac{1}{2}\nu,\tfrac{2}{3},\tfrac{1}{2}\nu,x_6), \end{aligned}\right\} \tag{14}$$

where

$$\left.\begin{aligned} x_2 &= 4x_3(1-x_3) = \frac{1}{4x_1(1-x_1)}, \\ x_5 &= 4x_4(1-x_4) = \frac{1}{4x_6(1-x_6)}, \\ x_4 &= \frac{(x_3+\omega^2)^3}{3(\omega-\omega^2)x_3(1-x_3)}, \quad (1-x_4) = \frac{(x_3+\omega)^3}{3(\omega^2-\omega)x_3(1-x_3)}. \end{aligned}\right\} \tag{15}$$

To each of these we can apply the general linear transformation.

Repeated Quadratic Transformation. If one exponent-difference is $\tfrac{1}{2}$ and the other two are equal, we have the equivalent schemes given by Riemann

$$\left.\begin{aligned} &P(\nu,\nu,\tfrac{1}{2},x_2), \quad P(\nu,2\nu,\nu,x_1), \\ &P(\tfrac{1}{4},\nu,\tfrac{1}{2},x_3), \quad P(\tfrac{1}{4},2\nu,\tfrac{1}{4},x_4), \end{aligned}\right\} \tag{16}$$

where $\quad x_2 = 4x_1(1-x_1), \quad x_3 = 4x_4(1-x_4) = \tfrac{1}{4}\!\left(2-x_2-\dfrac{1}{x_2}\right). \tag{17}$

If two exponent-differences are $\tfrac{1}{2}$, we have an elementary function, since

$$P\begin{pmatrix} 0 & 0 & \tfrac{1}{2}\nu & \\ \tfrac{1}{2} & \tfrac{1}{2} & -\tfrac{1}{2}\nu & x^2 \end{pmatrix} = P\begin{Bmatrix} \tfrac{1}{2}\nu & 0 & \tfrac{1}{2}\nu & \\ -\tfrac{1}{2}\nu & 1 & -\tfrac{1}{2}\nu & \left(\dfrac{1+x}{2}\right) \end{Bmatrix}$$

$$= \left(\frac{1+x}{1-x}\right)^{\tfrac{1}{2}\nu} P\begin{pmatrix} 0 & 0 & 0 & \dfrac{1+x}{2} \\ -\nu & 1 & \nu & \end{pmatrix}$$

$$= A\left(\frac{1+x}{1-x}\right)^{\tfrac{1}{2}\nu} + B\left(\frac{1-x}{1+x}\right)^{\tfrac{1}{2}\nu}. \tag{18}$$

26. Continuation of the Hypergeometric Series

Convergence on the Unit Circle. The simple ratio test suffices to prove the convergence of $F(a,b;c;x) \equiv \sum c_n x^n$ in the interior of the unit circle, but fails on the circumference. Now we have

$$\frac{c_n}{c_{n+1}} = \frac{(n+1)(c+n)}{(a+n)(b+n)} = 1 + \frac{1+c-a-b}{n} + O\!\left(\frac{1}{n^2}\right),$$

and
$$\left|\frac{c_n}{c_{n+1}}\right| = 1 + \frac{1+R(c-a-b)}{n} + O\!\left(\frac{1}{n^2}\right); \tag{1}$$

hence, by Raabe's test or by direct comparison with $\sum n^{-k}$, the series is absolutely convergent on the unit circle if $R(c-a-b) > 0$. We observe the significant role of the exponent-difference at $x = 1$; for if $R(c-a-b) > 0$, both the principal branches remain bounded as $x \to 1$, with any fixed amplitude of $(x-1)$.

If $R(c-a-b) \leqslant -1$, the general term does not tend to zero, so that the series is definitely divergent. A still more delicate test is required in the doubtful range of values $0 \geqslant R(c-a-b) > -1$. We observe that in this case $|c_n| \to 0$; provided that $x \neq 1$, we may make the transformation
$$(1-x)\sum_{n=0}^{\infty} c_n x^n = c_0 + \sum_{n=1}^{\infty}(c_n - c_{n-1})x^n, \tag{2}$$
and this is expressible in terms of two contiguous series
$$c(1-x)F = x(b-c)F_{c+} + cF_{a-}, \tag{3}$$
both of which are absolutely convergent. Hence the series converges (though not absolutely) at all points of the unit circle except $x = 1$, when $0 \geqslant R(c-a-b) > -1$.

To consider $F(a, b; c; 1)$, we use the gamma product
$$\prod_{r=0}^{n-1}\left[\frac{(a+r)(b+r)}{(c+r)(a+b-c+r)}\right] = \frac{\Gamma(c)\Gamma(a+b-c)}{\Gamma(a)\Gamma(b)}\left[1 + O\!\left(\frac{1}{n}\right)\right], \tag{4}$$
which enables us to compare the series with one involving binomial coefficients
$$\sum c_n^* = \sum (-)^n \binom{c-a-b}{n}\left[1 + O\!\left(\frac{1}{n}\right)\right]. \tag{5}$$

The series of error terms is absolutely convergent, by Raabe's test, if $R(c-a-b) > -1$. The sum of the first $(n+1)$ coefficients of the series $(1-x)^{c-a-b}$ is the coefficient of x^n in the series $(1-x)^{c-a-b-1}$, and this can be verified by elementary methods. This tends to infinity, and so $F(a, b; c; 1)$ diverges, when $R(c-a-b) \leqslant 0$. If $(c-a-b) = 0$, both sides of (4) become infinite. But we can remove the unwanted factor $(a+b-c)$ in the denominator, and so obtain the logarithmic comparison series $\sum x^n/n$, from which the same result follows.

Gauss's Evaluation of $F(a, b; c; 1)$. Consider the relation between contiguous series
$$(c-a)(c-b)xF_{c+} - c(c-a-b)xF = c(c-1)(1-x)(F_{c-} - F). \tag{6}$$
If $R(c-a-b) > 0$, both F and F_{c+} converge absolutely in the closed

interval ($0 \leqslant x \leqslant 1$). The series F_{c-} need not converge at $x = 1$; but at any rate its coefficients tend to zero. If we apply to it the transformation (2) and put $x = 1$, the sum of the series on the right is $\lim_{n \to \infty} c_n = 0$; by Abel's theorem on the continuity of a power-series we have therefore $\lim_{x \to 1}[(1-x)F_{c-}] = 0$, and hence

$$c(c-a-b)F(a,b;c;1) = (c-a)(c-b)F(a,b;c+1;1), \qquad (7)$$

and accordingly

$$\frac{F(a,b;c;1)}{F(a,b;c+n;1)} = \prod_{r=0}^{n-1} \left[\frac{(c-a+r)(c-b+r)}{(c+r)(c-a-b+r)}\right]. \qquad (8)$$

If a, b, c are fixed, the series $F(a,b;c+n;1)$ converges uniformly with respect to n, when $n > n_0$ say; since each term except the first tends to zero, we have

$$\lim_{n \to \infty} F(a,b;c+n;1) = 1. \qquad (9)$$

The terms grouped in square brackets in the product (8) give an expression of the type $\left[1+O\!\left(\frac{1}{r^2}\right)\right]$; so that the product converges, and may be evaluated like (4) by means of gamma functions. We thus have finally

$$F(a,b;c;1) = \frac{\Gamma(c)\Gamma(c-a-b)}{\Gamma(c-a)\Gamma(c-b)}. \qquad (10)$$

It is convenient, even when the hypergeometric series is divergent, to write

$$\phi(a,b,c) \equiv \frac{\Gamma(c)\Gamma(c-a-b)}{\Gamma(c-a)\Gamma(c-b)}. \qquad (11)$$

This remains finite unless the argument of one of the gamma functions is zero or a negative integer.

Kummer's Continuation Formulae. We can now evaluate directly the coefficients of the continuation formulae in § 23; but we must first remove all ambiguity regarding the branches of the hypergeometric function. We shall make a cut along the entire real axis from $x = -\infty$ to $x = \infty$, and consider the branches defined in the upper half-plane by the following conditions.

$$\left.\begin{array}{ll} P^{(\alpha)} \to 1, \quad P^{(\alpha')} \sim x^{1-c} & \text{(as } x \to 0\text{),} \\ P^{(\beta)} \sim (1-x)^{-a}, \quad P^{(\beta')} \sim (1-x)^{-b} & \text{(as } x \to \infty\text{),} \\ P^{(\gamma)} \to 1, \quad P^{(\gamma')} \sim (1-x)^{c-a-b} & \text{(as } x \to 1\text{),} \\ 0 \leqslant \operatorname{am}(x) \leqslant \pi, \quad -\pi \leqslant \operatorname{am}(1-x) \leqslant 0. & \end{array}\right\} \qquad (12)$$

THE HYPERGEOMETRIC EQUATION

The coefficients of the formulae
$$P^{(\alpha)} = \alpha_\gamma P^{(\gamma)} + \alpha_{\gamma'} P^{(\gamma')}, \quad P^{(\alpha')} = \alpha'_\gamma P^{(\gamma)} + \alpha'_{\gamma'} P^{(\gamma')}, \tag{13}$$
will be found by making $x \to 1$ and $x \to 0$.

(i) If $R(c-a-b) > 0$, the first expansions in the table of Kummer's series give, when $x \to 1$,
$$\left. \begin{aligned} P^{(\alpha)} &\to \phi(a,b,c), \quad P^{(\alpha')} \to \phi(a-c+1, b-c+1, 2-c), \\ P^{(\gamma)} &\to 1, \quad P^{(\gamma')} \to 0, \end{aligned} \right\} \tag{14}$$
and hence
$$\alpha_\gamma = \phi(a,b,c), \quad \alpha'_\gamma = \phi(a-c+1, b-c+1, 2-c). \tag{15a}$$

(ii) If $R(c-a-b) < 0$, the second Kummer series of each function gives, as $x \to 1$,
$$\left. \begin{aligned} P^{(\alpha)} &\sim (1-x)^{c-a-b}\phi(c-a, c-b, c), \\ P^{(\alpha')} &\sim (1-x)^{c-a-b}\phi(1-a, 1-b, 2-c), \\ P^{(\gamma)} &\to 1, \quad P^{(\gamma')} \sim (1-x)^{c-a-b}, \end{aligned} \right\} \tag{16}$$
and hence
$$\alpha_{\gamma'} = \phi(c-a, c-b, c), \quad \alpha'_{\gamma'} = \phi(1-a, 1-b, 2-c). \tag{15b}$$

(iii) If $R(1-c) > 0$, we have also, as $x \to 0$,
$$\left. \begin{aligned} P^{(\alpha)} &\to 1, \quad P^{(\alpha')} \to 0, \\ P^{(\gamma)} \to \phi(a,b,a+b-c+1), \quad & P^{(\gamma')} \to \phi(c-a, c-b, c-a-b+1). \end{aligned} \right\} \tag{17}$$
The relations now give
$$\left. \begin{aligned} \alpha_\gamma \phi(a,b,a+b-c+1) + \alpha_{\gamma'}\phi(c-a,c-b,c-a-b+1) &= 1, \\ \alpha'_\gamma \phi(a,b,a+b-c+1) + \alpha'_{\gamma'}\phi(c-a,c-b,c-a-b+1) &= 0. \end{aligned} \right\} \tag{18a}$$

(iv) If $R(1-c) < 0$, we have, as $x \to 0$,
$$\left. \begin{aligned} P^{(\alpha)} &\to 1, \quad P^{(\alpha')} \sim x^{1-c}, \\ P^{(\gamma)} &\sim x^{1-c}\phi(a-c+1, b-c+1, a+b-c+1), \\ P^{(\gamma')} &\sim x^{1-c}\phi(1-a, 1-b, c-a-b+1), \end{aligned} \right\} \tag{19}$$
and these now give
$$\left. \begin{aligned} \alpha_\gamma \phi(a-c+1,b-c+1,a+b-c+1) + & \\ +\alpha_{\gamma'}\phi(1-a,1-b,c-a-b+1) &= 0, \\ \alpha'_\gamma \phi(a-c+1,b-c+1,a+b-c+1) + & \\ +\alpha'_{\gamma'}\phi(1-a,1-b,c-a-b+1) &= 1. \end{aligned} \right\} \tag{18b}$$

In any particular case, two coefficients are given directly by (15a) or (15b), and the others are found from (18a) or (18b). But it is easy to verify, by means of the definition (11) and of the relation

$$\Gamma(z)\Gamma(1-z) = \pi \operatorname{cosec} \pi z,$$

that all the eight relations are consistent. The connexion between the solutions at $x = 0$ and $x = \infty$ are similarly obtained from the expansions in powers of $x/(x-1)$ and of $1/(1-x)$. We thus have always

$$\begin{pmatrix} \alpha_\gamma \, \alpha_{\gamma'} \\ \alpha'_\gamma \, \alpha'_{\gamma'} \end{pmatrix} = \begin{pmatrix} \phi(a,b,c) & \phi(c-a,c-b,c) \\ \phi(a-c+1,b-c+1,2-c) & \phi(1-a,1-b,2-c) \end{pmatrix}$$

$$\begin{pmatrix} \alpha_\beta \, \alpha_{\beta'} \\ \alpha'_\beta \, \alpha'_{\beta'} \end{pmatrix} = \begin{pmatrix} \phi(a,c-b,c) & \phi(b,c-a,c) \\ e^{i\pi(1-c)}\phi(a-c+1,1-b,2-c) & e^{i\pi(1-c)}\phi(b-c+1,1-a,2-c) \end{pmatrix}$$

(20)

27. Hypergeometric Integrals[†]

Euler's Transformation. An effective method of continuing the hypergeometric series beyond its circle of convergence is to represent it by a definite integral, where the variable x appears as a parameter. The oldest method of representation is due to Euler and has been developed by many writers. Euler's integral was transformed by Wirtinger[‡] into one involving theta functions, but we shall confine ourselves to the classical form. Equivalent results were obtained by Pincherle, Mellin and Barnes[||], using integrals with gamma functions; but we shall omit these, as they are well known and readily accessible to English students.

A linear differential equation of order n, whose coefficients are polynomials of degree m, can in general be satisfied by integrals of the type

$$y = \int_C (u-x)^{\xi-1} \phi(u) \, du, \qquad (1)$$

along a suitable path C, where ξ is a constant to be determined and where the auxiliary function $\phi(u)$ satisfies a differential equation of order $(m+n)$. If, however, the coefficient $p_r(x)$ of $D^{n-r}y$ is of degree $(n-r)$, the equation for $\phi(u)$ is of order n; and in a very special case it is only of the first order, so that the equation becomes soluble by

---

[†] B. Riemann, *Werke*, 81–3; (*Nachträge*) 69–75; C. Jordan, *Cours d'analyse*, **3**, 251–63; F. Klein, *Hypergeometrische Funktion* (1933), 88–111; L. Schlesinger, *Handbuch*, **2** (1), 405–524.

[‡] W. Wirtinger, *Wiener Berichte*, **3** (1902), 894–900; A. L. Dixon, *Quart. J. of Math. (Oxford)*, **1** (1930), 175–8.

[||] E. W. Barnes, *Proc. London Math. Soc.* (2), **6** (1907), 141–77; E. T. Whittaker and G. N. Watson, *Modern Analysis*, 280–5.

quadratures. Let $Q(x)$ and $R(x)$ be polynomials of degree n and $(n-1)$ respectively, and consider the equation

$$\sum_{r=0}^{n} \binom{n-\xi}{r} Q^{(r)}(x) D^{n-r} y - \sum_{r=0}^{n-1} \binom{n-\xi-1}{r} R^{(r)}(x) D^{n-r-1} y = 0. \quad (2)$$

On introducing the expression (1) for y and removing the constant factors $(n-\xi-1)(n-\xi-2)\ldots(1-\xi)$, we find

$$(n-\xi)\int_C (u-x)^{\xi-n-1}\left[\sum_{r=0}^{n} \frac{(u-x)^r}{r!} Q^{(r)}(x)\right]\phi(u)\,du -$$

$$-\int_C (u-x)^{\xi-n}\left[\sum_{r=0}^{n-1} \frac{(u-x)^r}{r!} R^{(r)}(x)\right]\phi(u)\,du = 0, \quad (3)$$

or, by Taylor's theorem,

$$\int_C [(n-\xi)(u-x)^{\xi-n-1}Q(u)-(u-x)^{\xi-n}R(u)]\phi(u)\,du = 0. \quad (4)$$

The integrand is an exact derivative, if $\phi(u)$ is so chosen that

$$\frac{d}{du}[Q(u)\phi(u)] = R(u)\phi(u), \quad (5)$$

or

$$\phi(u) = \frac{1}{Q(u)}\exp\left[\int \frac{R(u)}{Q(u)}\,du\right]. \quad (6)$$

We then complete the integration and determine the path C from the condition

$$[(u-x)^{\xi-n}Q(u)\phi(u)]_C = 0. \quad (7)$$

Hypergeometric Integrals. To identify the hypergeometric equation with the standard form (2), we must have

$$\left.\begin{array}{r}Q(x) = x(1-x),\\ (\xi-2)Q'(x)+R(x) = (a+b+1)x-c,\\ \tfrac{1}{2}(\xi-1)(\xi-2)Q''(x)+(\xi-1)R'(x) = -ab.\end{array}\right\} \quad (8)$$

On eliminating the polynomials, we have a quadratic for the parameter ξ, namely

$$(\xi+a-1)(\xi+b-1) = 0, \quad (9)$$

and the root $\xi = (1-a)$ gives

$$R(x) = (a-c+1)-(a-b+1)x. \quad (10)$$

From (6) we now get

$$\phi(u) = A u^{a-c}(1-u)^{c-b-1}, \quad (11)$$

and the path must be chosen so that

$$[u^{a-c+1}(1-u)^{c-b}(u-x)^{-a-1}]_C = 0, \quad (12)$$

with corresponding results for the second root $\xi = (1-b)$. The integrand of the expression

$$y = A \int_C u^{a-c}(1-u)^{c-b-1}(u-x)^{-a}\,du \qquad (13)$$

has in general four singularities $u = 0, \infty, 1, x$.

It may be shown that the six pairs of singularities each yield a solution corresponding to one of the six principal branches of the hypergeometric function (any three being of course linearly connected). If the expression (12) vanishes at both the critical points, an ordinary integral between these limits may be taken; such integrals were considered by Jacobi and by Goursat. In the most unfavourable circumstances it is necessary to consider double-loop integrals interlacing each pair of singularities; these were introduced by Jordan and afterwards rediscovered by Pochhammer. Let P be any convenient point in the u-plane, and let a definite initial value be assigned to the integrand. The three simple loop integrals about the three finite singularities may be denoted by

$$L_0 = \int_P^{(0+)}, \quad L_1 = \int_P^{(1+)}, \quad L_x = \int_P^{(x+)}. \qquad (14)$$

A positive loop about $u = \infty$ is topographically equivalent to a sequence of negative loops about $u = 0, 1, x$ (taken in the proper order).

If the integrand is *holomorphic* at one of the points $u = 0, 1, x$, the simple loop integral vanishes, but the point can be taken as an end point of an ordinary line integral or of a simple loop integral about one of the other singularities. If it is a *pole* of the integrand, the simple loop satisfies the condition (12) and furnishes a solution. In general, the expressions (14) are not themselves solutions; but any double loop circuit satisfies the condition (12) and the corresponding integral can be written in terms of the simple loop integrals (14), for example

$$\int_P^{(0+,1+,0-,1-)} = \int_P^{(0+)} + e^{2i\pi(a-c)}\int_P^{(1+)} + e^{2i\pi(a-b)}\int_P^{(0-)} + e^{2i\pi(c-b)}\int_P^{(1-)}$$

$$= \int_P^{(0+)} + e^{2i\pi(a-c)}\int_P^{(1+)} - e^{2i\pi(c-b)}\int_P^{(0+)} - \int_P^{(1+)}$$

$$= [1-e^{2i\pi(c-b)}]L_0 - [1-e^{2i\pi(a-c)}]L_1. \qquad (15)$$

The Forty-eight Eulerian Forms.

If we put $u = 1/t$, we obtain Euler's own forms

$$F(a,b;c;x) = \frac{\Gamma(c)}{\Gamma(b)\Gamma(c-b)} \int_0^1 t^{b-1}(1-t)^{c-b-1}(1-xt)^{-a} \, dt,$$

$$F(a,b;c;x) = \frac{\Gamma(c)}{\Gamma(a)\Gamma(c-a)} \int_0^1 t^{a-1}(1-t)^{c-a-1}(1-xt)^{-b} \, dt. \quad (16)$$

These are respectively valid if $(b, c-b)$ or $(a, c-a)$ have positive real parts and if x has any value not lying in the real interval $1 \leqslant x \leqslant \infty$; when the conditions are not satisfied, an appropriate double or simple loop contour must be substituted.

Each of the integrals (16) belongs to a set of twenty-four, obtainable by linear transformations of the integrand interchanging the points $(t = 0, \infty, 1, 1/x)$ in all possible ways. For example, the involutions

$$\left. \begin{array}{r} xtt' - t - t' + 1 = 0, \\ xtt' - xt - xt' + 1 = 0, \\ xtt' - 1 = 0, \end{array} \right\} \quad (17)$$

interchange the points in pairs and give four forms equivalent to the first of the pair (16), namely

$$\left. \begin{array}{c} \displaystyle\int_0^1 t^{b-1}(1-t)^{c-b-1}(1-xt)^{-a} \, dt, \\[2mm] (1-x)^{c-a-b} \displaystyle\int_0^1 t^{c-b-1}(1-t)^{b-1}(1-xt)^{a-c} \, dt, \\[2mm] x^{1-c}(1-x)^{c-a-b} \displaystyle\int_{1/x}^\infty t^{-a}(t-1)^{a-c}(xt-1)^{b-1} \, dt, \\[2mm] x^{1-c} \displaystyle\int_{1/x}^\infty t^{a-c}(t-1)^{-a}(xt-1)^{c-b-1} \, dt. \end{array} \right\} \quad (18)$$

The other transformations lead to Eulerian integrals where x is replaced by one of the expressions

$$x' = x, \quad \frac{1}{x}, \quad 1-x, \quad \frac{1}{1-x}, \quad \frac{x-1}{x}, \quad \frac{x}{x-1}. \quad (19)$$

But it was pointed out by Riemann that there is no elementary method of transforming the two integrals (16) into one another; he suggested that both could be derived from the same multiple integral, evaluated in different ways, and the required integral was afterwards found by Wirtinger.

Deformation of Contours. An expression for the hypergeometric function which is valid in the entire plane, cut along the real axis from $x = 1$ to $x = +\infty$, is given by

$$F(a,b;c;x) = \frac{\int_C t^{b-1}(1-t)^{c-b-1}(1-xt)^{-a}\,dt}{\int_C t^{b-1}(1-t)^{c-b-1}\,dt}. \qquad (20)$$

The contour C is in general a double loop interlacing $t = 0$ and $t = 1$; but in special circumstances it may be replaced by a simple loop, or a figure of eight, or an ordinary line of integration between the critical points.

Suppose now that the point x crosses the cut in the plane of x, and describes a complete circuit about $x = 1$. In the plane of t, the singular point $t = 1/x$ describes a circuit in the same sense about $t = 1$. The new form of the solution (20) is found by deforming the contour of integration so that it never passes through a singular point of the integrand. To fix ideas, let the starting-point P lie on the real axis in the interval $(0 < t < 1)$, and let the initial amplitudes of t, $(1-t)$, and $(1-xt)$ all tend to zero, as $t \to 0$ in that interval. We consider the contour integral

$$\int_P^{(1+,0+,1-,0-)} = [1-e^{2i\pi b}] \int_P^{(1+)} - [1-e^{2i\pi(c-b)}] \int_P^{(0+)}. \qquad (21)$$

After the point $t = 1/x$ has described a circuit counter-clockwise about $t = 1$, the simple loop integral $\int_P^{(0+)}$ is unchanged. But the deformation of the contour of the other integral is indicated by the figure

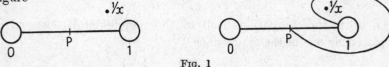

Fig. 1

and the new value of the integral is

$$\int_P^{(1+,1/x+,1+,1/x-,1-)} = \int_P^{(1+)} + e^{2i\pi(c-b)} \int_P^{(1/x+,1+,1/x-,1-)}. \qquad (22)$$

Hence the integral (21) becomes

$$\int_P^{(1+,0+,1-,0-)} + [e^{2i\pi(c-b)} - e^{2i\pi c}] \int_P^{(1/x+,1+,1/x-,1-)}, \qquad (23)$$

and the second double loop integral may be shown to be a multiple of $x^{1-c}(1-x)^{c-a-b}F(1-a, 1-b; c-a-b+1; 1-x)$.

EXAMPLES. VI

1. Verify that
$$(1+x)^n + (1-x)^n = 2F(-\tfrac{1}{2}n, -\tfrac{1}{2}n + \tfrac{1}{2}; \tfrac{1}{2}; x^2),$$
$$(1+x)^n - (1-x)^n = 2nxF(-\tfrac{1}{2}n + \tfrac{1}{2}, -\tfrac{1}{2}n + 1; \tfrac{3}{2}; x^2),$$
$$(1+x)^n = F(-n, 1; 1; -x),$$
$$(1+x)^n - 1 = nxF(1-n, 1; 2; -x),$$
$$\log(1+x) = xF(1, 1; 2; -x). \qquad \text{[GAUSS.]}$$

2. Verify that
$$1 + (1+x)^n = \lim_{b \to 0} 2F(-n, b; 2b; -x),$$
$$e^x = \lim_{b \to \infty} F(1, b; 1; x/b),$$
$$\cosh x = \lim_{a,b \to \infty} F(a, b; \tfrac{1}{2}; x^2/4ab),$$
$$\sinh x = \lim_{a,b \to \infty} xF(a, b; \tfrac{3}{2}; x^2/4ab). \qquad \text{[GAUSS.]}$$

3. Verify that
$${}_1F_1(a; c; x) = \lim_{b \to \infty} F(a, b; c; x/b),$$
$${}_0F_1(c; x) = \lim_{a,b \to \infty} F(a, b; c; x/ab). \qquad \text{[KUMMER.]}$$

4. By means of the equation $D^2 y + n^2 y = 0$, show that
$$\frac{\sin nx}{n \sin x} = F(\tfrac{1}{2}n + \tfrac{1}{2}, -\tfrac{1}{2}n + \tfrac{1}{2}; \tfrac{3}{2}; \sin^2 x),$$
$$= \cos x\, F(\tfrac{1}{2}n + 1, -\tfrac{1}{2}n + 1; \tfrac{3}{2}; \sin^2 x),$$
$$= \cos^{n-1} x\, F(-\tfrac{1}{2}n + \tfrac{1}{2}, -\tfrac{1}{2}n + 1; \tfrac{3}{2}; -\tan^2 x),$$
$$= \cos^{-n-1} x\, F(\tfrac{1}{2}n + \tfrac{1}{2}, \tfrac{1}{2}n + 1; \tfrac{3}{2}; -\tan^2 x),$$
and obtain corresponding forms for $\cos nx$. [GAUSS.]

5. By making $n \to 0$ in the above, show that
$$x = \sin x\, F(\tfrac{1}{2}, \tfrac{1}{2}; \tfrac{3}{2}; \sin^2 x),$$
$$= \sin x \cos x\, F(1, 1; \tfrac{3}{2}; \sin^2 x),$$
$$= \tan x\, F(\tfrac{1}{2}, 1; \tfrac{3}{2}; -\tan^2 x). \qquad \text{[GAUSS.]}$$

6. Obtain another solution of the hypergeometric equation satisfied by $F(a, b; 1; x)$ by evaluating
$$\lim_{\epsilon \to 0} \frac{1}{\epsilon}[F(a, b; 1-\epsilon; x) - x^\epsilon F(a+\epsilon, b+\epsilon; 1+\epsilon; x)].$$

7. Show that
$$a_n(c-b)_n(x-1)^{a-1}F(a+n, b; c+n; x) = c_n D^n[(x-1)^{a+n-1} F(a, b; c; x)],$$
$$(c-n)_n x^{c-n-1}(1-x)^{b-c-n+1} F(a-n, b; c-n; x)$$
$$= D^n[x^{c-1}(1-x)^{b-c+1} F(a, b; c; x)]. \qquad \text{[T. W. CHAUNDY.]}$$

8. **Hypergeometric Polynomials.** (i) If a is a positive integer,
$$(2-c)_{a-1} F(1-a, 1-b; 2-c; x) = x^{c-1}(1-x)^{a+b-c} D^{a-1}[x^{a-c}(1-x)^{c-b-1}].$$
(ii) If $(a-c)$ is a positive integer,
$$c_{a-c} F(c-a, c-b; c; x) = x^{1-c}(1-x)^{a+b-c} D^{a-c}[x^{a-1}(1-x)^{-b}].$$
(iii) If $(c-a)$ is a positive integer,
$$(2-c)_{c-a-1} F(a-c+1, b-c+1; 2-c; x)$$
$$= x^{c-1}(1-x)^{c-a-b} D^{c-a-1}[x^{-a}(1-x)^{b-1}].$$

9. **Legendre Polynomials.** (i) If n is a positive integer, the equation
$$(1-x^2)D^2 y - 2xDy + n(n+1)y = 0$$
is satisfied by $P_n(x) = D^n(x^2-1)^n / 2^n n!$.

(ii) By operating with $(xD-n+2)$ or $(xD+n+3)$ on
$$D^2 P_n(x) - (xD+n+1)(xD-n)P_n(x) = 0,$$
show that
$$DP_{n-1}(x) = (xD-n)P_n(x),$$
$$DP_{n+1}(x) = (xD+n+1)P_n(x),$$
$$(2n+1)P_n(x) = DP_{n+1}(x) - DP_{n-1}(x).$$

(iii) Obtain by integration the relations
$$nP_{n-1}(x) = (1-x^2)DP_n(x) + nxP_n(x),$$
$$(n+1)P_{n+1}(x) = (x^2-1)DP_n(x) + (n+1)xP_n(x),$$
$$(n+1)P_{n+1}(x) - (2n+1)xP_n(x) + nP_{n-1}(x) = 0.$$
Verify that these agree with the recurrence formulae of $P_n^{(0,0)}(x)$.

10. (i) If n is an integer, Legendre's equation is satisfied by
$$Q_n(x) = P_n(x) \int_{\infty}^{x} \frac{dx}{(1-x^2)P_n^2(x)}.$$

(ii) By examining the partial fractions corresponding to the zeros of $P_n(x)$, show that
$$Q_n(x) = \tfrac{1}{2} P_n(x) \log\left(\frac{x+1}{x-1}\right) - W_{n-1}(x),$$
where $W_{n-1}(x)$ is a polynomial expressible in the form
$$W_{n-1}(x) = \frac{2n-1}{1 \cdot n} P_{n-1}(x) + \frac{2n-5}{3(n-1)} P_{n-3}(x) + \frac{2n-9}{5(n-2)} P_{n-5}(x) + \ldots.$$

11. If n is a positive integer, show that
$$P_n(x) = F\left(n+1, -n; 1; \frac{1-x}{2}\right),$$
$$P_n(x) = \frac{(2n)!}{2^n n! n!} x^n F\left(-\frac{n}{2}, \frac{1-n}{2}; \tfrac{1}{2}-n; \frac{1}{x^2}\right),$$
$$Q_n(x) = \frac{2^n n! n!}{(2n+1)!} x^{-n-1} F\left(\frac{n+1}{2}, \frac{n+2}{2}; n+\tfrac{3}{2}; \frac{1}{x^2}\right),$$
$$(n+1)Q_{n+1}(x) - (2n+1)xQ_n(x) + nQ_{n-1}(x) = 0.$$

12. **Associated Legendre Equation.** By means of the scheme of § 25, (4), show that the associated Legendre equation is soluble in finite terms if n or

$(m\pm n)$ is an integer; by the quadratic transformation show that it is also soluble if $(m+\tfrac{1}{2})$ is an integer.

13. (i) Rodrigues's Lemma. If $(n\pm m)$ are positive integers, show that the associated Legendre equation is satisfied by the equivalent expressions

$$\frac{(x^2-1)^{m/2}}{(n+m)!}D^{n+m}(x^2-1)^n = \frac{(x^2-1)^{-m/2}}{(n-m)!}D^{n-m}(x^2-1)^n.$$

(ii) **Schendel's Lemma.** If n is a positive integer, show that the associated Legendre equation is satisfied by the equivalent expressions

$$\left(\frac{x+1}{x-1}\right)^{m/2}D^n[(x+1)^{n-m}(x-1)^{n+m}] = \left(\frac{x-1}{x+1}\right)^{m/2}D^n[(x+1)^{n+m}(x-1)^{n-m}].$$

14. Tschebyscheff's Polynomials. If $T_n(\cos\theta) \equiv \cos n\theta$, show that

$$(1-x^2)D^2T_n(x) - xDT_n(x) + n^2T_n(x) = 0,$$

$$T_n(x) = \frac{(x^2-1)^{\tfrac{1}{2}}D^n(x^2-1)^{n-\tfrac{1}{2}}}{1.3....(2n-1)},$$

$$T_n(x) = \frac{n!\,\Gamma(\tfrac{1}{2})}{\Gamma(n+\tfrac{1}{2})}P_n^{(-\tfrac{1}{2},-\tfrac{1}{2})}(x).$$

15. Jacobi's Relations. If $U_n(\cos\theta) \equiv \dfrac{\sin(n+1)\theta}{\sin\theta}$, show that

$$U_n(x) = \frac{DT_{n+1}(x)}{(n+1)} = \frac{(n+1)(x^2-1)^{-\tfrac{1}{2}}D^n(x^2-1)^{n+\tfrac{1}{2}}}{1.3....(2n+1)},$$

$$U_n(x) = \frac{(n+1)!\,\Gamma(\tfrac{1}{2})}{\Gamma(n+\tfrac{3}{2})}P_n^{(\tfrac{1}{2},\tfrac{1}{2})}(x),$$

$$\frac{d^n\sin^{2n+1}\theta}{d(\cos\theta)^n} = (-)^n\frac{1.3....(2n+1)}{(n+1)}\sin(n+1)\theta,$$

$$\int_0^\pi f(\cos\theta)\cos(n+1)\theta\,d\theta = \int_0^\pi \frac{f^{(n+1)}(\cos\theta)\sin^{2n+2}\theta\,d\theta}{1.3....(2n+1)}.$$

16. If n is a positive integer and $(\alpha+\beta+n+1) = 0$, show that

$$D^n[(x-1)^{\alpha+n}(x+1)^{\beta+n}] = 2^n\frac{\Gamma(\alpha+n+1)}{\Gamma(\alpha+1)}(x-1)^\alpha(x+1)^\beta.$$

17. Recurrence Formulae. Show that

$$2(n+1)(\alpha+\beta+n+1)P_{n+1}^{(\alpha,\beta)}(x)$$
$$= (\alpha+\beta+2n+2)(x^2-1)DP_n^{(\alpha,\beta)}(x) +$$
$$+ (\alpha+\beta+n+1)[(\alpha+\beta+2n+2)x+(\alpha-\beta)]P_n^{(\alpha,\beta)}(x),$$

$$2n(\beta+n)P_{n-1}^{(\alpha,\beta)}(x)$$
$$= (\alpha+\beta+2n)(1-x^2)DP_n^{(\alpha,\beta)}(x) +$$
$$+ n[(\alpha+\beta+2n)x-(\alpha-\beta)]P_n^{(\alpha,\beta)}(x);$$

by eliminating $DP_n^{(\alpha,\beta)}(x)$, obtain a recurrence formula connecting three successive Jacobi polynomials.

18. **Kummer's Quadratic Transformations.** Show that

$F(\alpha, \alpha+\tfrac{1}{2}; \gamma; x)$

$= \left(\dfrac{1+(1-x)^{\frac{1}{2}}}{2}\right)^{-2\alpha} F\left(2\alpha, 2\alpha-\gamma+1; \gamma; \dfrac{1-(1-x)^{\frac{1}{2}}}{1+(1-x)^{\frac{1}{2}}}\right)$

$= (1+x^{\frac{1}{2}})^{-2\alpha} F\left(2\alpha, \gamma-\tfrac{1}{2}; 2\gamma-1; \dfrac{2x^{\frac{1}{2}}}{1+x^{\frac{1}{2}}}\right)$

$= (1-x)^{-\alpha} F\left(2\alpha, 2\gamma-2\alpha-1; \gamma; \dfrac{(1-x)^{\frac{1}{2}}-1}{2(1-x)^{\frac{1}{2}}}\right);$

$F(\alpha, \beta; 2\beta; x)$

$= \left(\dfrac{1+(1-x)^{\frac{1}{2}}}{2}\right)^{-2\alpha} F\left\{\alpha, \alpha-\beta+\tfrac{1}{2}; \beta+\tfrac{1}{2}; \left(\dfrac{1-(1-x)^{\frac{1}{2}}}{1+(1-x)^{\frac{1}{2}}}\right)^2\right\}$

$= (1-x)^{-\alpha/2} F\left(\alpha, 2\beta-\alpha; \beta+\tfrac{1}{2}; \dfrac{\{1-(1-x)^{\frac{1}{2}}\}^2}{-4(1-x)^{\frac{1}{2}}}\right);$

$F(\alpha, \beta; \alpha+\beta+\tfrac{1}{2}; x)$

$= \left(\dfrac{1+(1-x)^{\frac{1}{2}}}{2}\right)^{-2\alpha} F\left(2\alpha, \alpha-\beta+\tfrac{1}{2}; \alpha+\beta+\tfrac{1}{2}; \dfrac{(1-x)^{\frac{1}{2}}-1}{(1-x)^{\frac{1}{2}}+1}\right)$

$= (-1)^{\alpha}\{(x-1)^{\frac{1}{2}}+x^{\frac{1}{2}}\}^{-2\alpha} F\left(2\alpha, \alpha+\beta; 2\alpha+2\beta; \dfrac{2x^{\frac{1}{2}}}{(x-1)^{\frac{1}{2}}+x^{\frac{1}{2}}}\right)$

$= F\left(2\alpha, 2\beta; \alpha+\beta+\tfrac{1}{2}; \dfrac{1-(1-x)^{\frac{1}{2}}}{2}\right).$

19. Show that

$F(\alpha, \beta; \alpha-\beta+1; x)$

$= (1+x)^{-\alpha} F\{\tfrac{1}{2}\alpha, \tfrac{1}{2}\alpha+\tfrac{1}{2}; \alpha-\beta+1; 4x/(1+x)^2\}$

$= (1+x^{\frac{1}{2}})^{-2\alpha} F\{\alpha, \alpha-\beta+\tfrac{1}{2}; 2\alpha-2\beta+1; 4x^{\frac{1}{2}}/(1+x^{\frac{1}{2}})^2\}$

$= (1-x)^{-\alpha} F\{\tfrac{1}{2}\alpha, \tfrac{1}{2}\alpha-\beta+\tfrac{1}{2}; \alpha-\beta+1; -4x/(1-x)^2\};$

$F(\alpha, \beta; 2\beta; x)$

$= (1-x)^{-\frac{1}{2}\alpha} F\{\tfrac{1}{2}\alpha, \beta-\tfrac{1}{2}\alpha; \beta+\tfrac{1}{2}; x^2/4(x-1)\}$

$= (1-\tfrac{1}{2}x)^{-\alpha} F\{\tfrac{1}{2}\alpha, \tfrac{1}{2}\alpha+\tfrac{1}{2}; \beta+\tfrac{1}{2}; x^2/(2-x)^2\};$

$F(\alpha, \beta; \tfrac{1}{2}\alpha+\tfrac{1}{2}\beta+\tfrac{1}{2}; x)$

$= (1-2x)^{-\alpha} F\{\tfrac{1}{2}\alpha, \tfrac{1}{2}\alpha+\tfrac{1}{2}; \tfrac{1}{2}\alpha+\tfrac{1}{2}\beta+\tfrac{1}{2}; (4x^2-4x)/(1-2x)^2\}$

$= \{(1-x)^{\frac{1}{2}}+(-x)^{\frac{1}{2}}\}^{-2\alpha} F[\alpha, \tfrac{1}{2}\alpha+\tfrac{1}{2}\beta; \alpha+\beta; 4(x^2-x)^{\frac{1}{2}}/\{(1-x)^{\frac{1}{2}}+(-x)^{\frac{1}{2}}\}^2]$

$= F\{\tfrac{1}{2}\alpha, \tfrac{1}{2}\beta; \tfrac{1}{2}\alpha+\tfrac{1}{2}\beta+\tfrac{1}{2}; 4x(1-x)\}.$

20. **Dihedral Equations.** If $\mu \neq 0$, show that

$$P\begin{pmatrix} 0 & \tfrac{1}{2}\mu & 0 \\ \tfrac{1}{2} & -\tfrac{1}{2}\mu & \tfrac{1}{2} \end{pmatrix} x = P\begin{pmatrix} 0 & \mu & 0 & \dfrac{1+\sqrt{x}}{2} \\ \tfrac{1}{2} & -\mu & \tfrac{1}{2} & \end{pmatrix}$$

$$= P\begin{pmatrix} \tfrac{1}{2}\mu & 0 & \tfrac{1}{2}\mu & \dfrac{\sqrt{x}+\sqrt{(x-1)}}{2\sqrt{(x-1)}} \\ -\tfrac{1}{2}\mu & 1 & -\tfrac{1}{2}\mu & \end{pmatrix}$$

$= A\{\sqrt{x}+\sqrt{(x-1)}\}^\mu + B\{\sqrt{x}-\sqrt{(x-1)}\}^\mu$

$= Ae^{i\mu\theta} + Be^{-i\mu\theta} \quad (x = \cos^2\theta).$

If two exponent-differences are halves of odd integers, the P-function can be written in terms of $P(\tfrac{1}{2}, \mu, \tfrac{1}{2}, x)$ and its derivatives.

THE HYPERGEOMETRIC EQUATION

21. If p is an integer, Elliot's limiting form of Lamé's equation
$$\frac{d^2u}{d\theta^2}+[k^2-p(p+1)\operatorname{cosec}^2\theta]u = 0$$
has the finite solutions
$$e^{ik\theta}F(p+1,-p;1+k;\tfrac{1}{2}-\tfrac{1}{2}i\cot\theta),\quad e^{-ik\theta}F(p+1,-p;1-k;\tfrac{1}{2}-\tfrac{1}{2}i\cot\theta).$$
The associated Legendre equation can be reduced to the above by either of the transformations

(i) $x = i\cot\theta,\quad y = u,\quad m^2 = k^2,\quad n = p$;

(ii) $x = \cos\theta,\quad y = u\operatorname{cosec}^{\frac{1}{2}}\theta,\quad m = p+\tfrac{1}{2},\quad (n+\tfrac{1}{2})^2 = k^2$.

[Z. V. Elliot, *Acta Mathematica*, **2** (1883), 233–60. See also *Journal London Math. Soc.* **5** (1930), 189–91.]

22. (i) Show that
$$u = \sin^m\theta\cos^n\theta F\{\tfrac{1}{2}(m+n+k),\tfrac{1}{2}(m+n-k);(m+\tfrac{1}{2});\sin^2\theta\}$$
satisfies the equation
$$\frac{d^2u}{d\theta^2}+[k^2-m(m-1)\operatorname{cosec}^2\theta-n(n-1)\sec^2\theta]u = 0,$$
and obtain other solutions by interchanging $(m, 1-m)$ or $(n, 1-n)$.

(ii) If $u_{m,n}$ is a general solution, verify the recurrence formulae
$$u_{m+1,n+1} = \frac{d}{d\theta}u_{m,n}+(n\tan\theta-m\cot\theta)u_{m,n},$$
$$u_{m,1} = u_{m,0},\qquad u_{1,n} = u_{0,n},$$
$$u_{0,0} = Ae^{ik\theta}+Be^{-ik\theta}.$$

(iii) If $k \neq 0$, show that
$$u_{m,n} = Ae^{ik\theta}F(\cot\theta)+Be^{-ik\theta}F(-\cot\theta),$$
where $F(z)$ is a rational function of the type
$$F(z) = A_{m-1}z^{m-1}+A_{m-2}z^{m-2}+\ldots+A_0+\ldots+A_{2-n}z^{2-n}+A_{1-n}z^{1-n}.$$
[G. Darboux, *Théorie générale des surfaces*, ii. 207–13.]

23. If the exponent-differences at $x = 0$ and $x = 1$ are halves of odd integers, the hypergeometric function has two solutions whose product is a rational function of x. If in addition the exponent-difference at $x = \infty$ is rational and has the denominator N, there are N solutions whose product is merely multiplied by a constant after any closed circuit.

[Transform to Darboux's equation and consider the expressions
$$F(\cot\theta)F(-\cot\theta),\qquad e^{iNk\theta}F^N(\cot\theta)\pm e^{-iNk\theta}F^N(-\cot\theta).]$$

24. If the hypergeometric function has two distinct branches whose product is merely multiplied by a constant after any closed circuit, the branches become multiples of themselves or of one another after any circuit. Deduce that *either* (i) there are two branches expressible in a finite form, *or* (ii) two of the exponent-differences are halves of odd integers.

[A. A. Markoff, *Math. Ann.* **28** (1887), 586–93; **29** (1887), 247–58; **40** (1892), 313–16. *Comptes rendus*, **114** (1892), 54–5. E. B. van Vleck, *American J. of Math.* **21** (1899), 126–67. E. G. C. Poole, *Quart. J. of Math.* (Oxford), **1** (1930), 108–15; **2** (1931), 90–6.]

25. (i) Show, by means of the relation $\sum \alpha = 1$, that Riemann's equation cannot have two solutions expressible in finite terms if the exponent-differences are not integers.

(ii) If a, b are integers of opposite sign and c is not an integer, verify that the finite expressions $F(a,b;c;x)$ and $x^{1-c}(1-x)^{c-a-b}F(1-a,1-b;2-c;x)$ are branches belonging to the same exponent at infinity.

(iii) If a, b, c are integers and $0 > a \geqslant c > b$, the complete primitive is a polynomial. Show that any solution vanishing at $x = 0$ and $x = 1$ must vanish identically. Examine the equation
$$x(1-x)D^2y - (3-7x)Dy - 12y = 0.$$

26. LOGARITHMIC SINGULARITIES. Show by the methods of §26 that
$$\lim_{x \to 1} \frac{F(a,b;a+b;x)}{\log(1-x)^{-1}} = \frac{\Gamma(a+b)}{\Gamma(a)\Gamma(b)}.$$

27. GROUP OF LEGENDRE'S EQUATION. (i) When n is not an integer, two independent solutions of Legendre's equation are
$$y_1 = P_n(x) = F\left(n+1, -n; 1; \frac{1-x}{2}\right),$$
$$y_2 = P_n(-x) = F\left(n+1, -n; 1; \frac{1+x}{2}\right).$$

(ii) $\lim\limits_{x \to -1} \dfrac{P_n(x)}{\log(1+x)} = \dfrac{\sin \pi n}{\pi}.$

(iii) The transformations corresponding to positive circuits about $x = \pm 1$ are
$$\left.\begin{array}{l} Y_1 = y_1, \\ Y_2 = (2i\sin \pi n)y_1 + y_2, \end{array}\right\} \quad \left.\begin{array}{l} Y_1 = y_1 + (2i\sin \pi n)y_2, \\ Y_2 = y_2. \end{array}\right\}$$

(iv) The transformation corresponding to a negative circuit about $x = \infty$, or to positive circuits about $x = 1$ and $x = -1$ in turn, is
$$\left.\begin{array}{l} Y_1 = y_1 + (2i\sin \pi n)y_2, \\ Y_2 = (2i\sin \pi n)y_1 + (1 - 4\sin^2 \pi n)y_2. \end{array}\right\}$$

(v) The characteristic equation is
$$\begin{vmatrix} 1-\lambda, & 2i\sin \pi n \\ 2i\sin \pi n, & 1 - 4\sin^2 \pi n - \lambda \end{vmatrix} = 0,$$
and the multipliers $\lambda = \exp(\pm 2i\pi n)$ are unequal unless n is half an odd integer; in that case the invariant factors are $[(\lambda+1)^2, 1]$ and the singularity $x = \infty$ is logarithmic.

(vi) From the formulae of Ex. 10 above, write down the corresponding transformations of two linearly independent solutions, when n is an integer.

28. SCHLÄFLI'S INTEGRALS. Solve Legendre's equation by means of Euler's transformation, and hence show that, when $|1-x| < 2$, $P_n(x)$ can be expressed in either of the forms
$$P_n(x) = F\left(n+1, -n; 1; \frac{1-x}{2}\right),$$
$$= \frac{1}{2^{n+1}i\pi} \int_P^{(1+,x+)} (t^2-1)^n(t-x)^{-n-1}\, dt,$$

$$= \frac{2^n}{i\pi} \int_P^{(1+,x+)} (t^2-1)^{-n-1}(t-x)^n \, dt,$$

where P is a point $t = t_0 > 1$ on the real axis and initially $|\operatorname{am}(t-x)| < \pi$.

Show that the integrals can be evaluated as residues when n is an integer, and deduce Rodrigues's formula.

29. If $R(n+1) > 0$ and $|x| > 1$, show that

$$Q_n(x) = \frac{\Gamma(n+1)\Gamma(\tfrac{1}{2})}{2^{n+1}\Gamma(n+\tfrac{3}{2})} x^{-n-1} F\!\left(\frac{n+1}{2}, \frac{n+2}{2}; n+\tfrac{3}{2}; \frac{1}{x^2}\right)$$

$$= \frac{1}{2^{n+1}} \int_{-1}^{1} (1-t^2)^n (x-t)^{-n-1} \, dt.$$

30. If $P_n(x)$ is defined by the first integral in Ex. 28, show that, after x has described a positive circuit about $x = -1$, the integral $\int_P^{(1+,x+)}$ becomes $\int_P^{(1+,x+,-1+,x+,-1-,x-)}$; and show that the latter gives the value

$$P_n(x) + \frac{e^{i\pi n} - e^{-i\pi n}}{2^{n+1} i\pi} \int_0^{(-1+,x+)} (1-t^2)^n (t-x)^{-n-1} \, dt = P_n(x) + (2i \sin \pi n) P_n(-x),$$

in agreement with Ex. 27.

31. **Associated Legendre Equation.** Show that branches of the associated Legendre function are represented by the Eulerian integrals along suitable contours

$$(x^2-1)^{m/2} \int_C (t^2-1)^n (t-x)^{-m-n-1} \, dt,$$

$$(x+1)^{m/2}(x-1)^{-m/2} \int_C (t-1)^{n+m}(t+1)^{n-m}(t-x)^{-n-1} \, dt.$$

Show that alternative forms are obtained by interchanging $(n+1, -n)$ or $(m, -m)$. When can these be evaluated as residues in finite terms?

[Cf. E. W. Hobson, *Spherical and Ellipsoidal Harmonics* (1931), ch. v.]

32. **Complete Elliptic Integrals.** If $k'^2 = (1-k^2)$, show that the complete elliptic integrals of the first kind

$$K = \int_0^{\pi/2} (1-k^2 \sin^2\theta)^{-\tfrac{1}{2}} \, d\theta = \tfrac{1}{2}\pi F(\tfrac{1}{2}, \tfrac{1}{2}; 1; k^2),$$

$$K' = \int_0^{\pi/2} (1-k'^2 \sin^2\theta)^{-\tfrac{1}{2}} \, d\theta = \tfrac{1}{2}\pi F(\tfrac{1}{2}, \tfrac{1}{2}; 1; k'^2),$$

are both branches of $P\!\begin{pmatrix} 0 & \tfrac{1}{2} & 0 \\ 0 & \tfrac{1}{2} & 0 \end{pmatrix} k^2$. Show that this is equivalent to a Legendre function of order $-\tfrac{1}{2}$. By putting $t^2 = \sin^2\theta + k^{-2}\cos^2\theta$, show that

$$K' = \int_1^{1/k} (t^2-1)^{-\tfrac{1}{2}}(1-k^2 t^2)^{-\tfrac{1}{2}} \, dt.$$

33. The complete elliptic integrals of the second kind are given by

$$E = \int_0^{\pi/2} (1-k^2\sin^2\theta)^{\frac{1}{2}} d\theta = \tfrac{1}{2}\pi F(\tfrac{1}{2}, -\tfrac{1}{2}; 1; k^2),$$

$$E' = \int_0^{\pi/2} (1-k'^2\sin^2\theta)^{\frac{1}{2}} d\theta = \tfrac{1}{2}\pi F(\tfrac{1}{2}, -\tfrac{1}{2}; 1; k'^2).$$

By means of the relations between contiguous functions, show that

$$E = kk'^2\frac{dK}{dk} + k'^2 K, \qquad E' = k'k^2\frac{dK'}{dk'} + k^2 K'.$$

By means of Abel's relation, show that the expression

$$EK' + E'K - KK' = 2x(1-x)\left(K'\frac{dK}{dx} - K\frac{dK'}{dx}\right),$$

where $x = k^2$, is a numerical constant. By examining the asymptotic form of K, K' as $x \to 1$, show that

$$\lim_{x\to 1} 2x(1-x)\left(K'\frac{dK}{dx} - K\frac{dK'}{dx}\right) = \pi \lim_{x\to 1} x(1-x)\frac{dK}{dx} = \tfrac{1}{2}\pi.$$

Hence obtain Legendre's relation

$$EK' + E'K - KK' = \tfrac{1}{2}\pi.$$

34. (i) Two independent solutions of the equation of the periods in elliptic functions
$$x(1-x)D^2 y + (1-2x)Dy - \tfrac{1}{4}y = 0$$
are
$$y_1 = f(x) = K \quad \text{and} \quad y_2 = if(1-x) = iK',$$

where

$$K = \tfrac{1}{2}\int_0^1 t^{-\frac{1}{2}}(1-t)^{-\frac{1}{2}}(1-xt)^{-\frac{1}{2}} dt,$$

$$K' = \tfrac{1}{2}\int_1^{1/x} t^{-\frac{1}{2}}(t-1)^{-\frac{1}{2}}(1-xt)^{-\frac{1}{2}} dt.$$

(ii) Show, both by the method of power-series and by the method of deformation of contours, that the transformations corresponding to positive circuits about $x = 0$ and $x = 1$ are

$$\left.\begin{array}{l} Y_1 = y_1, \\ Y_2 = -2y_1 + y_2, \end{array}\right\} \qquad \left.\begin{array}{l} Y_1 = y_1 - 2y_2, \\ Y_2 = y_2. \end{array}\right\}$$

(iii) If $z \equiv y_2/y_1$, show that after any closed circuit z undergoes a transformation of the *modular group*

$$Z = \frac{az+b}{cz+d},$$

where a, d are *odd* and b, c *even* integers, and $(ad-bc) = 1$.

[J. Tannery and J. Molk, *Fonctions elliptiques*, iii. 188–214; E. Picard, *Traité d'analyse*, iii. 358–64.]

35. LOGARITHMIC SINGULARITIES. (i) Any hypergeometric function with a logarithmic singularity at $x = 0$ is expressible in terms of the derivatives of

$$P\begin{pmatrix} 0 & a & 0 \\ 0 & b & 1-a-b \end{pmatrix} x\bigg).$$

THE HYPERGEOMETRIC EQUATION

(ii) The branch $F(a,b;1;x)$ is a multiple of the Eulerian integrals

$$\int_P^{(0+,x+)} u^{a-1}(1-u)^{-b}(u-x)^{-a}\,du,$$

or

$$\int_P^{(0+,x+)} u^{b-1}(1-u)^{-a}(u-x)^{-b}\,du.$$

(iii) The integrals $\int_P^{(0+,1+,0-,1-)}$ or their degenerate forms (when they vanish identically) give the branches

$$x^{-a}F(a,a;a-b;1/x) \quad \text{and} \quad x^{-b}F(b,b;b-a;1/x).$$

[Cf. W. L. Ferrar, *Proc. Edinburgh Math. Soc.* **43** (1925), 39–47. The notation has been altered to conform with the text.]

VII

CONFORMAL REPRESENTATION

28. Schwarz's Problem†

P-Functions whose Group is Finite. THE cases where the hypergeometric function is algebraic, or has only a finite number of branches, were enumerated by Schwarz in 1873. His methods and some of his results had been anticipated by Riemann in his lectures in 1858–9, but these remained unpublished until 1902.

If only a *particular branch* is algebraic, the complete primitive being transcendental, that branch must be merely multiplied by a constant after any closed circuit; for otherwise we should have two independent algebraic branches and the complete primitive would be algebraic. The particular branch is accordingly the product of $x^\alpha(1-x)^\gamma$ and a hypergeometric polynomial, and the equation can be solved completely in finite terms by means of one quadrature, in accordance with Jacobi's formula.

If *every* branch is algebraic, the exponents must be rational. For after m circuits about $x = 0$, say, any branch $\{AP^{(\alpha)} + A'P^{(\alpha')}\}$ is transformed into $\{Ae^{2i\pi m\alpha}P^{(\alpha)} + A'e^{2i\pi m\alpha'}P^{(\alpha')}\}$; and, if (α, α') were irrational, the number of distinct determinations would be infinite.

We can easily dispose of cases where one of the exponent-differences is an integer. For we know how to determine whether the corresponding singularity does or does not involve logarithms. If it does, the solution is certainly transcendental; if it does not, we can make a transformation of the Riemann equation which reduces the singularity at $x = 1$ (say) to an apparent singularity, where every branch is holomorphic. The two principal branches at $x = 0$ either resume their initial values, or are multiplied by constants, after any closed circuit (supposing that $(\alpha' - \alpha)$ is not also an integer). Accordingly each of them is the product of a power of x and a hypergeometric polynomial. If two exponent-differences are integers and the solution is free from logarithms, we can arrange that $x = 0$ and $x = 1$ shall both be apparent singularities, and the complete primitive will then be a rational integral function.

† H. A. Schwarz, *Gesammelte mathematische Abhandlungen*, ii. 211–59 (= *J. für Math.* **75** (1873), 292–335); B. Riemann, *Gesammelte mathematische Werke* (*Nachträge*), Leipzig (1902), 67–93.

Reduced Sets of Exponents. Let λ, μ, ν be any real rational numbers, but not integers; and let l, m, n be any integers whose sum is even. Then we saw in § 23 that all the Riemann functions

$$P(l\pm\lambda, m\pm\mu, n\pm\nu, x)$$

have the same group. To pick out the simplest function of the group, we first choose exponents satisfying the conditions

$$\left.\begin{array}{l}\lambda_1 \equiv \lambda, \quad \mu_1 \equiv \mu, \quad \nu_1 \equiv \nu \pmod{2}, \\ -1 < \lambda_1, \mu_1, \nu_1 < 1;\end{array}\right\} \quad (1)$$

and, because $P(\pm\lambda_1, \pm\mu_1, \pm\nu_1, x)$ all have the same group, we may arrange without loss of generality that

$$0 < \lambda_1, \mu_1, \nu_1 < 1, \quad (2)$$

integer values being excluded. We now consider the four associated functions

$$\left.\begin{array}{l}P(\lambda_1, \mu_1, \nu_1, x), \\ P(\lambda_1, 1-\mu_1, 1-\nu_1, x), \\ P(1-\lambda_1, \mu_1, 1-\nu_1, x), \\ P(1-\lambda_1, 1-\mu_1, \nu_1, x),\end{array}\right\} \quad (3)$$

and we can readily verify that at least one satisfies the conditions

$$\left.\begin{array}{l}0 < \lambda_0, \mu_0, \nu_0 < 1, \\ 0 < \mu_0+\nu_0, \nu_0+\lambda_0, \lambda_0+\mu_0 \leqslant 1.\end{array}\right\} \quad (4)$$

A set of exponent-differences satisfying these more stringent conditions is called a *reduced set*.

We can reduce all the associated P-functions to hypergeometric functions in such a manner that the branches whose exponents are zero at $x = 0$ and at $x = 1$ shall correspond to one another. If $F(a, b; c; x)$ is the hypergeometric function with the reduced exponent-differences, all the others can be expressed linearly in terms of branches of $F(a, b; c; x)$ and $DF = (ab/c)F(a+1, b+1; c+1; x)$, multiplied by rational functions of x.

Quotients of Solutions. Let us now consider the function defined by the quotient of two branches of a given P-function, together with its inverse function

$$z = \frac{y_1(x)}{y_2(x)} = f(x), \qquad x = \phi(z). \quad (5)$$

After the description of any closed circuit, z is transformed into

$$Z = \frac{ay_1+by_2}{cy_1+dy_2} = \frac{az+b}{cz+d} \quad (ad-bc \neq 0), \quad (6)$$

where (a, b, c, d) are certain constants, while the inverse function has the automorphic property

$$\phi\left(\frac{az+b}{cz+d}\right) = \phi(z),$$

for the same sets of constants. The relations expressing the periodicity of the circular and elliptic functions are particular cases of such automorphisms.

If (Z, z) are connected by a linear relation with constant coefficients of the form (6), we have

$$DZ = \frac{(ad-bc)Dz}{(cz+d)^2},$$

$$\frac{D^2Z}{DZ} = \frac{D^2z}{Dz} - \frac{2cDz}{(cz+d)},$$

$$\frac{D^3Z}{DZ} - \left(\frac{D^2Z}{DZ}\right)^2 = \frac{D^3z}{Dz} - \left(\frac{D^2z}{Dz}\right)^2 - \frac{2cD^2z}{(cz+d)} + \frac{2c^2(Dz)^2}{(cz+d)^2},$$

and hence
$$\frac{D^3Z}{DZ} - \frac{3}{2}\left(\frac{D^2Z}{DZ}\right)^2 = \frac{D^3z}{Dz} - \frac{3}{2}\left(\frac{D^2z}{Dz}\right)^2. \tag{7}$$

The expression
$$\{z, x\} \equiv \frac{D^3z}{Dz} - \frac{3}{2}\left(\frac{D^2z}{Dz}\right)^2 \tag{8}$$

was called by Cayley the *Schwarzian derivative*. As Schwarz pointed out, it had been used by Lagrange and Kummer, and it appears also in Riemann's lectures.

The property (7) may be written

$$\left\{\frac{az+b}{cz+d}, x\right\} = \{z, x\} \quad (ad-bc \neq 0), \tag{9}$$

and we have in particular

$$\left\{\frac{ax+b}{cx+d}, x\right\} = \{x, x\} = 0. \tag{10}$$

Any function $z = \phi(x)$, which undergoes a linear transformation with constant coefficients when x describes any closed circuit, satisfies an equation of the form $\{z, x\} = F(x)$, whose right-hand side is a uniform function of x. Moreover, if we put $(ad-bc) = 1$, we can write the relations between $Z = \dfrac{az+b}{cz+d}$ and z in Riemann's form

$$\left.\begin{aligned}(DZ)^{-\frac{1}{2}} &= (Dz)^{-\frac{1}{2}}(cz+d), \\ Z(DZ)^{-\frac{1}{2}} &= (Dz)^{-\frac{1}{2}}(az+b);\end{aligned}\right\} \tag{11}$$

CONFORMAL REPRESENTATION

these show that $(Dz)^{-\frac{1}{2}}$ and $z(Dz)^{-\frac{1}{2}}$ undergo a homogeneous linear transformation after any circuit, and therefore satisfy an equation of the second order with uniform coefficients.

The Schwarzian Equation. Before considering the quotient of two solutions of $D^2y+p_1Dy+p_2y = 0$, it is convenient to reduce the equation to the invariant form $D^2\bar{y}+I\bar{y} = 0$ ($I \equiv p_2-\frac{1}{2}Dp_1-\frac{1}{4}p_1^2$) by putting $y = \bar{y}\exp[-\frac{1}{2}\int p_1(x)\,dx]$. We shall suppose this to have been done. This implies that Riemann's scheme must be taken in the form

$$P\begin{pmatrix} \frac{1}{2}(1-\lambda) & -\frac{1}{2}(1+\mu) & \frac{1}{2}(1-\nu) \\ \frac{1}{2}(1+\lambda) & -\frac{1}{2}(1-\mu) & \frac{1}{2}(1+\nu) \end{pmatrix} x \Bigg), \tag{12}$$

with the equation

$$D^2y + \left\{\frac{1-\lambda^2}{4x^2} + \frac{1-\nu^2}{4(x-1)^2} + \frac{\lambda^2-\mu^2+\nu^2-1}{4x(x-1)}\right\}y = 0. \tag{13}$$

Now if we have

$$z = \frac{y_1}{y_2}, \qquad D^2y_i+Iy_i = 0 \quad (i = 1, 2), \tag{14}$$

we find

$$Dz = (y_2Dy_1-y_1Dy_2)/y_2^2, \tag{15}$$

which reduces, by Abel's relation, to

$$Dz = Cy_2^{-2}. \tag{16}$$

Hence $(Dz)^{-\frac{1}{2}}$ is a solution of (14), and so we have

$$D^2(Dz)^{-\frac{1}{2}}+I(Dz)^{-\frac{1}{2}} = 0, \tag{17}$$

or

$$\{z,x\} = 2I, \tag{18}$$

as is easily verified.

We can solve completely the equation $D^2y+Iy = 0$, if we know any solution of (18). The case $I \equiv 0$ being trivial, the solution z cannot be a constant; hence, using (9), we see that $-z^{-1}$ is a distinct solution, and so we have two solutions of (14) given by

$$y_1 = (-Dz^{-1})^{-\frac{1}{2}} = z(Dz)^{-\frac{1}{2}}, \qquad y_2 = (Dz)^{-\frac{1}{2}}, \tag{19}$$

which are evidently linearly independent.

Change of Variables. If we put in $D^2y+Iy = 0$ the forms

$$x = \psi(t), \qquad y = u[\psi'(t)]^{\frac{1}{2}}, \tag{20}$$

we get the equation

$$\frac{d^2u}{dt^2}+I^*u = 0, \tag{21}$$

where

$$I^* = I[\psi'(t)]^2 + \frac{1}{2}\frac{\psi'''(t)}{\psi'(t)} - \frac{3}{4}\left[\frac{\psi''(t)}{\psi'(t)}\right]^2. \tag{22}$$

Since $z = y_1/y_2 = u_1/u_2$, we have
$$\{z,x\} = 2I, \qquad \{z,t\} = 2I^*, \tag{23}$$
and so, from (22) and (23), we have the relation
$$\{z,t\} = \{z,x\}\left(\frac{dx}{dt}\right)^2 + \{x,t\}. \tag{24}$$

This identity was given explicitly by Cayley, but was used implicitly in Kummer's investigation of the transformation of the hypergeometric equation. By putting $t \equiv z$ in (24), we have
$$\{z,x\} = -\{x,z\}\left(\frac{dz}{dx}\right)^2. \tag{25}$$

By means of (24), we can prove that, if z is a particular solution of $\{z,x\} = 2I$, then the most general solution is
$$Z = (az+b)/(cz+d) \quad (ad-bc \neq 0),$$
the solution suggested by (9). For we have the relations
$$\left.\begin{aligned}\{Z,x\} &= \{Z,z\}\left(\frac{dz}{dx}\right)^2 + \{z,x\}, \\ \{Z,x\} &= 2I = \{z,x\};\end{aligned}\right\} \tag{26}$$
hence
$$\{Z,z\} = 0,$$
or
$$\frac{d^2}{dz^2}\left(\frac{dZ}{dz}\right)^{-\frac{1}{2}} = 0; \tag{27}$$
hence
$$\frac{dZ}{dz} = (cz+d)^{-2}, \tag{28}$$
and so
$$Z = \frac{az+b}{cz+d} \quad (ad-bc = 1). \tag{29}$$

From these relations and (19) we see that the solution of the equations $D^2 y + Iy = 0$ and $\{z,x\} = 2I$ are equivalent problems.

Every equation of the second order is formally equivalent to $d^2w/dz^2 = 0$; for if $\{\phi_1(x), \phi_2(x)\}$ are known independent solutions, we have only to put $y = \phi_1(x)w$ and $z = \phi_2(x)/\phi_1(x)$.

29. The Reduced Curvilinear Triangle

Conformal Representation. We shall now examine the mapping of the upper half-plane of x by the quotient of two distinct branches of $P\begin{pmatrix} \alpha & \beta & \gamma \\ \alpha' & \beta' & \gamma' \end{pmatrix} x$, say $z = y_1/y_2 = f(x)$.

(i) If the exponents are subject only to the general condition $\sum \alpha = 1$, we can prove that the relation between z and x is locally one-to-one (*schlicht*), except at the points $(x = 0, \infty, 1)$. For the branches $[y_1(x), y_2(x)]$ are holomorphic and have at most a simple zero at any ordinary point, and they cannot vanish together. Thus either z or $1/z$ is holomorphic; without loss of generality, let us suppose that $y_2 \neq 0$, so that z is holomorphic. Then, by the Abel-Liouville formula, we have

$$Dz = \frac{y_2 Dy_1 - y_1 Dy_2}{y_2^2} = \frac{Cx^{\alpha+\alpha'-1}(1-x)^{\gamma+\gamma'-1}}{y_2^2} \neq 0. \qquad (1)$$

Hence, at an ordinary point $(x = \xi, z = \zeta)$ we have, by the implicit function theorem,

$$x - \xi = c_1(z-\zeta) + c_2(z-\zeta)^2 + \dots \quad (y_2 \neq 0), \qquad (2\,\text{a})$$

or else, when z has a pole,

$$x - \xi = c_1/z + c_2/z^2 + \dots \quad (y_2 = 0). \qquad (2\,\text{b})$$

(ii) We next restrict the exponents to *real values*, and consider more particularly the points corresponding to real values of x other than the singular points. In the typical interval $(0 < x < 1)$ the principal branches are expressions of the type $Ax^\alpha(1-x)^\gamma F$, where F is a real hypergeometric series, and we can select two branches (y_1, y_2) which remain real in the entire interval. Thus z is also real, and the formula (1) shows that it is monotonic (say increasing). If the denominator vanishes at any point, z passes discontinuously from ∞ to $-\infty$ and begins to increase again as x increases, and it must pass through the value $z = 0$ before it can again become infinite, so that the zeros of (y_1, y_2) must separate one another.

This particular quotient z therefore travels steadily along the real axis of the z-plane, but it may pass over the same point more than once; the general form $Z = (az+b)/(cz+d)$ gives a point which travels steadily around a circle, but the arc corresponding to the interval $(0 < x < 1)$ may overlap itself.

(iii) At the typical singularity $x = 0$, we have in particular

$$z_0 = \frac{P^{(\alpha')}}{P^{(\alpha)}} = \frac{A'x^\lambda F\{\tfrac{1}{2}(1+\lambda-\mu-\nu), \tfrac{1}{2}(1+\lambda+\mu-\nu); (1+\lambda); x\}}{A F\{\tfrac{1}{2}(1-\lambda-\mu-\nu), \tfrac{1}{2}(1-\lambda+\mu-\nu); (1-\lambda); x\}}$$
$$= Mx^\lambda[1 + c_1 x + c_2 x^2 + \dots]. \qquad (3)$$

If (λ, μ, ν) are real, and x is real and sufficiently small, the series in brackets converges and takes real values. Hence as x describes the segments $-\infty < x < 0$ and $0 < x < 1$ of the boundary of the upper half-plane of x, the quotient z_0 describes segments of two straight lines enclosing an interior angle $\pi\lambda$. The most general form of quotient $z = y_1/y_2$ thus maps the upper half-plane of x upon a domain bounded by three arcs of circles (which may overlap themselves) enclosing interior angles $(\pi\lambda, \pi\mu, \pi\nu)$ at the points corresponding to $x = 0, \infty, 1$.

Reduced Exponent-Differences. When we have

$$\left. \begin{array}{l} 0 < \lambda, \mu, \nu < 1, \\ 0 < \mu+\nu, \nu+\lambda, \lambda+\mu \leqslant 1, \\ 0 < \lambda+\mu+\nu \leqslant \tfrac{3}{2}, \end{array} \right\} \quad (4)$$

the parameters of the hypergeometric series

$$a = \tfrac{1}{2}(1-\lambda-\mu-\nu), \quad b = \tfrac{1}{2}(1-\lambda+\mu-\nu), \quad c = (1-\lambda) \quad (5)$$

are found to satisfy the corresponding conditions

$$-\tfrac{1}{4} \leqslant a \leqslant \tfrac{1}{2}, \quad 0 < b, c-b < 1. \quad (6)$$

Following Wirtinger,† we choose as denominator the branch

$$F(a,b;c;x) = \frac{\Gamma(c)}{\Gamma(b)\Gamma(c-b)} \int_0^1 t^{b-1}(1-t)^{c-b-1}(1-xt)^{-a}\,dt, \quad (7)$$

and we show that this does not vanish in the interval $(0 < x < 1)$, nor in the plane of x, cut from $x = 1$ to $x = +\infty$. For the integral converges when the value of x is not real and greater than unity. If we determine the amplitude of $(1-xt)^{-a}$ in the cut plane by assigning the value zero at $x = 0$, we have, in the upper half-plane either

$$\left. \begin{array}{ll} 0 \leqslant \operatorname{am}(1-xt)^{-a} \leqslant \tfrac{1}{2}\pi & (0 \leqslant a \leqslant \tfrac{1}{2}), \\ 0 \geqslant \operatorname{am}(1-xt)^{-a} \geqslant -\tfrac{1}{4}\pi & (0 \geqslant a \geqslant -\tfrac{1}{4}). \end{array} \right\} \quad (8)$$

In either case the real and imaginary parts of the integrand (7) do not change sign, and so $y_2 \equiv F(a,b;c;x) \neq 0$. Hence the arcs corresponding to the segments $(-\infty < x < 0)$ and $(0 < x < 1)$ do not overlap themselves; and, by considering another special case, we can show that the arc corresponding to $(1 < x < \infty)$ does not overlap itself either. Thus as x describes the real axis from $-\infty$ to ∞, with indentations above the singularities $x = 0, 1$, z describes once counterclockwise the boundary of a curvilinear triangle with interior angles $(\pi\lambda, \pi\mu, \pi\nu)$, whose sides do not overlap. We can now prove, by a classical argument, that there is one-to-one correspondence of the

† Summarized by O. Haupt in his new edition of F. Klein, *Hypergeometrische Funktion* (1933), 326–30.

two domains. For consider the increment of

$$(2\pi)^{-1}\mathrm{am}(z-\zeta) \equiv (2\pi)^{-1}\mathrm{am}\{f(x)-\zeta\}, \tag{9}$$

as the points z and x each describe once counter-clockwise the boundaries of their respective domains.

The expression on the left increases by unity or zero, according as ζ does or does not lie within the triangle. The function $f(x) = y_1/y_2$ has no poles in the upper half-plane, on account of the way in which the denominator has been chosen; hence the increment of the expression on the right is equal to the number of zeros of $\{f(x)-\zeta\}$ in the upper half-plane. Accordingly it has exactly one zero when ζ lies within the triangle, and none when it lies outside.

The Rectilinear Triangle. If $(\lambda+\mu+\nu) = 1$, we have $a = 0$, and then Riemann's equation is soluble by quadratures,

$$P\begin{pmatrix} 0 & 0 & 0 \\ \lambda & \mu & \nu \end{pmatrix} x = A \int x^{\lambda-1}(1-x)^{\nu-1}\, dx + B. \tag{10}$$

The branch $y_2 \equiv 1$ fulfils the condition of not vanishing in the upper half-plane of x; and the quotient

$$z = \int x^{\lambda-1}(1-x)^{\nu-1}\, dx \tag{11}$$

is the ordinary Schwarz-Christoffel formula giving the conformal representation of the upper half-plane of x upon a triangle $(\pi\lambda, \pi\mu, \pi\nu)$.

30. Symmetrical Continuation†

Schwarz's Principle of Symmetry. If an analytic function $z = f(x)$ is holomorphic in a domain intersected by the real axis of x and real along the segment of the axis, it takes conjugate complex values $z, \bar{z}$ at conjugate points $x, \bar{x}$ in the domain. If we make linear transformations of both planes

$$Z = \frac{az+b}{cz+d}, \qquad X = \frac{a'x+b'}{c'x+d'}, \tag{1}$$

we find that if $f(x)$ is holomorphic in a domain intersected by a circle, and if points on the circumference lying in the domain correspond to points lying on a circle in the z-plane, then inverse points with respect to the circle in the x-plane correspond to inverse points with respect to the circle in the z-plane. This is obvious if we remember that inverse points are common points of a family of circles orthogonal

---

† H. A. Schwarz, *Ges. math. Abhandlungen*, ii. 65–83; or *J. für Math.* **70** (1869), 105–20. See, for example, E. C. Titchmarsh, *Theory of Functions* (1932), 155.

to a given circle, and that the transformations (1) convert circles or straight lines into circles or straight lines.

Now the function $z = f(x)$, which maps the upper half-plane of x on a triangle of circular arcs ABC, is holomorphic except at the points $x = 0, \infty, 1$ corresponding to the corners. Hence continuation into the lower half-plane across the segment $(-\infty < x < 0)$ gives a new triangle of circular arcs ABC', by inversion in AB; and a return to the upper half-plane across $(0 < x < 1)$, which completes a positive circuit about $x = 0$, gives a fresh triangle $AB''C'$ by inversion in AC'. The two successive inversions give a linear transformation $Z = (az+b)/(cz+d)$.

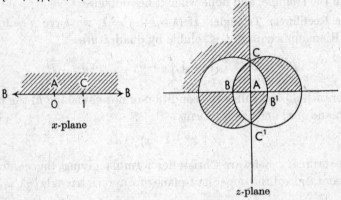

Fig. 2. Dihedral configuration ($n = 3$).

Similarly, if we had left the original domain across $(0 < x < 1)$ or $(1 < x < \infty)$, we should have found different representations $AB'C$ or $A'BC$ of the lower half-plane, by inversion with respect to CA or BC respectively. Continuing this process, we obtain a pattern of (say) black and white triangles, corresponding respectively to the upper and lower half-planes of x. The pattern may or may not overlap; it may cover the whole or only part of the z-plane; and in certain cases the number of triangles may be finite.

Types of Reduced Triangles. The circles BC, CA, AB have a unique radical centre O (which may lie at infinity if the centres are collinear). If O is exterior to the three circles, they have a real common orthogonal circle with O as centre. The *reduced* triangle ABC is found to lie wholly on one side of this orthogonal circle. We may invert AB, AC into straight lines, and then the arc BC will be convex to A, and the sum of the angles is less than two right angles, that is to say $(\lambda+\mu+\nu) < 1$. We now find that all the successive images are in the interior of the fixed orthogonal circle, which can never be filled by any finite number of repetitions. Since the number of values of z corresponding to any given x is infinite, the relation $z = f(x)$ cannot be algebraic.

Similarly, if $(\lambda+\mu+\nu) = 1$, the triangle can be inverted into a rectilinear one, having an infinite number of distinct repetitions.

A necessary condition for a finite number of repetitions is accordingly that $(\lambda+\mu+\nu) > 1$. In this case the radical centre O is interior to the circles. The pattern can be more easily visualized by an inversion in three dimensions converting BC, CA, AB into great circles on a sphere. The centre of inversion V must lie on the normal at O to the plane of the figure, and $-OV^2$ must be the power of the radical centre with respect to each circle, so that OV is the geometric mean of the segments of any chord through O. It can then be shown by elementary geometry that the circles BC, CA, AB are inverted into great circles on a sphere passing through V. Inverse points with respect to the circle BC, in the plane figure, are common points of a family of circles orthogonal to BC. They therefore become the common points of a family of small circles cutting orthogonally the great circle BC, in the corresponding spherical figure. The operation of inversion in the plane thus corresponds to reflection in the plane of the great circle of the sphere.

Steiner's Problem. We must now find all spherical triangles having only a finite number of distinct repetitions on the sphere. Every side lies in a plane of symmetry, and the number of such planes must be finite, because two different planes of symmetry give different reflections of a figure. These planes cut out a finite number of spherical triangles. Let PQR be a triangle of minimum area. If the angle at P is $\pi\theta$, and θ is irrational, the number of planes of symmetry passing through the diameter at P is infinite. If θ is a fraction in its lowest terms $\theta = p'/p$, there are p such planes passing through the diameter at P; and if $p' > 1$ we can choose one of them cutting off a spherical triangle of smaller area than PQR. For a triangle of minimum area with a finite set of repetitions the angles must be aliquot parts of π, say $(\pi/p, \pi/q, \pi/r)$, where (p, q, r) are integers greater than unity, and

$$\frac{1}{p}+\frac{1}{q}+\frac{1}{r} > 1. \qquad (2)$$

The number of solutions is limited, and each solution gives the planes of symmetry of a regular solid. Arranging the numbers in ascending order, we find the following sets.

$$\left.\begin{array}{rl} \textit{Bipyramid:} & (2,\ 2,\ r \text{ arbitrary}) \\ \textit{Tetrahedron:} & (2,\ 3,\ 3) \\ \textit{Cube and Octahedron:} & (2,\ 3,\ 4) \\ \textit{Dodecahedron and Icosahedron:} & (2,\ 3,\ 5) \end{array}\right\} \quad (3)$$

By picking out *all* the reduced triangles cut out by the planes of each configuration, Schwarz enumerated fifteen cases.

Table of Schwarz's Reduced Triangles

No.	λ, μ, ν	Spherical Excess $/\pi$	Configuration
I	$\frac{1}{2}, \frac{1}{2}, \frac{1}{n}$	$\frac{1}{n}$	Regular Bipyramid
II	$\frac{1}{2}, \frac{1}{3}, \frac{1}{3}$	$\frac{1}{6} = A$	Tetrahedron
III	$\frac{2}{3}, \frac{1}{3}, \frac{1}{3}$	$\frac{1}{3} = 2A$	
IV	$\frac{1}{2}, \frac{1}{3}, \frac{1}{4}$	$\frac{1}{12} = B$	Cube and Octahedron
V	$\frac{2}{3}, \frac{1}{4}, \frac{1}{4}$	$\frac{1}{6} = 2B$	
VI	$\frac{1}{2}, \frac{1}{3}, \frac{1}{5}$	$\frac{1}{30} = C$	
VII	$\frac{2}{5}, \frac{1}{3}, \frac{1}{3}$	$\frac{1}{15} = 2C$	
VIII	$\frac{2}{3}, \frac{1}{5}, \frac{1}{5}$	$\frac{1}{15} = 2C$	
IX	$\frac{1}{2}, \frac{2}{5}, \frac{1}{5}$	$\frac{1}{10} = 3C$	
X	$\frac{3}{5}, \frac{1}{3}, \frac{1}{5}$	$\frac{2}{15} = 4C$	Dodecahedron and Icosahedron
XI	$\frac{2}{5}, \frac{2}{5}, \frac{2}{5}$	$\frac{1}{5} = 6C$	
XII	$\frac{2}{3}, \frac{1}{3}, \frac{1}{5}$	$\frac{1}{5} = 6C$	
XIII	$\frac{4}{5}, \frac{1}{5}, \frac{1}{5}$	$\frac{1}{5} = 6C$	
XIV	$\frac{1}{2}, \frac{2}{5}, \frac{1}{3}$	$\frac{7}{30} = 7C$	
XV	$\frac{3}{5}, \frac{2}{5}, \frac{1}{3}$	$\frac{1}{3} = 10C$	

Uniform Schwarzian Functions. In general, when z is the quotient of two branches of $P(\lambda, \mu, \nu, x)$, the pattern of triangles covers the z-domain with an overlapping Riemann surface of many sheets. A given complex number z corresponds to points differently situated in different overlapping triangles; and so the inverse function $x = \phi(z)$ is many-valued. If, however, λ is the reciprocal of an integer, equal numbers of black and white triangles fit exactly around the points corresponding to corners $\pi\lambda$, and a unique value of x corresponds to $1/\lambda$ values of z near such a point, so that $x = \phi(z)$ is locally uniform. If (λ, μ, ν) are all reciprocals of integers, $x = \phi(z)$ will be uniform everywhere. There are a limited number of cases where $(\lambda + \mu + \nu) \geqslant 1$, giving the polyhedral triangles and space-filling recti-

Chap. VII, § 30   CONFORMAL REPRESENTATION   129

linear triangles, covering the entire z-plane. There are an infinite number of uniform Schwarzian functions where $(\lambda+\mu+\nu) < 1$, the example $(\frac{1}{5}, \frac{1}{4}, \frac{1}{2})$ being illustrated in Schwarz's memoir. In these cases the sides of every triangle are orthogonal to a fixed circle or line, forming a natural boundary beyond which continuation is impossible.

31. Some Special Cases

The Dihedral Equation. If the reduced exponent-differences are reciprocals of integers, and their sum is also greater than unity, the function $x = \phi(z)$ is both uniform and algebraic, and therefore is a rational function. These are the cases I, II, IV, VI of Schwarz's table.

The first case $P\left(\frac{1}{2}, \frac{1}{2}, \frac{1}{n}, x\right)$ is soluble by elementary methods; for the quadratic transformation

$$P\left\{\begin{array}{ccc} 0 & \infty & 1 \\ 0 & 0 & \frac{1}{2n} \\ \frac{1}{2} & \frac{1}{2} & -\frac{1}{2n} \end{array} x\right\} = P\left\{\begin{array}{ccc} -1 & \infty & 1 \\ \frac{1}{2n} & 0 & \frac{1}{2n} \\ -\frac{1}{2n} & 1 & -\frac{1}{2n} \end{array} \sqrt{x}\right\}$$

$$= P\left\{\begin{array}{cccc} 0 & \infty & 1 & \\ \frac{1}{2n} & \frac{1}{2n} & 0 & \frac{1+\sqrt{x}}{1-\sqrt{x}} \\ -\frac{1}{2n} & -\frac{1}{2n} & 1 & \end{array}\right\} \quad (1)$$

gives an equation with only *two* regular singularities, which is equivalent to Euler's homogeneous equation. The general solution being

$$y = A\left(\frac{1+\sqrt{x}}{1-\sqrt{x}}\right)^{\frac{1}{2n}} + B\left(\frac{1-\sqrt{x}}{1+\sqrt{x}}\right)^{\frac{1}{2n}}, \quad (2)$$

we can choose the quotient

$$z = \left(\frac{1+\sqrt{x}}{1-\sqrt{x}}\right)^{\frac{1}{n}}, \quad (3)$$

giving the inverse function

$$x = \left(\frac{z^n-1}{z^n+1}\right)^2. \quad (4)$$

The z-sphere is divided by the equator and n complete meridians into $2n$ pairs of triangles, whose corners form a bipyramid. In general, $2n$ distinct values of z correspond to each value of x; but these become united in pairs at the corners $\tfrac{1}{2}\pi$, along the equator, as $x \to 0$ or $x \to \infty$; and as $x \to 1$ they are united in two sets of n at the poles.

The Octahedral Equation. A cube and a regular octahedron are inscribed in a unit sphere, so that the edges of the cube are parallel to the diagonals of the octahedron. If the pole of coordinates is at a vertex of the octahedron, the point $z = e^{i\phi}\cot\tfrac{1}{2}\theta$ is the stereographic projection on the plane of the complex variable z of the point (θ,ϕ) on the unit sphere. The corners of the octahedron are then given by
$$z = 0, \infty, \pm 1, \pm i. \tag{5}$$
At the corners of the cube $\cos\theta = \pm 1/\sqrt{3}$, and their stereographic projections are found to satisfy the equation
$$z^8 + 14z^4 + 1 = 0. \tag{6}$$
The middle points of the edges correspond to the affixes
$$(z^4+1)(z^4 - \tan^4\pi/8)(z^4 - \cot^4\pi/8) = 0,$$
or $\quad (z^4+1)(z^4 - 17 + 12\sqrt{2})(z^4 - 17 - 12\sqrt{2}) = 0,$
or $\quad (z^{12} - 33z^8 - 33z^4 + 1) = 0. \tag{7}$

The three fundamental polynomials satisfy the identity
$$108 z^4 (z^4-1)^4 - (z^8+14z^4+1)^3 + (z^{12}-33z^8-33z^4+1)^2 \equiv 0. \tag{8}$$

Consider the function $x = \phi(z)$ defined by the hypergeometric equation with exponent-differences $(\tfrac{1}{3}, \tfrac{1}{2}, \tfrac{1}{4})$. The fundamental spherical triangle with angles $(\tfrac{1}{3}\pi, \tfrac{1}{2}\pi, \tfrac{1}{4}\pi)$ is one-sixth of an octant, or one forty-eighth of a sphere, and is bounded by planes of symmetry of the octahedral configuration. Since each half-plane of x is represented on twenty-four different triangles, the equation $x = \phi(z)$ is of degree 24 in z. When $x = 0$, the points on the sphere are united in threes at the corners of the cube; when $x = \infty$, they are united in pairs at the middle points of the edges; when $x = 1$, they are united in fours at the corners of the octahedron. Hence we must have
$$x = \phi(z) = \frac{A(z^8+14z^4+1)^3}{(z^{12}-33z^8-33z^4+1)^2}. \tag{9}$$

In order that this may have the root $z = 0$ when $x = 1$, we have to put $A = 1$, and then the identity (8) shows that we can write
$$\frac{(z^8+14z^4+1)^3}{x} = \frac{(z^{12}-33z^8-33z^4+1)^2}{1} = \frac{108z^4(z^4-1)^4}{x-1}. \tag{10}$$

The Tetrahedral Equation. Two adjacent octahedral triangles $(\frac{1}{3}\pi, \frac{1}{2}\pi, \frac{1}{4}\pi)$ joined along the sides opposite to the angles $\frac{1}{3}\pi$, make up one tetrahedral triangle $(\frac{1}{3}\pi, \frac{1}{2}\pi, \frac{1}{3}\pi)$. There are twenty-four such triangles on the sphere, each of which gives a complete representation of the x-plane in the octahedral relation (10), but only half of the plane of x_1 corresponding to the tetrahedral relation with exponent-differences $(\frac{1}{3}, \frac{1}{2}, \frac{1}{3})$. Two adjacent tetrahedral triangles contain four complete octahedral triangles, and give a single representation of the plane of x_1, and a double representation of the plane of x. Thus to every value of x there correspond two values of x_1, and it is found that we can pass to the tetrahedral equation by putting

$$x/(x-1) = 4x_1(1-x_1). \tag{11}$$

The transformation of Riemann's equation is as follows

$$P\begin{Bmatrix} 0 & \infty & 1 & \\ 0 & \frac{5}{12} & 0 & x_1 \\ \frac{1}{3} & -\frac{1}{12} & \frac{1}{3} & \end{Bmatrix} = P\begin{Bmatrix} -1 & \infty & 1 & \\ 0 & \frac{5}{12} & 0 & 2x_1-1 \\ \frac{1}{3} & -\frac{1}{12} & \frac{1}{3} & \end{Bmatrix}$$

$$= P\begin{Bmatrix} 0 & \infty & 1 & \\ 0 & \frac{5}{24} & 0 & (2x_1-1)^2 \\ \frac{1}{2} & -\frac{1}{24} & \frac{1}{3} & \end{Bmatrix}$$

$$= P\begin{Bmatrix} 0 & \infty & 1 & \\ 0 & \frac{5}{24} & 0 & 4x_1(1-x_1) \\ \frac{1}{3} & -\frac{1}{24} & \frac{1}{2} & \end{Bmatrix}. \tag{12}$$

From (10), we have algebraically

$$(1-x)^{\frac{1}{2}} = (2x_1-1)^{-1} = \pm \frac{6i\sqrt{3}z^2(z^4-1)^2}{z^{12}-33z^8-33z^4+1}, \tag{13}$$

and on taking the upper sign, we get the tetrahedral relation

$$\frac{(z^4+2i\sqrt{3}z^2+1)^3}{x_1} = \frac{(z^4-2i\sqrt{3}z^2+1)^3}{x_1-1} = \frac{12i\sqrt{3}z^2(z^4-1)^2}{1}. \tag{14}$$

We have the identity

$$(z^8+14z^4+1) \equiv (z^4+2i\sqrt{3}z^2+1)(z^4-2i\sqrt{3}z^2+1), \tag{15}$$

which corresponds to the division of the vertices of the cube into those of two desmic regular tetrahedra, whose edges are the diagonals of the faces of the cube. The corners of one tetrahedron correspond to the centres of the faces of the other, and the middle points of the edges of either lie at the corners of the octahedron. When $x_1 = 0$, the twelve roots of the tetrahedral equation $x_1 = \phi_1(z)$ are united in threes at the corners of one tetrahedron; when $x_1 = 1$, they are united in threes at the corners of the other; when $x_1 = \infty$, they are united in pairs at the middle points of the edges. The algebraic identity corresponding to (10) is

$$(z^4+2i\sqrt{3}z^2+1)^3-(z^4-2i\sqrt{3}z^2+1)^3-12i\sqrt{3}z^2(z^4-1)^2 \equiv 0. \quad (16)$$

The Icosahedral Equation. The triangle $(\tfrac{1}{2}\pi, \tfrac{1}{3}\pi, \tfrac{1}{5}\pi)$ corresponds to the sixth of a face of the regular icosahedron, and to the tenth of a face of the regular dodecahedron. As there are 120 such triangles on the sphere, the equation $x = \phi(z)$ is of degree 60 in z, with quintuple roots at the corners of the icosahedron (say $x = \infty$), triple roots at those of the dodecahedron (say $x = 0$), and double roots at the midpoints of the edges. With a suitable orientation of the figure, the equation may be written

$$\frac{1728z^5(z^{10}+11z^5-1)^5}{1} = \frac{(z^{20}-228z^{15}+494z^{10}+228z^5+1)^3}{-x}$$
$$= \frac{(z^{30}+522z^{25}-10005z^{20}-10005z^{10}-522z^5+1)^2}{1-x}. \quad (17)$$

For full details, reference should be made to Schwarz's memoir or the writings of Klein.†

Composite and Associated Triangles. The remaining cases of reduced triangles giving algebraic functions can be solved by the method of adjunction of domains, used above to obtain the tetrahedral function from the octahedral. The method was used in Riemann's lectures (posthumously published in 1902), but the fundamental theorem was more explicitly established by Burnside;‡ it has been applied in numerous problems by Hodgkinson.∥ When the reduced equation is solved, it is possible to solve the associated non-reduced equations having the same group.

The Modular Function. The limiting case $P(0, 0, 0, x)$ gives a uniform transcendental Schwarzian function $x = \phi(z)$, which is obtained by inverting the quotient of the elliptic integrals $z = iK'/K$, where $x = k^2$. The initial triangle is bounded by two parallel lines

---

† F. Klein, *Vorlesungen über das Ikosaeder* (1884); *Vorlesungen über die hypergeometrische Funktion* (reprint 1933, edited by O. Haupt). See also A. R. Forsyth, *Theory of Functions*, ch. xx; *Theory of Differential Equations*, iv. 174–90.

‡ W. Burnside, *Proc. London Math. Soc.* (1) **24** (1893), 187–206.

∥ J. Hodgkinson, ibid. (2) **15** (1916), 166–81; **17** (1918), 17–24; **18** (1920), 268–73; **24** (1926), 71–82.

perpendicular to the real axis of z and a semicircle touching them on the axis. The generating transformations of the modular group

$$Z = z-2, \qquad Z = \frac{z}{-2z+1}, \tag{18}$$

correspond to circuits about $x = 0$ and $x = 1$. The function $\phi(z)$ has an automorphic property analogous to those of the simply and doubly periodic functions,

$$\phi\left(\frac{az+b}{cz+d}\right) = \phi(z), \tag{19}$$

for all linear transformations with integer coefficients such that $(ad-bc) = 1$ and $a \equiv d \equiv 1$, $b \equiv c \equiv 0 \pmod{2}$. The function is the subject of a large literature.†

EXAMPLES. VII

1. CIRCLES ON A SPHERE. (i) If two circles on a sphere cut orthogonally, their planes are conjugate with respect to the sphere.

(ii) Families of planes through two fixed lines in space, which are conjugate with respect to a sphere, cut the sphere in two families of orthogonal circles, and conversely.

(iii) If two points of a sphere are collinear with the pole of a circle on the sphere, their stereographic projections are mutually inverse with respect to the stereographic projection of the circle.

2. ROTATION. If the stereographic projection of points on a sphere is given by the relation $z = e^{i\phi}\cot\tfrac{1}{2}\theta$, show that the linear transformation

$$\left[\frac{Z-e^{i\phi}\cot\tfrac{1}{2}\theta}{Z+e^{i\phi}\tan\tfrac{1}{2}\theta}\right] = e^{-i\psi}\left[\frac{z-e^{i\phi}\cot\tfrac{1}{2}\theta}{z+e^{i\phi}\tan\tfrac{1}{2}\theta}\right]$$

corresponds to a rotation of the sphere.

3. KLEIN'S PARAMETERS. (i) The above rotation may be written

$$z' = \frac{(d+ic)z-(b-ia)}{(b+ia)z+(d-ic)},$$

where

$(a:b:c:d) = (\sin\tfrac{1}{2}\psi\sin\theta\cos\phi: \sin\tfrac{1}{2}\psi\sin\theta\sin\phi: \sin\tfrac{1}{2}\psi\cos\theta: \cos\tfrac{1}{2}\psi).$

(ii) The rotation (a, b, c, d) followed by the rotation (a', b', c', d') is equivalent to the rotation (A, B, C, D), where

$$A = ad'+a'd-bc'+b'c,$$
$$B = bd'+b'd-ca'+c'a,$$
$$C = cd'+c'd-ab'+a'b,$$
$$D = -aa'-bb'-cc'+dd'.$$

† See, for instance, W. Burnside, *Theory of Groups* (1911), 372–427; H. Weber, *Algebra* (1908), iii; F. Klein and R. Fricke, *Elliptische Modulfunktionen* (1890–2), i–ii.

(iii) Verify that the resultant rotation is given by the quaternion product
$$Ai+Bj+Ck+D = (a'i+b'j+c'k+d')(ai+bj+ck+d),$$
where $\quad ij = k = -ji, \quad jk = i = -kj, \quad ki = j = -ik.$

4. Rectilinear Space-filling Triangles. The only rectilinear triangles whose symmetric repetitions fill the plane without overlapping are those with angles
$$(\tfrac{1}{2}\pi, \tfrac{1}{2}\pi, 0), \quad (\tfrac{1}{2}\pi, \tfrac{1}{4}\pi, \tfrac{1}{4}\pi), \quad (\tfrac{1}{2}\pi, \tfrac{1}{3}\pi, \tfrac{1}{6}\pi), \quad (\tfrac{1}{3}\pi, \tfrac{1}{3}\pi, \tfrac{1}{3}\pi).$$
The triangle with angles $(\tfrac{2}{3}\pi, \tfrac{1}{6}\pi, \tfrac{1}{6}\pi)$ gives a double covering of the plane.

[Schwarz.]

5. (i) The relation $z = \arcsin x = xF(\tfrac{1}{2}, \tfrac{1}{2}; \tfrac{3}{2}; x^2)$ maps each quadrant of the x-plane conformally on a semi-infinite rectangular strip of breadth $\tfrac{1}{2}\pi$.

(ii) By the method of symmetric continuation, show that $x = \sin z$ is a uniform analytic function, and obtain geometrically the relations
$$\sin(z+\pi) = -\sin z, \qquad \sin(-z) = -\sin z.$$

6. The conformal representation of an isosceles right-angled triangle on a half-plane is effected by the relation
$$z = \int_0^x [4x(1-x)]^{-\tfrac{3}{4}}\, dx.$$
This is equivalent to the relation
$$4x(1-x) = 1/\wp^2(z),$$
where $\wp(z)$ is the Weierstrassian elliptic function satisfying the equation
$$\wp'^2(z) = 4\wp^3(z) - 4\wp(z).$$
The relations $\quad [\wp(z)-e_\alpha]^{\tfrac{1}{2}} = \dfrac{e^{\eta_\alpha z}\sigma(z-\omega_\alpha)}{\sigma(z)\sigma(\omega_\alpha)} \quad (\alpha = 1, 2, 3)$

show that x is a uniform function of z.

[A. E. H. Love, *American J. of Mathematics*, **11** (1889), 158–71.]

7. The triangle $(\tfrac{1}{2}\pi, \tfrac{1}{3}\pi, \tfrac{1}{6}\pi)$ is conformally represented on a half-plane by the relation
$$z = \frac{1}{6i}\int_0^x x^{-\tfrac{5}{6}}(x-1)^{-\tfrac{2}{3}}\, dx.$$
This is satisfied by putting
$$x = 1/[1-\wp^3(z)], \quad \text{where } \wp'^2(z) = 4\wp^3(z)-4. \qquad \text{[Love.]}$$

8. Obtain the representation of the equilateral triangle from the above by putting $x = x_1^2$, or $x_1 = 2i/\wp'(z)$. Obtain the representation of the isosceles triangle with a vertical angle $2\pi/3$ by putting $x = 1-x_2^2$, and verify that $x_2 = 2\wp^{\tfrac{3}{2}}(z)/\wp'(z)$ is *not* a uniform function. Obtain also the Schwarz-Christoffel formulae expressing z in terms of x_1 and x_2. [Love.]

9. Regular Polygon. (i) The triangle whose angles are $\{\pi/2,\ \pi/n,\ \pi(n-2)/2n\}$ is conformally represented on a half-plane by the relation
$$z = C\int x^{-\tfrac{1}{2}}(x-1)^{(1-n)/n}\, dx.$$

(ii) The representation of the same triangle on a semicircle is given by putting $x = (1+t)^2/(1-t)^2$, or
$$z = C'\int t^{(1-n)/n}(t-1)^{-2/n}\, dt.$$

(iii) The representation of the interior of the regular polygon with n sides on a circle is given by putting $t = w^n$, or
$$z = C'' \int (w^n - 1)^{-2/n}\, dw.$$

[H. A. Schwarz, *Gesammelte mathematische Abhandlungen*, ii. 65–83
(= *J. für Math.* **70** (1869), 105–20).]

10. **The Icosahedral Triangle.** In the equilateral spherical triangle with angles $2\pi/5$, show that
$$\tan a = 2, \quad \tan R = 3 - \sqrt{5}, \quad \sin 2r = \tfrac{2}{3}, \quad \tan p = \tfrac{1}{2}(\sqrt{5}+1),$$
$$p = R + r = \tfrac{1}{2}(\pi - a), \quad \tan(R+a) = -3 - \sqrt{5}.$$

The angles a and $2r$ are those subtended at the centre by edges of the regular icosahedron and dodecahedron.

11. Verify that $\tan \tfrac{1}{2}a = 2\cos 2\pi/5 = (\sqrt{5}-1)/2$. Hence show that the corners of the regular icosahedron can be stereographically projected into the points
$$z = 0, \infty, \quad \epsilon^\nu(\epsilon + \epsilon^4), \quad \epsilon^\nu(\epsilon^2 + \epsilon^3) \quad (\nu = 0, 1, 2, 3, 4),$$
where $\epsilon = \exp(2i\pi/5)$. Verify that these values (except $z = \infty$) satisfy the equation
$$z(z^{10} + 11z^5 - 1) = 0.$$

12. Verify by spherical trigonometry the expressions given for the stereographic projections of the corners of the regular dodecahedron and for the middle points of the edges.

VIII
LAPLACE'S TRANSFORMATION

32. Laplace's Linear Equation†

Form and Singularities. THE equation

$$[P(D)+xQ(D)]y \equiv \sum_{r=0}^{n}(a_r+b_r x)D^{n-r}y = 0 \qquad (1)$$

is a generalization of the equation with constant coefficients, and is completely soluble by an artifice of Laplace. If $b_0 = 0$, we may take unity as the coefficient of $D^n y$, and the only singular point is seen to be $x = \infty$, which is irregular. If $b_0 \neq 0$, we shift the origin and write

$$xD^n y+(a_1+b_1 x)D^{n-1}y+\ldots+(a_n+b_n x)y = 0. \qquad (2)$$

The equation now has one regular singularity at $x = 0$ and one irregular one at $x = \infty$. The exponents at $x = 0$ are

$$(0, 1, 2,\ldots, n-2, n-1-a_1).$$

If a_1 is not an integer, we find $(n-1)$ holomorphic solutions forming one Hamburger set, and one regular solution belonging to the exponent $(n-1-a_1)$ forming another. If a_1 is an integer, all the exponents belong to the same set. It is easily verified that there are always at least $(n-1)$ solutions which are free from logarithms and holomorphic. Since there is only one winding point in the finite part of the plane, and since $(n-1)$ independent solutions are uniform, the group properties of the equation are trivial. The feature of interest is the behaviour of the solutions at the irregular singularity $x = \infty$.

Laplace's equation furnishes a simple illustration of an interesting, but difficult, theorem due to Perron.‡ If $p_0(x)$ is a polynomial of degree $s < n$, and $\{p_r(x)\}$ are integral functions of x, the equation

$$p_0(x)D^n y+p_1(x)D^{n-1}y+\ldots+p_n(x)y = 0 \qquad (3)$$

has at least $(n-s)$ linearly independent solutions, which are integral functions of x.

Solution by Definite Integrals. Laplace uses the transformation

$$y = \int_C e^{xt}f(t)\,dt, \qquad (4)$$

---

† C. Jordan, *Cours d'analyse*, iii. 251–65; E. Picard, *Traité d'analyse*, iii. 394–402.
‡ O. Perron, *Math. Ann.* **70** (1911), 1–32.

which may be regarded as a limiting form of the Eulerian transformation of § 27 as $\xi \to \infty$. The equation

$$\int_C e^{xt}[P(t)+xQ(t)]f(t)\,dt = 0 \tag{5}$$

can be integrated exactly if $f(t)$ is chosen so that

$$\frac{d}{dt}[Q(t)f(t)] = [P(t)f(t)], \tag{6}$$

or
$$Q(t)f(t) = \exp\left[\int \frac{P(t)}{Q(t)}\,dt\right]. \tag{7}$$

We must now find n contours giving linearly independent solutions and satisfying the condition

$$[e^{xt}Q(t)f(t)] = 0. \tag{8}$$

Suppose first that $b_0 \neq 0$, so that we can use the form (2), where $Q(t)$ is of degree n and $P(t)$ of degree $(n-1)$. We have in general, if the roots (β_r) are unequal,

$$\left. \begin{array}{c} Q(t) = \prod_{r=1}^{n}(t-\beta_r), \qquad \dfrac{P(t)}{Q(t)} = \sum_{r=1}^{n}\dfrac{\alpha_r}{t-\beta_r}, \\[1em] f(t) = A\prod_{r=1}^{n}(t-\beta_r)^{\alpha_r-1}. \end{array} \right\} \tag{9}$$

Let I denote a point at infinity, taken in such a direction that xt is real and negative. Then the condition (8) is satisfied by the simple loop integrals

$$L_r = \int_I^{(\beta_r+)} e^{xt}f(t)\,dt \quad (r = 1, 2, ..., n). \tag{10}$$

If α_r is a positive integer, $L_r \equiv 0$; but the condition is then satisfied by the definite integral

$$L_r^* = \int_I^{\beta_r} e^{xt}f(t)\,dt. \tag{11}$$

If α_r is zero or a negative integer, $\text{am}[f(t)]$ returns to its initial value and the contour may be shrunk to a small circle about $t = \beta_r$. The integral can be evaluated as a Cauchy residue. If we expand $f(t)$ and $e^{xt} = e^{x\beta_r}e^{x(t-\beta_r)}$ in ascending powers of $(t-\beta_r)$ and evaluate, we obtain an integral function which is the product of $e^{x\beta_r}$ and a polynomial.

If x is very large and if the contour consists of a small circle about $t = \beta_r$, together with a straight line to infinity described twice, it may be shown that the dominant term in the integral is given by the portion of the contour near $t = \beta_r$. Thus we find $L_r \sim C_r e^{\beta_r x} x^{-\alpha_r}$; and since (β_r) are supposed unequal, the n solutions are linearly independent.

If α_r, α_s are not integers, consider the double-loop integral

$$L_{rs} = \int_I^{(\beta_r+,\beta_s+,\beta_r-,\beta_s-)} e^{xt} f(t)\, dt$$

$$= (1-e^{2i\pi\alpha_s})L_r - (1-e^{2i\pi\alpha_r})L_s. \qquad (12)$$

Since the double circuit restores the initial amplitude of $f(t)$, we may contract this path to a finite contour interlacing $t = \beta_r, \beta_s$, and clear of all singularities of the integrand. Since $f(t)$ is bounded and the contour is of finite length, we can expand e^{xt} and evaluate term by term; the resulting expression is a power-series converging for all finite values of x, i.e. an integral function. Putting $r = 1$, $s = 2, 3, \ldots, n$, we have $(n-1)$ independent integral functions.

If α_r or α_s is a positive integer, $L_{rs} \equiv 0$; but we can then satisfy the conditions by a simple loop from one singularity about the other, or by an ordinary definite integral between $t = \beta_r$ and $t = \beta_s$.

For the last solution, we take an infinite simple loop Ω enclosing *all* the singularities $t = \beta_r$. If we enlarge the loop sufficiently, we can write along the contour

$$f(t) = t^{a_1-n} \sum_{r=0}^{\infty} A_r t^{-r}, \qquad (13)$$

since $\sum \alpha_r = a_1$. We now integrate term by term, using Hankel's integral

$$\frac{1}{\Gamma(z)} = \frac{1}{2\pi i} \int_{-\infty}^{(0+)} e^t t^{-z}\, dt, \qquad (14)$$

and find the solution

$$\int_\Omega e^{xt} f(t)\, dt = 2\pi i \sum_{r=0}^{\infty} A_r x^{n+r-a_1-1}/\Gamma(n+r-a_1). \qquad (15)$$

If a_1 is not an integer, this is regular and belongs to the exponent $(n-a_1-1)$ at the origin.

The expression (15) cannot give a solution belonging to a negative integral exponent, nor a solution of logarithmic type. In fact the method fails when a_1 is an integer, because the set of solutions (12) and (15) are not linearly independent. But we may retain *one* simple loop integral (10), or *one* definite integral (11), and the logarithmic property can be exhibited by deformation of the path as $\operatorname{am}(x)$

increases by 2π. It appears that the integral in question is augmented by a multiple of the solution (15).

Exceptional Cases. The appropriate paths when $Q(t)$ is of lower degree than $P(t)$, or when it has multiple zeros, have been indicated by Jordan. If $Q(t)$ is of degree $(n-\lambda)$, we have

$$\begin{aligned}\frac{P(t)}{Q(t)} &= \sum_{r=0}^{\lambda} A_r t^r + \sum_{s=1}^{n-\lambda} \frac{\alpha_s}{(t-\beta_s)}, \\ f(t) &= A\exp\left[\sum_{r=0}^{\lambda}\frac{A_r t^{r+1}}{(r+1)}\right] \prod_{s=1}^{n-\lambda}(t-\beta_s)^{\alpha_s-1}.\end{aligned} \qquad (16)$$

The space at $t = \infty$ is divided into $(2\lambda+2)$ sectors where $f(t)$ is alternately very large and very small. We can make a festoon of $(\lambda+1)$ paths, beginning and ending in sectors where $f(t)$ is small and crossing the intervening sectors in the finite part of the plane. These and the $(n-\lambda-1)$ double loop circuits interlacing (β_r) make up the n required paths.

Similarly, if $t = \beta$ is a zero of multiplicity λ, we have

$$\log f(t) \sim B(t-\beta)^{1-\lambda}, \quad \text{as } t \to \beta.$$

The space about $t = \beta$ is now divided into $(2\lambda-2)$ sectors where $f(t)$ is alternately large and small. We can construct small loops entering and leaving $t = \beta$ in sectors where $f(t)$ is small, and crossing at a finite distance the intervening ones where it is large. This set of $(\lambda-1)$ loops just compensates for the loss of $(\lambda-1)$ simple zeros of $Q(t)$.

33. The Confluent Hypergeometric Equation[†]

Canonical Forms. Just as Riemann's equation can be transformed in twenty-four ways into a hypergeometric one, so the equation

$$x^2 D^2 y + (A_0 + A_1 x) x Dy + (B_0 + B_1 x + B_2 x^2) y = 0 \qquad (1)$$

can be transformed in four ways into *Kummer's first confluent hypergeometric equation*

$$xD^2 y + (c-x)Dy - ay = 0. \qquad (2)$$

We write $\qquad y = x^\rho e^{\beta x} y', \qquad x' = (\beta'-\beta)x, \qquad (3)$

where ρ is either root of the indicial equation

$$\rho^2 + (A_0-1)\rho + B_0 = 0, \qquad (4)$$

[†] E. T. Whittaker and G. N. Watson, *Modern Analysis*, ch. xvi; G. N. Watson, *Bessel Functions* (1922), 100–5, 188–93; H. Bateman, *Differential Equations* (1918), 75–9, 110–15.

and (β, β') are the roots of
$$\beta^2 + A_1\beta + B_2 = 0. \tag{5}$$

If $\beta = \beta'$, we obtain similarly *Kummer's second equation*
$$xD^2y + cDy - y = 0, \tag{6}$$
or an elementary one of Euler's homogeneous type. The typical solutions of (2) and (6) are respectively
$$\left.\begin{aligned}{}_1F_1(a;c;x) &= \lim_{b\to\infty} F\!\left(a,b;c;\frac{x}{b}\right) = 1 + \frac{a}{1!\,c}x + \frac{a(a+1)}{2!\,c(c+1)}x^2 + \dots, \\ {}_0F_1(c;x) &= \lim_{a,b\to\infty} F\!\left(a,b;c;\frac{x}{ab}\right) = 1 + \frac{x}{1!\,c} + \frac{x^2}{2!\,c(c+1)} + \dots.\end{aligned}\right\} \tag{7}$$

Again, the equation
$$D^2y + (A_0 + A_1 x)Dy + (B_0 + B_1 x + B_2 x^2)y = 0 \tag{8}$$
can be reduced by putting
$$y = e^{\beta x + \gamma x^2} y', \qquad x' = kx + l, \tag{9}$$
to one of the forms
$$D^2y + (2n + 1 - x^2)y = 0, \tag{10}$$
$$D^2y - 9xy = 0, \tag{11}$$
or to an elementary equation with constant coefficients. The equation (10) was found by Weber on transforming the equation of the potential $\nabla^2 V = 0$ to parabolic cylindrical coordinates
$$(x + iy) = \tfrac{1}{2}(\xi + i\eta)^2.$$
It is reduced to Hermite's equation
$$D^2y - 2xDy + 2ny = 0 \tag{12}$$
by putting $y = e^{-\frac{1}{2}x^2} y'$; and then to an equation of Kummer's first type
$$z\frac{d^2y}{dz^2} + (\tfrac{1}{2} - z)\frac{dy}{dz} + \tfrac{1}{2}ny = 0, \tag{13}$$
by the change of variable $z = x^2$.

The equation (11) is satisfied by a definite integral considered by Airy, and on changing the variable to $z = x^3$ it is reduced to Kummer's second type
$$z\frac{d^2y}{dz^2} + \frac{2}{3}\frac{dy}{dz} - y = 0. \tag{14}$$

Finally, Kummer's second equation is transformed by the change of variable $z = \pm 4\sqrt{x}$ and the substitution $y = e^{-\frac{1}{2}z} y'$ into one of his first type
$$z\frac{d^2y}{dz^2} + (2c - 1 - z)\frac{dy}{dz} - (c - \tfrac{1}{2})y = 0. \tag{15}$$

The equations (2), (6), (11), (12) are soluble as they stand by Laplace's method, and their solutions can also be expressed in terms of the standard power-series (7). But it is convenient to regard (2) as the ultimate canonical form. The analogue of Schwarz's form of the hypergeometric equation is obtained by removing the middle term. We thus find *Whittaker's equation*

$$D^2y + \left[-\frac{1}{4} + \frac{k}{x} + \frac{\frac{1}{4}-m^2}{x^2}\right]y = 0, \qquad (16)$$

exhibiting the exponents $(\frac{1}{2} \pm m)$ at the origin. The solutions of (16) are written $M_{k,\pm m}(x)$, where

$$M_{k,m}(x) = x^{\frac{1}{2}+m} e^{-\frac{1}{2}x} {}_1F_1(\tfrac{1}{2}+m-k; 2m+1; x). \qquad (17)$$

If $k = 0$, the equation is invariant when x is replaced by $-x$ and is reducible to Kummer's second form.

Solutions in Power-Series.† Kummer's first equation (2) is transformed into another of the same type by putting

$$\left.\begin{array}{llll} y = e^x y', & x = -x', & a = c'-a', & c = c'; \\ \text{or}\quad y = x^{1-c} y', & x = x', & a = a'-c'+1, & c = 2-c'; \\ \text{or}\quad y = e^x x^{1-c} y', & x = -x', & a = 1-a', & c = 2-c'. \end{array}\right\} \quad (18)$$

From the standard solution (7) we can write down two alternative expressions of each principal branch, when c is not an integer. We thus have Kummer's linear transformation

$$\left.\begin{array}{l} y_1 \equiv {}_1F_1(a;c;x) \equiv e^x {}_1F_1(c-a;c;-x), \\ y_2 \equiv x^{1-c} {}_1F_1(a-c+1;2-c;x) \equiv e^x x^{1-c} {}_1F_1(1-a;2-c;-x). \end{array}\right\} \quad (19)$$

These four expressions correspond to Kummer's twenty-four hypergeometric series; and the relation is the limiting form of Euler's identity

$$F(a,b;c;x) \equiv (1-x)^{c-a-b} F(c-a, c-b; 2-c; x), \qquad (20)$$

when x is replaced by x/b, and $b \to \infty$.

The principal solutions of (6) are

$$y_1 = {}_0F_1(c;x), \qquad y_2 = x^{1-c} {}_0F_1(2-c;x), \qquad (21)$$

when c is not an integer. Kummer's transformation of (6) into (15) is the analogue of Riemann's quadratic transformation and gives the identities

$${}_0F_1\left(p+\tfrac{1}{2}; \frac{z^2}{16}\right) = e^{-\frac{1}{2}z} {}_1F_1(p; 2p; z) = e^{\frac{1}{2}z} {}_1F_1(p; 2p; -z), \qquad (22)$$

where $p = (c-\tfrac{1}{2})$.

† Cf. Ex. V, 4–7.

Bessel's Equation. If $A_1 = 0 = B_1$, the equation (1) is invariant when x is replaced by $-x$. It may then be reduced to Bessel's equation
$$x^2D^2y + xDy + (x^2-k^2)y = 0, \tag{23}$$
or to an elementary equation of Euler's homogeneous type. This property is analogous to the symmetry of the associated Legendre equation
$$\frac{d}{d\mu}\left[(1-\mu^2)\frac{dw}{d\mu}\right] + \left[n(n+1) - \frac{m^2}{1-\mu^2}\right]w = 0, \tag{24}$$
with the scheme
$$P\left\{\begin{array}{ccc} -1 & \infty & 1 \\ \tfrac{1}{2}m & n+1 & \tfrac{1}{2}m \quad \mu \\ -\tfrac{1}{2}m & -n & -\tfrac{1}{2}m \end{array}\right\}. \tag{25}$$

If we write $\mu = m/ix$ and let $m \to \infty$, we get an equation
$$x^2\frac{d^2w}{dx^2} + [x^2 - n(n+1)]w = 0, \tag{26}$$
having a regular singularity at $x = 0$ with exponents $(n+1, -n)$ and an irregular singularity at infinity. If we now put $w = x^{\frac{1}{2}}y$, the equation is reduced to Bessel's equation of order $k = (n+\tfrac{1}{2})$, with the exponents $\pm k$ at the regular singularity $x = 0$.

When k is not an integer, the solutions of Bessel's equation are $J_k(x)$ and $J_{-k}(x)$, where
$$J_k(x) \equiv \frac{x^k}{2^k\Gamma(k+1)}\,_0F_1\!\left(k+1; -\frac{x^2}{4}\right). \tag{27}$$

Let n be positive and let m tend to infinity by real positive values. We have with Hobson's notation
$$Q_n^m(\mu) = \frac{e^{m\pi i}}{2^{n+1}}\frac{\Gamma(n+m+1)\Gamma(\tfrac{1}{2})}{\Gamma(n+\tfrac{3}{2})} \times$$
$$\times \mu^{-n-1}\left(1 - \frac{1}{\mu^2}\right)^{m/2} F\!\left(\frac{n+m+1}{2}, \frac{n+m+2}{2}; n+\tfrac{3}{2}; \frac{1}{\mu^2}\right). \tag{28}$$

As $m \to \infty$, we have by Stirling's formula $\Gamma(n+m+1) \sim m^{n+1}\Gamma(m)$, and so
$$Q_n^m\!\left(\frac{m}{ix}\right) \sim \frac{e^{m\pi i}\Gamma(m)\Gamma(\tfrac{1}{2})}{2^{n+1}\Gamma(n+\tfrac{3}{2})}(ix)^{n+1}S, \tag{29}$$
where
$$S = \lim_{m\to\infty} F\!\left(\frac{n+m+1}{2}, \frac{n+m+2}{2}; n+\tfrac{3}{2}; -\frac{x^2}{m^2}\right). \tag{30}$$

The series being convergent uniformly with respect to m for a fixed value of x, we may evaluate the limit term by term and so get $S = {}_0F_1\left(n+\tfrac{3}{2}; -\tfrac{x^2}{4}\right)$.

Hence, with $k = (n+\tfrac{1}{2})$, we have finally

$$\lim_{m \to \infty}\left[e^{-m\pi i}Q_n^m\left(\frac{m}{ix}\right)\bigg/\Gamma(m)\right] = i^{n+1}\left(\frac{\pi x}{2}\right)^{\frac{1}{2}} J_{n+\frac{1}{2}}(x). \tag{31}$$

Finite Solutions. From Kummer's relations (19), we see that one of the four series is finite, if a or $(a-c)$ is an integer. As a typical case, let $a = -n$. As in the proof of Jacobi's formula in § 24, we find for any solution of (2) the relation

$$D^k[x^{c+k-1}e^{-x}D^k y] = a_k x^{c-1} e^{-x} y, \tag{32}$$

where $a_k \equiv a(a+1)...(a+k-1)$.

In particular, if $y \equiv {}_1F_1(a;c;x)$, we get

$$D^k[x^{c+k-1}e^{-x}{}_1F_1(a+k;c+k;x)] = c_k x^{c-1} e^{-x} {}_1F_1(a;c;x). \tag{33}$$

If now n is a positive integer, and $a = -n$, $k = n$, we get the expression

$$c_n {}_1F_1(-n;c;x) = x^{1-c} e^x D^n[x^{c+n-1} e^{-x}]. \tag{34}$$

The standard polynomials of this class are known as *Sonine polynomials*.

34. Integral Representations of Kummer's Series

Standard Forms. To solve Kummer's first equation, we write in § 32, (4)–(8),

$$\left.\begin{array}{c} P(t) \equiv ct-a, \qquad Q(t) \equiv t^2 - t, \\[4pt] f(t) = \dfrac{1}{Q(t)} \exp\left[\int \dfrac{P(t)}{Q(t)}\, dt\right] = A t^{a-1}(1-t)^{c-a-1}, \end{array}\right\} \tag{1}$$

and obtain the expression

$$y = \int_C e^{xt} t^{a-1}(1-t)^{c-a-1}\, dt, \tag{2}$$

where the contour C must satisfy the condition

$$[e^{xt} t^a (1-t)^{c-a}]_C = 0. \tag{3}$$

Independent solutions are in general given by the simple loop integrals $L_0 = \int_I^{(0+)}$ and $L_1 = \int_I^{(1+)}$, where I is the point $xt = -\infty$. From these we construct the combinations

$$\int_\Phi \equiv \int_I^{(0+,1+,0-,1-)} = [1-e^{2i\pi(c-a)}]L_0-[1-e^{2i\pi a}]L_1,$$
$$\int_\Omega \equiv \int_I^{(0+,1+)} = L_0+e^{2i\pi a}L_1, \qquad (4)$$

which are linearly independent, provided that $e^{2i\pi c} \neq 1$ and that L_0, L_1 do not vanish identically.

The contour Φ may be deformed into a fixed double loop interlacing $t = 0, 1$, and lying entirely in the finite part of the plane. On expanding e^{xt} under the integral sign and evaluating term by term, we get

$$\int_\Phi e^{xt}t^{a-1}(1-t)^{c-a-1}\,dt = {}_1F_1(a;c;x)\int_\Phi t^{a-1}(1-t)^{c-a-1}\,dt, \qquad (5)$$

the integral on the right being a generalized beta function.

The contour Ω may be enlarged so that $|t| > 1$ everywhere; on expanding $f(t)$ in descending powers of t, and evaluating term by term by means of Hankel's integral for the gamma function, we have

$$\int_\Omega e^{xt}t^{a-1}(t-1)^{c-a-1}\,dt = x^{1-c}\int_{-\infty}^{(0+)} e^u u^{c-2}\left(1-\frac{x}{u}\right)^{c-a-1}du$$
$$= x^{1-c}\sum_{n=0}^\infty \binom{c-a-1}{n}(-x)^n \int_{-\infty}^{(0+)} e^u u^{c-2-n}\,du$$
$$= \frac{2i\pi x^{1-c}}{\Gamma(2-c)}{}_1F_1(a-c+1;2-c;x). \qquad (6)$$

If a or $(c-a)$ is an integer, the contours must be modified. If a is a positive integer, $L_0 \equiv 0$; but $t = 0$ may now be taken as a limit of integration. The holomorphic solution is obtained as in (5), except that the double loop is replaced by a simple one from $t = 0$ about $t = 1$. The other solution is

$$\int_\Omega e^{xt}t^{a-1}(t-1)^{c-a-1}\,dt = D^{a-1}\left[e^x\int_\Omega e^{x(t-1)}(t-1)^{c-a-1}\,dt\right],$$
$$= D^{a-1}\left[e^x x^{a-c}\int_{-\infty}^{(0+)} e^u u^{c-a-1}\,du\right],$$
$$= \frac{2i\pi}{\Gamma(a-c+1)}D^{a-1}[e^x x^{a-c}],$$

$$= \frac{2i\pi}{\Gamma(2-c)} x^{1-c} F(a-c+1; 2-c; x),$$

$$= \frac{2i\pi}{\Gamma(2-c)} x^{1-c} e^x F(1-a; 2-c; -x), \qquad (7)$$

as may be verified by elementary methods.

If a is zero or a negative integer, L_0 can be evaluated as a finite series of Cauchy residues, and the contour Φ gives a numerical multiple of L_0.

$$\int\limits^{(0+)} e^{xt} t^{a-1}(1-t)^{c-a-1} dt = \sum_{n=0}^{-a} \frac{x^n}{n!} \int\limits^{(0+)} t^{a+n-1}(1-t)^{c-a-1} dt$$

$$= 2i\pi \sum_{n=0}^{-a} \frac{x^n}{n!} (-)^{-a-n} \binom{c-a-1}{-a-n}$$

$$= 2i\pi \frac{\Gamma(c-a)}{\Gamma(c)\Gamma(1-a)} {}_1F_1(a; c; x). \qquad (8)$$

Similar considerations apply to L_1 when $(c-a)$ is an integer, and so we obtain integral representations of the Sonine polynomials. When $a, (c-a)$ are integers of the same sign, $x = 0$ is an *apparent singularity* or reducible thereto, and there are *two* solutions expressible in finite terms.

If c is any integer and a not one of the integers lying between zero and c, $x = 0$ is a logarithmic singularity, and the solutions (4) are not independent, because $e^{2i\pi c} = 1$. A simple circuit enclosing $t = 0$ and $t = 1$, and lying in the finite part of the plane, gives the holomorphic solution $y_1 \equiv (L_0 + e^{2i\pi a} L_1)$. The solution $y_2 \equiv L_0$ is logarithmic.

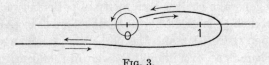

Fig. 3.

For as $\operatorname{am}(x)$ increases from zero to 2π, $\operatorname{am}(t)$ at I decreases from π to $-\pi$. As the path of integration swings round and bends to avoid the singularities, we get the new branch

$$Y_2 = \int\limits_I^{(1+, 0+, 1-)} = L_1 + e^{-2i\pi a} L_0 - e^{2i\pi a} L_1$$

$$= L_0 + (e^{-2i\pi a} - 1)(L_0 + e^{2i\pi a} L_1) = y_2 + (e^{-2i\pi a} - 1) y_1, \qquad (9)$$

whose form shows the presence of a logarithmic term.

Alternative Forms. According as we operate on Kummer's equation as it stands or on one of the equivalent forms, we have the four Laplace integrals

$$\left.\begin{array}{rl} \int e^{xt} t^{a-1}(1-t)^{c-a-1}\, dt, & \text{(A)} \\[4pt] e^x \int e^{-xt} t^{c-a-1}(1-t)^{a-1}\, dt, & \text{(B)} \\[4pt] x^{1-c} \int e^{xt} t^{a-c}(1-t)^{-a}\, dt, & \text{(C)} \\[4pt] e^x x^{1-c} \int e^{-xt} t^{-a}(1-t)^{a-c}\, dt, & \text{(D)} \end{array}\right\} \quad (10)$$

which correspond to the four Eulerian hypergeometric integrals

$$\left.\begin{array}{rl} \int_0^1 (1-xt)^{-b} t^{a-1}(1-t)^{c-a-1}\, dt, & \text{(A)} \\[4pt] (1-x)^{c-a-b}\int_0^1 (1-xt)^{b-c} t^{c-a-1}(1-t)^{a-1}\, dt, & \text{(B)} \\[4pt] x^{1-c}\int_{1/x}^{\infty} (1-xt)^{c-b-1} t^{a-c}(1-t)^{-a}\, dt, & \text{(C)} \\[4pt] x^{1-c}(1-x)^{c-a-b}\int_{1/x}^{\infty} (1-xt)^{b-1} t^{-a}(1-t)^{a-c}\, dt. & \text{(D)} \end{array}\right\} \quad (11)$$

The latter all converge and give multiples of $F(a,b;c;x)$, if $1 > c > a > 0$. In each set, the pairs (AB) and (CD) belong to the same sets. Kummer's linear relations § 33 (19) are given by putting $t' = (1-t)$ in (10); and Euler's relation § 33 (20) is similarly given by putting $t' = (1-t)/(1-xt)$ in (11).

There is in general a curious reciprocity of paths between the two pairs. The branch y_1 is given by the double loop Φ in (AB) and by the infinite loop Ω in (CD), and the paths are interchanged for the branch y_2. There is an exception, however, in the logarithmic case, where we saw that Φ and Ω both reduce to the same branch. In that case only, the two types can be transformed into one another by elementary methods, so that the *same* path gives the *same* branch for each. Without loss of generality, we suppose that c is a positive integer. We take the representation (A) and integrate by parts $(c-1)$ times, the integrated parts vanishing around any admissible contour,

$$\int_C e^{xt} t^{a-1}(1-t)^{c-a-1}\, dt = (-x)^{1-c}\int_C e^{xt}\left[\left(\frac{d}{dt}\right)^{c-1}\{t^{a-1}(1-t)^{c-a-1}\}\right] dt. \quad (12)$$

Now in Jacobi's identity

$$\left(\frac{d}{dt}\right)^n \{t^{n+\alpha}(1-t)^{n+\beta}\} = \frac{\Gamma(n+1+\alpha)}{\Gamma(1+\alpha)} t^\alpha (1-t)^\beta F(-n, n+1+\alpha+\beta; 1+\alpha; t), \tag{13}$$

let us write $\alpha = (a-c), \beta = -a, n = (c-1)$; then we have

$$(n+1+\alpha+\beta) = 0,$$

and so
$$\left(\frac{d}{dt}\right)^{c-1} \{t^{a-1}(1-t)^{c-a-1}\} = \frac{\Gamma(a)}{\Gamma(a-c+1)} t^{a-c}(1-t)^{-a}. \tag{14}$$

Hence

$$\int_C e^{xt} t^{a-1}(1-t)^{c-a-1}\,dt = \frac{\Gamma(a)}{\Gamma(a-c+1)}(-x)^{1-c}\int_C e^{xt} t^{a-c}(1-t)^{-a}\,dt. \tag{15}$$

Both sides vanish around any closed contour if

$$(a-1)(a-2)\ldots(a-c+1) = 0, \tag{16}$$

these values of a giving an apparent singularity when c is a positive integer.

Kummer's Second Equation. The equation of § 33 (6), which is satisfied by $_0F_1(c; x)$, also admits two alternative solutions of Laplace's type

$$\left.\begin{aligned}\int e^{xt+1/t} t^{c-2}\,dt,\\ x^{1-c}\int e^{xt+1/t} t^{-c}\,dt.\end{aligned}\right\} \tag{17}$$

But these can be transformed into one another by putting $t' = 1/xt$. The most interesting feature of this solution is the choice of contours. One solution is given by a loop from $xt = -\infty$ about $t = 0$, and another by a small heart-shaped Jordan loop entering and leaving the origin on the side where $R(t)$ is negative, so that the condition

$$[e^{xt+1/t} t^c]_C = 0 \tag{18}$$

is satisfied. The infinite contour and the small loop are interchanged when we pass from one representation to the other.

35. Bessel's Equation†

Integrals of Poisson's Type. Bessel's equation

$$\delta^2 y + (x^2 - k^2) y = 0 \tag{1}$$

is reduced to an auxiliary equation of Laplace's type

$$xD^2w + (2k+1)Dw + xw = 0 \tag{2}$$

† G. N. Watson, *Bessel Functions* (1922).

by the substitution $y = x^k w$. On solving by the usual rule, we find
$$y = x^k \int_C e^{ixt}(1-t^2)^{k-\frac{1}{2}}\, dt, \tag{3}$$
where the contour C must satisfy the condition
$$[e^{ixt}(1-t^2)^{k+\frac{1}{2}}]_C = 0. \tag{4}$$

There is an alternative solution where k is everywhere replaced by $-k$. If k is half an odd integer, the solutions can be evaluated in an elementary form; in the one case they are given as residues by small circuits about $t = \pm 1$, and in the other as elementary definite integrals between limits $t = \pm 1$, $ixt = -\infty$.

In general, an infinite loop from $ixt = -\infty$ enclosing $t = \pm 1$ and a figure of eight interlacing these points give independent solutions $J_k(x)$ and $J_{-k}(x)$ [cf. § 33 (27)]. But when k is an integer,
$$J_{-k}(x) = (-)^k J_k(x).$$
Except when $2k$ is an odd positive integer, independent solutions are always given by Hankel's simple loop integrals
$$\left.\begin{aligned} H_k^{(1)}(x) &= \frac{\Gamma(\frac{1}{2}-k)(\frac{1}{2}x)^k}{i\pi\Gamma(\frac{1}{2})} \int_I^{(1+)} e^{ixt}(t^2-1)^{k-\frac{1}{2}}\, dt, \\ H_k^{(2)}(x) &= \frac{\Gamma(\frac{1}{2}-k)(\frac{1}{2}x)^k}{i\pi\Gamma(\frac{1}{2})} \int_I^{(-1-)} e^{ixt}(t^2-1)^{k-\frac{1}{2}}\, dt, \end{aligned}\right\} \tag{5}$$
the phases of these integrands being so adjusted that
$$J_k(x) = \tfrac{1}{2}\{H_k^{(1)}(x) + H_k^{(2)}(x)\} = \frac{\Gamma(\frac{1}{2}-k)(\frac{1}{2}x)^k}{2i\pi\Gamma(\frac{1}{2})} \int_I^{(1+,-1-)} e^{ixt}(t^2-1)^{k-\frac{1}{2}}\, dt. \tag{6}$$

If we choose k with the real part positive, these expressions may be replaced by convergent definite integrals. For example, if x is real and positive, we can show by putting $t = \left(\dfrac{iu}{x} \pm 1\right)$ that
$$\left.\begin{aligned} H_k^{(1)}(x) &= \left(\frac{2}{\pi x}\right)^{\frac{1}{2}} \frac{e^{i(x-\frac{1}{2}k\pi-\frac{1}{4}\pi)}}{\Gamma(k+\frac{1}{2})} \int_0^\infty e^{-u} u^{k-\frac{1}{2}} \left(1+\frac{iu}{2x}\right)^{k-\frac{1}{2}} du, \\ H_k^{(2)}(x) &= \left(\frac{2}{\pi x}\right)^{\frac{1}{2}} \frac{e^{-i(x-\frac{1}{2}k\pi-\frac{1}{4}\pi)}}{\Gamma(k+\frac{1}{2})} \int_0^\infty e^{-u} u^{k-\frac{1}{2}} \left(1-\frac{iu}{2x}\right)^{k-\frac{1}{2}} du. \end{aligned}\right\} \tag{7}$$

Asymptotic Expansions. From Hankel's integrals we obtain approximate values of the Bessel functions, when x is large with a given amplitude. We shall take x and k both real and positive. Then, using Cauchy's form of the remainder in Taylor's series, we write in (7)

$$\left(1+\frac{iu}{2x}\right)^{k-\frac{1}{2}} = \sum_{r=0}^{p-1} \frac{\Gamma(k+\frac{1}{2})}{\Gamma(k-r+\frac{1}{2})r!}\left(\frac{iu}{2x}\right)^r + R_p^{(1)}, \tag{8}$$

where

$$R_p^{(1)} = \frac{\Gamma(k+\frac{1}{2})}{\Gamma(k-p+\frac{1}{2})(p-1)!}\left(\frac{iu}{2x}\right)^p \int_0^1 (1-\theta)^{p-1}\left(1+\frac{iu\theta}{2x}\right)^{k-p-\frac{1}{2}} d\theta. \tag{9}$$

If $p > k$, $\left|1+\dfrac{iu\theta}{2x}\right|^{k-p-\frac{1}{2}} < 1$, since $u\theta/2x$ is real and positive. Hence the remainder is numerically smaller than the first term neglected. On integrating term by term, we have the *asymptotic representations*

$$\left. \begin{aligned} H_k^{(1)}(x) &\sim \left(\frac{2}{\pi x}\right)^{\frac{1}{2}} e^{i(x-\frac{1}{2}k\pi-\frac{1}{4}\pi)} \sum_{r=0}^{p-1} \frac{\Gamma(k+r+\frac{1}{2})}{\Gamma(k-r+\frac{1}{2})r!}\left(\frac{i}{2x}\right)^r, \\ H_k^{(2)}(x) &\sim \left(\frac{2}{\pi x}\right)^{\frac{1}{2}} e^{-i(x-\frac{1}{2}k\pi-\frac{1}{4}\pi)} \sum_{r=0}^{p-1} \frac{\Gamma(k+r+\frac{1}{2})}{\Gamma(k-r+\frac{1}{2})r!}\left(\frac{-i}{2x}\right)^r. \end{aligned} \right\} \tag{10}$$

The series continued to infinity would be divergent, the ratio of two successive terms being $\pm\dfrac{(k+r+\frac{1}{2})(k-r-\frac{1}{2})i}{2x(r+1)}$, which tends to infinity. But when x is large the terms at the beginning of the series decrease very rapidly, and the error committed by keeping p terms is numerically smaller than the $(p+1)$th. A very general theory of asymptotic solutions was developed by Poincaré.†

Bessel's Integral. Another way of using Laplace's transformation is to write $z = x^2$ in (2) and so obtain the auxiliary equation

$$z\frac{d^2w}{dz^2} + (k+1)\frac{dw}{dz} + \tfrac{1}{4}w = 0. \tag{11}$$

This is satisfied by

$$w = \int_C \exp\left(zt - \frac{1}{4t}\right) t^{k-1} dt, \tag{12}$$

along a contour chosen as for Kummer's second equation so that

$$\left[\exp\left(zt - \frac{1}{4t}\right) t^{k+1}\right]_C = 0. \tag{13}$$

† H. Poincaré, *American J. of Math.* 7 (1885), 203–58; *Acta Math.* 8 (1886), 295–344.

If we restore $y = x^n w$, $z = x^2$ and put $xt = \tfrac{1}{2}u$, we have the forms

$$\left.\begin{aligned} y &= \int \exp\!\left[\frac{x}{2}\!\left(u-\frac{1}{u}\right)\right] u^{k-1}\,du, \\ y &= \int \exp\!\left[\frac{x}{2}\!\left(u-\frac{1}{u}\right)\right] u^{-k-1}\,du, \end{aligned}\right\} \tag{14}$$

which can be obtained from one another by putting $u' = -1/u$. Loops from infinity give distinct solutions, except when k is an integer; we can then use a path from $u=0$ to $u=\infty$, if we approach the limits in such directions that the integrand tends to zero.

For integer values of k, we have

$$J_k(x) = (-)^k J_k(-x) = \frac{1}{2i\pi}\int^{(0+)} \exp\!\left[\frac{x}{2}\!\left(u-\frac{1}{u}\right)\right] u^{-k-1}\,du, \tag{15}$$

and so we find the generating function

$$\exp\!\left[\frac{x}{2}\!\left(u-\frac{1}{u}\right)\right] = \sum_{k=-\infty}^{\infty} u^k J_k(x). \tag{16}$$

On putting $u = e^{i\theta}$ in (15), we have

$$J_k(x) = \frac{1}{2\pi}\int_0^{2\pi} \exp(ix\sin\theta - ik\theta)\,d\theta, \tag{17}$$

and on putting $\theta' = (2\pi - \theta)$ and adding, we have Bessel's integral

$$J_k(x) = \frac{1}{2\pi}\int_0^{2\pi} \cos(x\sin\theta - k\theta)\,d\theta. \tag{18}$$

EXAMPLES. VIII

1. RECURRENCE FORMULAE. If $F \equiv {}_1F_1(a;c;x)$, show that
$$D^k[x^{a+k-1}F] = a_k\,x^{a-1}\,{}_1F_1(a+k;c;x),$$
$$D^k[x^{c-1}F] = (c-k)_k\,x^{c-k-1}\,{}_1F_1(a;c-k;x),$$
$$c_k\,D^k[e^{-x}F] = (-)^k(c-a)_k\,e^{-x}\,{}_1F_1(a;c+k;x),$$
$$D^k[e^{-x}x^{c-a+k-1}F] = (c-a)_k\,e^{-x}x^{c-a-1}\,{}_1F_1(a-k;c;x).$$

2. Show that Bessel's equation may be written in either of the forms
$$\left[D+\frac{k+1}{x}\right]\!\left[D-\frac{k}{x}\right]y + y = 0,$$
$$\left[D-\frac{k-1}{x}\right]\!\left[D+\frac{k}{x}\right]y + y = 0.$$

Hence prove that
$$J_{k+1}(x) = -\!\left[D-\frac{k}{x}\right]J_k(x),$$
$$J_{k-1}(x) = \left[D+\frac{k}{x}\right]J_k(x),$$

$$J_{k+1}(x)+J_{k-1}(x) = \frac{2k}{x} J_k(x),$$

$$J_{k+1}(x)-J_{k-1}(x) = -2J_k'(x).$$

3. If k is a positive integer, show that

$$J_k(x) = (-)^k x^k \left(\frac{d}{x\,dx}\right)^k J_0(x),$$

$$J_{m+k}(x) = (-)^k x^{m+k} \left(\frac{d}{x\,dx}\right)^k [x^{-m} J_m(x)],$$

$$J_{m-k}(x) = x^{k-m} \left(\frac{d}{x\,dx}\right)^k [x^m J_m(x)].$$

4. Show that

$$J_{\frac{1}{2}}(x) = \left(\frac{2}{\pi x}\right)^{\frac{1}{2}} \sin x, \qquad J_{-\frac{1}{2}}(x) = \left(\frac{2}{\pi x}\right)^{\frac{1}{2}} \cos x,$$

$$J_{k+\frac{1}{2}}(x) = (-)^k \left(\frac{2}{\pi}\right)^{\frac{1}{2}} x^{k+\frac{1}{2}} \left(\frac{d}{x\,dx}\right)^k \left(\frac{\sin x}{x}\right),$$

$$J_{-k-\frac{1}{2}}(x) = \left(\frac{2}{\pi}\right)^{\frac{1}{2}} x^{k+\frac{1}{2}} \left(\frac{d}{x\,dx}\right)^k \left(\frac{\cos x}{x}\right).$$

5. Show that, if k is a positive integer

$$J_{k+\frac{1}{2}}(x) = \frac{(\frac{1}{2}x)^{k+\frac{1}{2}}}{k!\,\pi^{\frac{1}{2}}} \int_{-1}^{1} e^{ixt}(1-t^2)^k\,dt = (-i)^k \left(\frac{x}{2\pi}\right)^{\frac{1}{2}} \int_{-1}^{1} e^{ixt} P_k(t)\,dt,$$

$$J_{-k-\frac{1}{2}}(x) = \frac{k!\,(\frac{1}{2}x)^{-k-\frac{1}{2}}}{2i\pi^{\frac{3}{2}}} \int_{0}^{(1+,-1-)} e^{ixt}(t^2-1)^{-k-1}\,dt.$$

6. Deduce Mehler's expression $J_0(x) = \lim_{n\to\infty} P_n\!\left(\cos\frac{x}{n}\right)$ from Laplace's integral

$$P_n(\cos\theta) = \frac{1}{\pi} \int_0^\pi (\cos\theta + i\sin\theta\cos\phi)^n\,d\phi.$$

7. If $R(k+1) > 0$, show that

$$(\alpha^2-\beta^2)\int_0^x x J_k(\alpha x) J_k(\beta x)\,dx = \beta x J_k(\alpha x) J_k'(\beta x) - \alpha x J_k'(\alpha x) J_k(\beta x),$$

$$2\int_0^x x J_k^2(\alpha x)\,dx = \left(x^2 - \frac{k^2}{\alpha^2}\right) J_k^2(\alpha x) + x^2 J_k'^2(\alpha x).$$

If $\alpha \neq \beta$ and $J_k(\alpha)/\alpha J'(\alpha) = J_k(\beta)/\beta J_k'(\beta)$, show that

$$\int_0^1 x J_k(\alpha x) J_k(\beta x)\,dx = 0.$$

8. If k is a positive integer, prove that Poisson's integral

$$J_k(x) = \frac{x^k}{1.3.5\ldots(2k-1)\pi} \int_0^\pi \cos(x\cos\theta)\sin^{2k}\theta\,d\theta$$

can be transformed into Bessel's integral

$$J_k(x) = \frac{1}{2\pi} \int_0^{2\pi} \exp(ix\sin\theta - ik\theta)\, d\theta$$

by integrating by parts and using Jacobi's formula, Ex. VI, 15. [JACOBI.]

9. If x, a, $(a-c+1)$ are all positive, show that

$$\int_{-\infty}^{(0+,x+)} du \int_0^\infty \exp\left\{u-v+\frac{xv}{u}\right\} u^{-c} v^{a-1}\, dv = \frac{2i\pi\Gamma(a)}{\Gamma(c)}\,{}_1F_1(a;c;x).$$

Show that the alternative representations of ${}_1F_1(a;c;x)$ as integrals of Laplace's type are obtained by evaluating the double integral in two ways.

10. SCHERK'S EQUATION. Show that the equation
$$D^n y - xy = 0$$
is satisfied by
$$y = \sum_{r=0}^n C_r \int_0^\infty \exp\left[\omega^r xt - \frac{t^{n+1}}{n+1}\right] dt,$$
where $\omega = \exp[2i\pi/(n+1)]$ and $\sum_{r=0}^n C_r = 0$.

[H. F. Scherk, *J. für Math.* **10** (1832), 92–7; C. G. J. Jacobi, *Werke*, iv. 33–4.]

11. HERMITE'S EQUATION. Show that the equation
$$D^2 y - 2xDy + 2ny = 0$$
is satisfied by
$$y_1 = {}_1F_1(-\tfrac{1}{2}n;\tfrac{1}{2};x^2) = e^{x^2}\,{}_1F_1(\tfrac{1}{2}+\tfrac{1}{2}n;\tfrac{1}{2};-x^2),$$
$$y_2 = x\,{}_1F_1(\tfrac{1}{2}-\tfrac{1}{2}n;\tfrac{3}{2};x^2) = xe^{x^2}\,{}_1F_1(1+\tfrac{1}{2}n;\tfrac{3}{2};-x^2).$$

Obtain integral representations, along appropriate contours, of the types

$$\int e^{2xt-t^2} t^{-n-1}\, dt, \qquad \int e^{x^2 s} s^{-1-\frac{1}{2}n}(1-s)^{-\frac{1}{2}+\frac{1}{2}n}\, ds, \qquad x\int e^{x^2 s} s^{-\frac{1}{2}-\frac{1}{2}n}(1-s)^{\frac{1}{2}n}\, ds.$$

[C. Hermite, *Œuvres*, ii. 293–308; E. T. Whittaker and G. N. Watson, *Modern Analysis*, 341–5; R. Courant and D. Hilbert, *Methoden der mathematischen Physik*, 76–7, 261, 294.]

12. HERMITE'S POLYNOMIALS. (i) Prove by Jacobi's method that the equation is satisfied, when n is a positive integer, by the polynomial
$$H_n(x) = (-)^n e^{x^2} D^n e^{-x^2},$$
and by $e^{x^2} H_{-n-1}(ix)$, when n is a negative integer.

(ii) Show that
$$H_{2n}(x) = (-)^n 2^n e^{x^2} x \left(\frac{d}{x\,dx}\right)^n (x^{2n-1} e^{-x^2}),$$
$$H_{2n+1}(x) = (-)^n 2^{n+1} e^{x^2} \left(\frac{d}{x\,dx}\right)^n (x^{2n+1} e^{-x^2}).$$
$$H_n'(x) = 2n H_{n-1}(x) = 2x H_n(x) - H_{n+1}(x).$$

(iii) Prove the orthogonal relations
$$\int_{-\infty}^\infty e^{-x^2} H_m(x) H_n(x)\, dx \begin{cases} = 0 & (m \neq n), \\ = 2^n n! \sqrt{\pi} & (m = n). \end{cases}$$

(iv) Show that $\quad \phi(x,t) = e^{2xt-t^2} = \sum_{n=0}^{\infty} \frac{t^n}{n!} H_n(x),$

and verify the above formulae from the relations

$$\frac{\partial \phi}{\partial x} = 2t\phi, \quad \frac{\partial \phi}{\partial t} = 2(x-t)\phi, \quad \int_{-\infty}^{\infty} e^{-x^2+(2xt-t^2)+(2xs-s^2)}\,dx = e^{2st}\sqrt{\pi}.$$

13. LAGUERRE'S POLYNOMIALS. (i) Find the integral representations of the polynomials
$$L_n(x) = e^x D^n[e^{-x}x^n] = n!\,{}_1F_1(-n;1;x).$$

(ii) Show that
$$\int_0^{\infty} e^{-x} L_m(x) L_n(x)\,dx \begin{cases} = 0 & (m \neq n), \\ = (n!)^2 & (m = n). \end{cases}$$

(iii) By means of Lagrange's series, show that
$$\exp\left(\frac{-vx}{1-v}\right)\frac{1}{(1-v)} = \sum_{n=0}^{\infty} \frac{v^n}{n!} L_n(x),$$
and hence or otherwise prove that
$$L_{n+1}(x) - (2n+1-x)L_n(x) + n^2 L_{n-1}(x) = 0.$$

(iv) Show that
$$D^m L_{m+n}(x) = D^m[e^x D^{m+n}(e^{-x} x^{m+n})]$$
$$= \frac{(m+n)!}{n!} e^x D^{m+n}(e^{-x} x^n)$$
$$= (-)^m \frac{(m+n)!}{n!} e^x x^{-m} D^n(e^{-x} x^{m+n}).$$

[Courant-Hilbert, loc. cit., 77–9; see also E. Schrödinger, *Abhandlungen zur Wellenmechanik* (1928), 131–6.]

14. SONINE POLYNOMIALS. A set of polynomials $\{\phi_n(x)\}$ are defined by the generating function
$$(1-v)^{-1-k}\exp\left[\frac{-xv}{1-v}\right] = \sum_{n=0}^{\infty} v^n \phi_n(x).$$

Show that

(i) $(k+n)\phi_{n-1}(x) = n\phi_n(x) - x\phi_n'(x);$

(ii) $(n+1)\phi_{n+1}(x) = (x+n+k+1)\phi_n(x) - x\phi_n'(x);$

(iii) $\phi_n(x) = \dfrac{e^x x^{-k}}{n!} D^n[e^{-x} x^{k+n}]$
$$= \frac{\Gamma(n+k+1)}{\Gamma(k+1)n!}\,{}_1F_1(-n;k+1;x);$$

(iv) $\displaystyle\int_0^{\infty} e^{-x} x^k \phi_m(x)\phi_n(x)\,dx \begin{cases} = 0 & (m \neq n), \\ = \Gamma(n+k+1)/n! & (m=n). \end{cases}$

[N. J. Sonine, *Math. Ann.* **16** (1880), 1–80; H. Bateman, *Partial Differential Equations of Mathematical Physics* (1932), 451–9; G. Pólya and G. Szegö, *Aufgaben und Lehrsätze*, ii. 94, 293–4.]

IX
LAMÉ'S EQUATION

36. Lamé Functions†

Ellipsoidal Harmonics. A SYSTEM of confocal quadrics being given by

$$\Theta \equiv \frac{x_1^2}{e_1-\theta}+\frac{x_2^2}{e_2-\theta}+\frac{x_3^2}{e_3-\theta}-1 = 0, \quad \bigg\} \quad (1)$$

where $\quad e_1+e_2+e_3 = 0 \quad (e_1 > e_2 > e_3),$

the analogy of spherical harmonics makes us look for polynomials in the Cartesian coordinates satisfying Laplace's equation $\nabla^2 V = 0$, whose nodal surfaces belong to the confocal system. Since even and odd terms must satisfy $\nabla^2 V = 0$ separately, we expect to find eight types of polynomials

$$V \equiv \begin{Bmatrix} & x_1 & x_2 x_3 & \\ 1 & x_2 & x_3 x_1 & x_1 x_2 x_3 \\ & x_3 & x_1 x_2 & \end{Bmatrix} \times \Theta_1 \Theta_2 \ldots \Theta_m, \quad (2)$$

where Θ_i is an expression of the form (1) with θ_i written for θ.

For any assigned point (x_1, x_2, x_3), the equation (1) in θ has three real roots (λ, μ, ν) lying in the intervals

$$e_1 > \lambda > e_2 > \mu > e_3 > \nu. \quad (3)$$

These correspond respectively to the hyperboloid of two sheets, the hyperboloid of one sheet, and the ellipsoid of the system passing through the point, and are called its *confocal coordinates*. From the identity

$$\Theta \equiv \sum_{i=1}^{3} \frac{x_i^2}{e_i-\theta}-1 \equiv \frac{(\theta-\lambda)(\theta-\mu)(\theta-\nu)}{(e_1-\theta)(e_2-\theta)(e_3-\theta)}, \quad (4)$$

we have by partial fractions

$$x_1^2 = \frac{(e_1-\lambda)(e_1-\mu)(e_1-\nu)}{(e_1-e_2)(e_1-e_3)}, \text{ etc.} \quad (5)$$

On substituting the expressions (4) and (5) for the several factors of V in (2), the expression V breaks up into a product of three similar factors

$$V \equiv C E(\lambda) E(\mu) E(\nu), \quad (6)$$

† G. H. Halphen, *Fonctions elliptiques*, ii. 457–531; E. T. Whittaker and G. N. Watson, *Modern Analysis*, 536–78; E. W. Hobson, *Spherical and Ellipsoidal Harmonics*, 454–96; P. Humbert, *Mémorial des sciences mathématiques*, x (1926); M. J. O. Strutt, *Ergebnisse der Mathematik*, I, **3** (1932).

where
$$E(\theta) \equiv \prod_{i=1}^{3} (\theta-e_i)^{\kappa_i} P(\theta) \quad (\kappa_i = 0 \text{ or } \tfrac{1}{2}),$$
$$P(\theta) \equiv (\theta-\theta_1)(\theta-\theta_2)\ldots(\theta-\theta_m). \tag{7}$$

We now make the well-known transformation of Laplace's equation to confocal coordinates,
$$2\frac{dx_i}{x_i} = \frac{d\lambda}{\lambda-e_i} + \frac{d\mu}{\mu-e_i} + \frac{d\nu}{\nu-e_i}, \tag{8}$$

$$ds^2 = \sum \frac{d\lambda^2}{4}\left[\sum_{i=1}^{3}\frac{x_i^2}{(e_i-\lambda)^2}\right] + \sum \frac{d\mu\,d\nu}{2}\left[\sum_{i=1}^{3}\frac{x_i^2}{(e_i-\mu)(e_i-\nu)}\right]$$
$$= \sum \frac{d\lambda^2}{4}\left[\frac{\partial\Theta}{\partial\theta}\right]_{\theta=\lambda} + \sum \frac{d\mu\,d\nu}{2(\mu-\nu)}\left[\sum_{i=1}^{3}\frac{x_i^2}{e_i-\mu} - \sum_{i=1}^{3}\frac{x_i^2}{e_i-\nu}\right]$$
$$= \sum \frac{(\lambda-\mu)(\nu-\lambda)d\lambda^2}{4(\lambda-e_1)(\lambda-e_2)(\lambda-e_3)} = \sum \frac{d\lambda^2}{h_1^2}. \tag{9}$$

$$\nabla^2 V \equiv h_1 h_2 h_3 \sum \frac{\partial}{\partial\lambda}\left[\frac{h_1}{h_2 h_3}\frac{\partial V}{\partial\lambda}\right]$$
$$= \sum \frac{\sqrt{f(\lambda)}}{(\lambda-\mu)(\nu-\lambda)}\frac{\partial}{\partial\lambda}\left[\sqrt{f(\lambda)}\frac{\partial V}{\partial\lambda}\right], \tag{10}$$

where $\quad f(\theta) \equiv 4(\theta-e_1)(\theta-e_2)(\theta-e_3) \equiv 4\theta^3 - g_2\theta - g_3. \tag{11}$

On dividing by V, we get from $\nabla^2 V = 0$ the equation
$$\sum (\mu-\nu)\left[\frac{f(\lambda)E''(\lambda) + \tfrac{1}{2}f'(\lambda)E'(\lambda)}{E(\lambda)}\right] = 0. \tag{12}$$

This implies that we can find (A, B) to satisfy the relations
$$\frac{f(\lambda)E''(\lambda) + \tfrac{1}{2}f'(\lambda)E'(\lambda)}{E(\lambda)} = A\lambda + B,$$
$$\frac{f(\mu)E''(\mu) + \tfrac{1}{2}f'(\mu)E'(\mu)}{E(\mu)} = A\mu + B, \tag{13}$$
$$\frac{f(\nu)E''(\nu) + \tfrac{1}{2}f'(\nu)E'(\nu)}{E(\nu)} = A\nu + B.$$

The last two equations show that (A, B) cannot depend on λ; and similarly they cannot depend on μ, ν, so that they are numerical constants. Thus $E(\theta)$ satisfies a linear differential equation of the second order
$$f(\theta)E''(\theta) + \tfrac{1}{2}f'(\theta)E'(\theta) - (A\theta + B)E(\theta) = 0, \tag{14}$$
which is called *Lamé's equation*.

This is an equation of the Fuchsian class having four regular singularities. At $\theta = e_i$ the exponents are $(0, \frac{1}{2})$, while at $\theta = \infty$ the indicial equation is

$$4\rho(\rho+1) - 6\rho - A = 0. \tag{15}$$

But, if n is the degree in the Cartesian coordinates of the expression V in (2), the corresponding solution (7) belongs to the exponent $-\frac{1}{2}n$ at infinity; and since this must satisfy (15), we must have

$$A = n(n+1), \tag{16}$$

where n is a positive integer.

If n is even, $E(\theta)$ must be a function of the first or third types, having (κ_i) all zero or two of them equal to $\frac{1}{2}$; if n is odd, $E(\theta)$ must be of the second or fourth types, having one or three of (κ_i) equal to $\frac{1}{2}$. In each case it is convenient (following Crawford)† to expand the polynomial $P(\theta)$ in descending powers of $(\theta - e_2)$.

Solutions of the First Type. The solution belonging to the exponent $-\frac{1}{2}n$ at $\theta = \infty$ can always be written $E(\theta) = \sum_{r=0}^{\infty} c_r (\theta - e_2)^{\frac{1}{2}n-r}$; but this will not be a polynomial unless n is an even integer and B is properly chosen. We write the equation in the form

$$4(\theta-e_2)^3 E'' + 6(\theta-e_2)^2 E' - n(n+1)(\theta-e_2)E +$$
$$+ 12e_2(\theta-e_2)^2 E'' + 12e_2(\theta-e_2)E' - [B+n(n+1)e_2]E +$$
$$+ f'(e_2)[(\theta-e_2)E'' + \tfrac{1}{2}E'] = 0, \tag{17}$$

and obtain the recurrence formulae

$$\left.\begin{array}{r} 2(2n-1)c_1 + [B + n(1-2n)e_2]c_0 = 0, \\ (2r+2)(2n-2r-1)c_{r+1} + [B + n(1-2n)e_2 + \\ + 12r(n-r)e_2]c_r - \tfrac{1}{4}f'(e_2)(n-2r+1)(n-2r+2)c_{r-1} = 0. \end{array}\right\} \tag{18}$$

The necessary and sufficient condition for $E(\theta)$ to be a polynomial of degree $m = \frac{1}{2}n$ is $c_{m+1} = 0$. For this automatically gives $c_{m+2} = 0$, $c_{m+3} = 0$, etc. Now if $c_0 = 1$, c_r is a polynomial of degree r in B; and so B must satisfy an algebraic equation of degree $(m+1)$. It was proved by Lamé, and then more simply by Liouville, that the values of B are real and unequal. For the relations (18), where $f'(e_2) < 0$, show that (c_r) is a Sturm sequence of polynomials in B, and that c_{r-1} and c_{r+1} take opposite signs when $c_r = 0$. No changes of sign are lost in the sequence as B varies, except when B passes through

---

† L. Crawford, *Quart. J. of Math.* **27** (1895), 93–8, **29** (1898), 196–201.

a zero of the last polynomial c_{m+1} of the set considered. But since the highest term of c_r is a positive numerical multiple of $(-)^r B^r$, we see by putting $B = \pm \infty$ that c_{m+1} must change sign for $(m+1)$ real values of B.

Other Types of Solutions. If an expression of the form (7) is substituted for $E(\theta)$ in Lamé's equation, the equation satisfied by $P(\theta)$ takes the form

$$4(\theta-e_2)^3 P'' + (4n-8m+6)(\theta-e_2)^2 P' - 2m(2n-2m+1)(\theta-e_2)P +$$

$$+ 12e_2(\theta-e_2)^2 P'' + \left[(8n-16m+12)e_2 + \sum_{i=1}^{3} 8\kappa_i e_i\right](\theta-e_2)P' - B^* P +$$

$$+ f'(e_2)[(\theta-e_2)P'' + (\tfrac{1}{2}+2\kappa_2)P'] = 0, \quad (19)$$

where $m = (\tfrac{1}{2}n - \kappa_1 - \kappa_2 - \kappa_3)$ and

$$B^* = B + n(n+1)e_2 + (2\kappa_2 + 2\kappa_3 + 8\kappa_2\kappa_3)(e_2-e_1) +$$
$$+ (2\kappa_1 + 2\kappa_2 + 8\kappa_1\kappa_2)(e_2-e_3). \quad (20)$$

If we expand in a descending series $P(\theta) = \sum_{r=0}^{\infty} c_r (\theta-e_2)^{m-r}$, we get the recurrence formulae

$$\left.\begin{aligned}
& 2(2n-1)c_1 + \left[B^* + m(4m-8n)e_2 - 8m\sum_{i=1}^{3}\kappa_i e_i\right]c_0 = 0, \\
& (2r+2)(2n-2r-1)c_{r+1} + \left[B^* + (m-r)(4m-4r-8n)e_2 - \right. \\
& \left. - 8(m-r)\sum_{i=1}^{3}\kappa_i e_i\right]c_r - f'(e_2)(m-r+1)(m-r+\tfrac{1}{2}+2\kappa_2)c_{r-1} = 0.
\end{aligned}\right\} \quad (21)$$

As before, $c_{m+1} = 0$ is the necessary and sufficient condition for $P(\theta)$ to reduce to a polynomial, and this is satisfied for $(m+1)$ distinct real values of B.

The Lamé functions of any given order and type, corresponding to distinct parameters (B_r), are linearly independent. For if we had $\sum A_r E_r(\theta) \equiv 0$, we should obtain also

$$\left[f(\theta)\frac{d^2}{d\theta^2} + \tfrac{1}{2}f'(\theta)\frac{d}{d\theta} - n(n+1)\theta\right]^k \sum A_r E_r(\theta) \equiv \sum A_r B_r^k E_r(\theta) \equiv 0, \quad (22)$$

for all positive integers k; and this would imply $A_r = 0$.

If n is even, we have $\tfrac{1}{2}(n+2)$ independent solutions of the first type and $\tfrac{3}{2}n$ of the third; if n is odd, we have $\tfrac{3}{2}(n+1)$ solutions of the second type and $\tfrac{1}{2}(n-1)$ of the fourth. This gives in either case $(2n+1)$ ellipsoidal harmonics, which is the same as the number of independent spherical harmonics of order n. These ellipsoidal harmonics are themselves linearly independent. For if we had an identity $\sum C_r V_r \equiv 0$, we could write it in the form

$$\sum C_r E_r(\lambda) E_r(\mu) E_r(\nu) \equiv 0;$$

and, if μ, ν are fixed, this is an identity between Lamé functions $\sum A_r E_r(\lambda)$, which has been proved to be impossible.

We note that the same value of B cannot give two Lamé functions of distinct types and of the same order n. For they would have to be independent principal branches, belonging to distinct exponents (κ_i) and $(\frac{1}{2}-\kappa_i)$ at each of the points $\theta = e_i$; and in the one case the positive integer n would be even, and in the other odd.

Zeros of Lamé Functions.† The polynomial $P(\theta)$ cannot have a double zero; for otherwise the differential equation would show that it must vanish identically. Accordingly, if $P(\theta_r) = 0$, we have $P'(\theta_r) \neq 0$; and so (19) can be written

$$\frac{P''(\theta_r)}{P'(\theta_r)} + \sum_{i=1}^{3} \frac{\frac{1}{2}+2\kappa_i}{\theta_r - e_i} = 0. \tag{23}$$

If $P(\theta) \equiv (\theta - \theta_r) Q_r(\theta)$, we have

$$\frac{P''(\theta_r)}{P'(\theta_r)} = \frac{2Q_r'(\theta_r)}{Q_r(\theta_r)} = \sum_{t \neq r} \frac{2}{\theta_r - \theta_t}; \tag{24}$$

and so (23) becomes

$$\sum_{t \neq r} \frac{1}{(\theta_r - \theta_t)} + \sum_{i=1}^{3} \frac{(\frac{1}{4}+\kappa_i)}{(\theta_r - e_i)} = 0 \quad (r = 1, 2, ..., m). \tag{25}$$

If any of the roots (θ_r) are complex, let θ_1 be the one with the numerically greatest imaginary part (say positive). Then every term of the relation (25) corresponding to $r = 1$ would have an imaginary part with the same sign (say negative); and, since this is impossible, none of the roots can be imaginary.

Again, suppose any of the roots are greater than e_1; if θ_1 now denotes the greatest root, every term of the relation corresponding to $r = 1$ would be positive; and, since this is impossible, no root can be greater than e_1, and similarly none can be less than e_3.

We shall show that, for each value $k = 0, 1, 2, ..., m$, there is one polynomial $P(\theta)$ whose roots are distributed in the following manner

$$e_1 > \theta_1 > \theta_2 > ... > \theta_k > e_2 > \theta_{k+1} > ... > \theta_m > e_3. \tag{26}$$

† E. Heine, *Kugelfunktionen* (1878), 382; F. Klein, *Math. Ann.* **18** (1881), 237–46; T. J. Stieltjes, *Acta Math.* **6** (1885), 321–6; G. Pólya and G. Szegö, *Aufgaben und Lehrsätze* (1925), ii. 57–9, 243–5.

For as (θ_r) vary in the intervals (26), the expression

$$\Pi = \prod_{r=1}^{m} \prod_{i=1}^{3} |\theta_r - e_i|^{\frac{1}{2}+\kappa_i} \cdot \prod_{1 \leqslant t < r \leqslant m} |\theta_r - \theta_t| \qquad (27)$$

is bounded and continuous. It therefore has a maximum, at which the conditions (25) are satisfied. The polynomial

$$f(\theta)\left[P''(\theta) + \sum_{i=1}^{3} \frac{(\frac{1}{2}+2\kappa_i)}{(\theta-e_i)} P'(\theta)\right] \qquad (28)$$

then vanishes at every zero of $P(\theta)$, and so must be of the form $(A\theta+B)P(\theta)$, and we get a Lamé function.

Stieltjes interprets (25) as the conditions of equilibrium of a system of collinear particles. Three of these have masses $(\frac{1}{4}+\kappa_i)$ and fixed coordinates (e_i), and m have unit mass and variable coordinates (θ_r). If they repel one another directly as their masses and inversely as their distances apart, there will clearly be a position of equilibrium where k of the movable particles lie in one interval and $(m-k)$ in the other.

37. Introduction of Elliptic Functions

Uniformization. We may now drop the physical interpretation, and consider as a purely analytical problem the solution of Lamé's equation

$$\left.\begin{aligned} \sqrt{\{f(x)\}} \frac{d}{dx}\left[\sqrt{\{f(x)\}} \frac{dy}{dx}\right] - [n(n+1)x+B]y &= 0, \\ f(x) \equiv 4\prod_{i=1}^{3}(x-e_i) \equiv 4x^3 - g_2 x - g_3 \quad (\textstyle\sum e_i = 0). & \end{aligned}\right\} \qquad (1)$$

Further progress depends on the introduction of elliptic functions, and we use the Weierstrassian form defined by

$$x = \wp(u), \qquad u = \int_{x}^{\infty} \frac{dx}{\sqrt{\{4(x-e_1)(x-e_2)(x-e_3)\}}}. \qquad (2)$$

Suppose first that (e_i) are real $(e_1 > e_2 > e_3)$. Then (2) is the Schwarz-Christoffel formula giving the conformal representation of the upper half-plane of x upon a rectangle in the plane of u. We construct a Riemann surface covering the x-domain, with winding points at $x = e_1, e_2, e_3, \infty$. When x describes a circuit about one of these points, the analytical continuation of $u \equiv u(x)$ is given by the principle of symmetry. We thus obtain a pattern of rectangles covering the u-plane without overlapping. The path in the u-plane corresponding to any given closed circuit in the x-plane can be determined without ambiguity; and, conversely, each value of u gives a unique value of x, so that $x = \wp(u)$ is a uniform analytic function. We make the

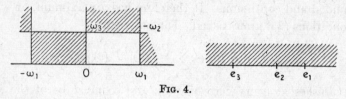

Fig. 4.

domains correspond as in the figure, ω_1 and ω_3 being the positive real and positive imaginary semi-periods of the elliptic functions, and $(\omega_1+\omega_2+\omega_3) = 0$. We have $\wp(\omega_i) = e_i$, and from (2) we have

$$u \sim x^{-\frac{1}{2}} \quad (x \to \infty), \qquad u-\omega_i = O[(x-e_i)^{\frac{1}{2}}] \quad (x \to e_i). \tag{3}$$

Thus $u = 0$ is a double pole of $\wp(u)$, and $u = \omega_i$ is a double zero of $\{\wp(u)-e_i\}$. We can express the radicals $(x-e_i)^{\frac{1}{2}}$ as uniform functions of u by the well-known formulae

$$[\wp(u)-e_i]^{\frac{1}{2}} = \frac{e^{-\eta_i u}\sigma(u+\omega_i)}{\sigma(u)\sigma(\omega_i)}. \tag{4}$$

These relations still hold when (e_i) are complex.† Each period parallelogram of sides $(2\omega_1, 2\omega_3)$ contains two complete pictures of

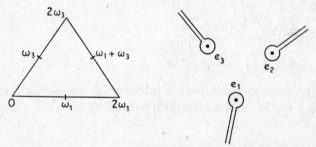

Fig. 5.

the x-plane. The triangle $(0, 2\omega_1, 2\omega_3)$ contains one complete representation of the plane bounded by three curvilinear cuts from $x = e_i$ to $x = \infty$. If we take any value of x and an initial value of u within this triangle, then when x describes a circuit about e_1, e_2, or e_3, u is transformed into $(2\omega_1-u)$, $(-2\omega_2-u)$, or $(2\omega_3-u)$, the reflections of the initial point in the mid-points of the sides of the triangle.

† For a detailed study see C. Jordan, *Cours d'analyse*, ii. 413–29.

Lamé's equation now takes the form

$$\frac{d^2y}{du^2} - [n(n+1)\wp(u) + B]y = 0. \tag{5}$$

The singularities $x = e_i$, with exponent-difference $\frac{1}{2}$, have become ordinary points, by virtue of the relations (4). The point $x = \infty$ corresponds to singularities at the lattice-points $u \equiv 0$ (*modulis* $2\omega_1, 2\omega_3$). At $u = 0$, we have $\wp(u) \sim u^{-2}$; the indicial equation is now $(\rho+n)(\rho-n-1) = 0$, the exponents being doubled. If n is an integer, the exponent-difference $(2n+1)$ is also an integer. But since (5) is invariant when we change the sign of u, one solution contains only odd and the other only even powers of u. Thus the complete primitive of (5) is free from logarithms everywhere and is a uniform analytic function of u.

Special Cases. Suppose first that B has a special value giving the Lamé function

$$\begin{aligned}\phi(u) &= E\{\wp(u)\} \\ &= \prod_{i=1}^{3} \{\wp(u)-e_i\}^{\kappa_i} \prod_{r=1}^{m} \{\wp(u)-\theta_r\} \\ &= \prod_{i=1}^{3} \left[\frac{e^{-\eta_i u}\sigma(u+\omega_i)}{\sigma(u)\sigma(\omega_i)}\right]^{2\kappa_i} \prod_{r=1}^{m}\left[\frac{\sigma(u_r+u)\sigma(u_r-u)}{\sigma^2(u)\sigma^2(u_r)}\right],\end{aligned} \tag{6}$$

where $\kappa_i = 0$ or $\frac{1}{2}$.

This is a uniform function of u having n-tuple poles at the lattice-points $u \equiv 0$ and n zeros in each period-parallelogram. The eight types are distinguished by whether they do or do not vanish at $u = \omega_i$, or by whether $(2\omega_i)$ are full periods or semi-periods. In all cases $\phi^2(u)$ is an even doubly-periodic function with periods $(2\omega_1, 2\omega_3)$. The second solution is now given by the usual formula

$$y = \phi(u) \int [\phi(u)]^{-2} \, du. \tag{7}$$

To effect the integration, we must resolve the integrand into its principal parts at the poles. Let $u = a$ be any zero of $\phi(u)$, so that $u = -a$ is also a zero. We must have $\phi'(a) \neq 0$, or $\phi(u)$ would vanish identically; and the differential equation (5) gives $\phi''(a) = 0$. From this it readily follows that the residue of $[\phi(u)]^{-2}$ is zero; and since $[\phi(0)]^{-2} = 0$ we get the expression

$$\begin{aligned}[\phi(u)]^{-2} &= \sum [\phi'(a)]^{-2}[\wp(u-a)-\wp(a)] \\ &= \tfrac{1}{2}\sum[\phi'(a)]^{-2}[\wp(u-a)+\wp(u+a)-2\wp(a)],\end{aligned} \tag{8}$$

summed over the n poles in a period-parallelogram. We now obtain by integration

$$y = -\tfrac{1}{2}\phi(u) \sum [\phi'(a)]^{-2}[\zeta(u-a)+\zeta(u+a)+2u\wp(a)]$$
$$= -\phi(u) \sum [\phi'(a)]^{-2}\left[\zeta(u)+u\wp(a)+\frac{\tfrac{1}{2}\wp'(u)}{\wp(u)-\wp(a)}\right]. \quad (9)$$

There are m pairs of identical terms arising from poles other than $u = \omega_i$.

As in the case of Legendre's equation, we know in general the *coefficients* of the Lamé function, but not its *factors*. But the second solution can be found by the method of undetermined coefficients, without knowing the roots of the first. For the expression (9) is of the form

$$y = \prod_{i=1}^{3}\{\wp(u)-e_i\}^{\kappa_i}P\{\wp(u)\}[L\zeta(u)+Mu]+$$
$$+ \prod_{i=1}^{3}\{\wp(u)-e_i\}^{\frac{1}{2}-\kappa_i}Q\{\wp(u)\}, \quad (10)$$

where P is a known polynomial of degree m, and Q an unknown one of degree $(n-m-1)$, and where L, M are constants. If we substitute the expression (10) in Lamé's equation and remove the factors $\prod \{\wp(u)-e_i\}^{\frac{1}{2}-\kappa_i}$, we have a polynomial of degree $(n-m)$ which must vanish identically. This gives $(n-m+1)$ conditions determining the ratios of L, M and the coefficients of Q.

Halphen's Transformation. Every Lamé function is an elliptic function of u, admitting $(2\omega_1, 2\omega_3)$ either as full periods or as semi-periods. If we put $u = 2v$, it is an elliptic function of v with periods $(2\omega_1, 2\omega_3)$ and is rationally expressible in terms of $\wp(v)$ and $\wp'(v)$. In fact the irrationals $(x-e_s)^{\frac{1}{2}}$ are removed by means of the identities of the type

$$[\wp(2v)-e_1]^{\frac{1}{2}} = [\wp^2(v)-2e_1\wp(v)-e_1^2-e_2e_3]/\wp'(v). \quad (11)$$

The Lamé function $y \equiv \phi(u) \equiv \phi(2v)$ is either an even or an odd function of v, according as n is even or odd. It has four poles of order n in each period-parallelogram, at the lattice-points $v \equiv 0, \omega_1, \omega_2, \omega_3$. If we multiply it by $[\wp'(v)]^n$, three of the poles are cancelled, and we have in all cases an even function of v with poles of order $4n$ at the lattice-points $v \equiv 0$. The resulting function must therefore be a polynomial of degree $2n$ in $\wp(v)$.

Accordingly Halphen puts in (4)

$$u = 2v, \qquad y = [\wp'(v)]^{-n}z, \quad (12)$$

and obtains for z the equation

$$\frac{d^2z}{dv^2} - 2n\frac{\wp''(v)}{\wp'(v)}\frac{dz}{dv} + \left[n(n+1)\left(\frac{\wp''(v)}{\wp'(v)}\right)^2 - 12n\wp(v)\right]z$$
$$= 4[n(n+1)\wp(2v) + B]z. \quad (13)$$

But we know that

$$\wp(2v) + 2\wp(v) = \frac{1}{4}\left(\frac{\wp''(v)}{\wp'(v)}\right)^2, \quad (14)$$

and so (13) becomes

$$\frac{d^2z}{dv^2} - 2n\frac{\wp''(v)}{\wp'(v)}\frac{dz}{dv} + 4[n(2n-1)\wp(v) - B]z = 0. \quad (15)$$

If now we put $\xi = \wp(v)$, we have (with our previous notation)

$$f(\xi)\frac{d^2z}{d\xi^2} + (\tfrac{1}{2}-n)f'(\xi)\frac{dz}{d\xi} + 4[n(2n-1)\xi - B]z = 0. \quad (16)$$

This is an equation of the Fuchsian class with four singularities; the exponents are $(0, n+\tfrac{1}{2})$ at each of the points $\xi = e_1, e_2, e_3$, and $(-2n, \tfrac{1}{2}-n)$ at $\xi = \infty$. Following Crawford's procedure, we may expand the solution belonging to the exponent $-2n$ at $\xi = \infty$ in the form

$$z = \sum_{r=0}^{\infty} c_r (\xi - e_2)^{2n-r}, \quad (17)$$

and obtain the recurrence formulae

$$\left.\begin{array}{r}(n-\tfrac{1}{2})c_1 + B^*c_0 = 0, \\ (r+1)(n-\tfrac{1}{2}-r)c_{r+1} + [B^* + 3r(2n-r)e_2]c_r - \\ -\tfrac{1}{4}f'(e_2)(2n+1-r)(n+\tfrac{1}{2}-r)c_{r-1} = 0 \quad (r = 1, 2, ...),\end{array}\right\} \quad (18)$$

where $B^* = B + n(1-2n)e_2$.

The necessary and sufficient condition that z should be a polynomial is $c_{2n+1} = 0$, which entails $c_r = 0$ $(r > 2n)$. If $c_0 = 1$, c_r is a polynomial in B of degree r, so that there are $(2n+1)$ special values of B corresponding to the Lamé functions of different types. If $0 < r < 2n+1$, and $c_r = 0$ for some value of B, then c_{r-1} and c_{r+1} have opposite signs, *except* when $r = n$, when c_{n-1} and c_{n+1} have the same sign. A Sturm sequence of polynomials is, however, formed by the set

$$+c_0, +c_1, +c_2, ..., +c_n, -c_{n+1}, +c_{n+2}, -c_{n+3}, ..., (-)^{n+1}c_{2n+1}, \quad (19)$$

and we can prove that the values of B are real and distinct.

Brioschi's Solutions. If n is half an odd integer, the exponent-difference $(n+\tfrac{1}{2})$ is an integer and the solution is in general logarithmic. But for certain special values of B we have algebraic solutions;

these were discovered by Brioschi, but are most simply exhibited by the analysis of Halphen and Crawford. Putting $n = (m+\tfrac{1}{2})$ in (18), we have the recurrence formulae

$$\left.\begin{array}{c} mc_1 + B^* c_0 = 0, \\ (r+1)(m-r)c_{r+1} + [B^* + 3r(2m+1-r)e_2]c_r - \\ -\tfrac{1}{4}f'(e_2)(2m+2-r)(m+1-r)c_{r-1} = 0 \quad (r=1,2,\dots), \end{array}\right\} \quad (20)$$

where $B^* = B - m(2m+1)e_2$.

The critical equation of the set is

$$[B^* + 3m(m+1)e_2]c_m - \tfrac{1}{4}f'(e_2)(m+2)c_{m-1} = 0, \qquad (21)$$

where c_{m+1} does not appear. If the solution is free from logarithms, the first $(m+1)$ equations must be compatible, so that B satisfies an equation of order $(m+1)$. When this is the case, c_0 and c_{m+1} may be arbitrarily assigned, and it turns out (owing to the vanishing of a certain determinant) that $c_r = 0$ $(r > 2m+1)$, so that the solution is a polynomial. The explanation of this is the fact that, if $\phi(u)$ is a solution of (4), then $\phi(u+2\omega_2)$ is another solution. In terms of v, these solutions are respectively

$$y_1 = [\wp'(v)]^{-n} \cdot \sum_{r=0}^{n-\tfrac{1}{2}} c_r [\wp(v) - e_2]^{2n-r}, \qquad (22)$$

$$y_2 = [\wp'(v+\omega_2)]^{-n} \cdot \sum_{r=0}^{n-\tfrac{1}{2}} c_r [\wp(v+\omega_2) - e_2]^{2n-r},$$

$$= (-)^n \left[\frac{(e_2-e_1)(e_2-e_3)\wp'(v)}{[\wp(v)-e_2]^2}\right]^{-n} \cdot \sum_{r=0}^{n-\tfrac{1}{2}} c_r \left[\frac{(e_2-e_1)(e_2-e_3)}{\wp(v)-e_2}\right]^{2n-r},$$

$$= C[\wp'(v)]^{-n} \cdot \sum_{r=0}^{n-\tfrac{1}{2}} c_r \left[\frac{\wp(v)-e_2}{(e_2-e_1)(e_2-e_3)}\right]^r. \qquad (23)$$

If we put $\quad c_r = (e_2-e_1)^r (e_2-e_3)^r c'_{2m+1-r} \quad (r \leqslant 2m+1),$

$\qquad\qquad c_r = 0 \qquad\qquad\qquad\qquad\qquad (r > 2m+1),$

in (20), it is found that the formulae for (c'_r) are of the same form as those for (c_r). Thus the one condition satisfied by B gives two finite and compatible sets of equations for $(c_0, c_1, \dots, c_{2m+1})$, and then

$$c_r = 0 \quad (r > 2m+1).$$

38. Oscillation and Comparison Theorems

A General Orthogonal Property. If V_1, V_2 are distinct solid ellipsoidal harmonics, we have by Green's theorem

$$\iint \left(V_1 \frac{dV_2}{dn} - V_2 \frac{dV_1}{dn}\right) dS = 0, \qquad (1)$$

the integral being over an ellipsoid of the confocal system and the derivatives along the outward normal. In confocal coordinates we have therefore

$$\iint \frac{h_3}{h_1 h_2}\left(V_1 \frac{\partial V_2}{\partial \nu} - V_2 \frac{\partial V_1}{\partial \nu}\right) d\lambda d\mu = 0; \qquad (2)$$

and on expressing the ellipsoidal harmonics as products of Lamé functions, we get (on the ellipsoid $\nu = $ constant)

$$\sqrt{\{f(\nu)\}}[E_1(\nu)E_2'(\nu) - E_2(\nu)E_1'(\nu)] \times$$
$$\times \iint (\lambda-\mu)E_1(\lambda)E_1(\mu)E_2(\lambda)E_2(\mu) \frac{d\lambda d\mu}{\sqrt{\{f(\lambda)f(\mu)\}}} = 0. \qquad (3)$$

If we now put $(\lambda, \mu, \nu) = \{\wp(u), \wp(v), \wp(w)\}$, we have $w = $ constant on the ellipsoid. The factor depending on w does not vanish identically unless $E_1(\nu) \equiv E_2(\nu)$, which is excluded. Hence we have the relation

$$\int_{\omega_1}^{\omega_1+4\omega_3} \int_{\omega_3}^{\omega_3+4\omega} [\wp(u)-\wp(v)]\phi_1(u)\phi_1(v)\phi_2(u)\phi_2(v) \, du dv = 0, \qquad (4)$$

between any two distinct Lamé functions. On a fixed ellipsoid we have $e_1 > \wp(u) > e_2 > \wp(v) > e_3$, and $\wp(u), \wp(v)$ will be real if we assign to u a fixed *real* part ω_1 and to v a fixed *imaginary* part ω_3. The range of integration covers the entire surface twice.

Functions of the Same Order. If the functions are of the same order n, but belong to unequal parameters B_1, B_2, we have

$$\left.\begin{aligned} \phi_1''(u) &= [n(n+1)\wp(u) + B_1]\phi_1(u), \\ \phi_2''(u) &= [n(n+1)\wp(u) + B_2]\phi_2(u), \end{aligned}\right\} \qquad (5)$$

and hence by cross-multiplication

$$(B_1 - B_2) \int_a^b \phi_1(u)\phi_2(u) \, du = [\phi_2(u)\phi_1'(u) - \phi_1(u)\phi_2'(u)]_a^b. \qquad (6)$$

Since Lamé functions of all types admit the periods $(4\omega_1, 4\omega_3)$, we get

$$\int_a^{a+4\omega_i} \phi_1(u)\phi_2(u) \, du = 0 \quad (B_1 \neq B_2). \qquad (7)$$

The eight types of Lamé functions are characterized by three boundary conditions of the form

$$\phi(\omega_i) = 0 \quad or \quad \phi'(\omega_i) = 0 \quad (i = 1, 2, 3), \qquad (8)$$

according as they do or do not vanish at the lattice-points $u \equiv \omega_i$. For two functions of the same order and the same type, we have from (6) the relation

$$\int_{\omega_i}^{\omega_j} \phi_1(u)\phi_2(u)\, du = 0 \quad (i,j = 1, 2, 3). \tag{9}$$

From this we infer that the critical values of B are real (as has been otherwise proved already). For if the two functions corresponded to conjugate imaginary values of B, their product would not change sign on the segment parallel to the real axis joining ω_3 and $(\omega_3+\omega_1)$; and the relation (9) would be impossible.

Oscillation Theorems.† Upper and lower bounds for the critical values of B, and further information about the zeros of the Lamé functions can be obtained very simply and directly by the methods of Sturm and Liouville. The parameters (e_i) being real and $\sum e_i = 0$, let $\phi(u)$ be any Lamé function and B the corresponding critical value of the parameter. We can choose a coefficient c so that $Y = c\phi(u)$ shall be real in the interval $\{e_3 < \wp(u) < e_2\}$, where the imaginary part of u remains fixed and equal to a half-period, say $I(u-\omega_3) = 0$; we put $X = (u-\omega_3)$ and consider the graph of a point whose Cartesian coordinates are (X, Y). If d^2Y/dX^2 and Y have the same sign, the curve is convex towards the axis of X; in any interval where $[n(n+1)\wp(u)+B]$ remains positive, the curve cannot cross the axis more than once, after which Y, dY/dX, d^2Y/dX^2 all retain the same sign and the curve moves steadily away from the axis. Such a curve cannot represent a real periodic function with the period $4\omega_1$, and so the minimum values of the expression in the interval must be negative, i.e. $[n(n+1)e_3+B] < 0$. But similar reasoning applied to the real curve traced by the point $[i(u-\omega_1), c'\phi(u)]$ in the interval $\{e_2 < \wp(u) < e_1\}$ shows that $[n(n+1)e_1+B] > 0$. Hence all the critical values lie in the interval

$$n(n+1)(e_2+e_3) < B < n(n+1)(e_1+e_2). \tag{10}$$

Now let $B_i > B_j$ be two of the $(2n+1)$ critical values, and $\{\phi_i(u), \phi_j(u)\}$ the corresponding Lamé functions. By means of (6) we can prove that the zeros of $\phi_j(u)$ separate those of $\phi_i(u)$ in the interval $\{e_2 > \wp(u) > e_3\}$, the situation being reversed in the interval

---

† M. Bôcher, *Les Méthodes de Sturm* (1917); *Die Reihenentwickelungen der Potentialtheorie* (1894); F. Klein, *Gesammelte mathematische Abhandlungen*, ii. 512–600.

$\{e_1 > \wp(u) > e_2\}$. For let $u = a, b$ be adjacent zeros of $\phi_i(u)$ in the range

$$0 < a-\omega_3 < b-\omega_3 < \omega_1, \qquad e_3 < \wp(a) < \wp(b) < e_2. \tag{11}$$

We can make $\phi_i(u)$ real and positive on this segment, and let us suppose (if possible) that $\phi_j(u)$ does not change sign in the interval, say $\phi_j(u) > 0$. Then the relation

$$(B_i - B_j) \int_a^b \phi_i(u)\phi_j(u)\, du = \phi_j(b)\phi_i'(b) - \phi_j(a)\phi_i'(a) \tag{12}$$

gives a contradiction. For the left-hand side is positive; but at the end-points $\phi_i'(a) > 0$, $\phi_i'(b) < 0$, $\phi_j(a) > 0$, $\phi_j(b) > 0$, so that the right-hand side is negative. Thus $\phi_j(u)$ must have a zero in the interval, and a similar argument applies to the segment where $\{e_2 < \wp(\omega) < e_1\}$.

By considering the two segments, we can arrange in order the $(2n+1)$ Lamé functions corresponding to the sequence

$$(B_1 > B_2 > \ldots > B_{2n+1}).$$

For example, if $n = 2$, we find two functions of the first type $[\wp(u) \pm \theta]$, where θ is the positive root of $(12\theta^2 - g_2) = 0$, and three of the third type $[\{\wp(u) - e_i\}\{\wp(u) - e_j\}]^{\frac{1}{2}}$, and the sequence is

$$\{\wp(u)-\theta\}, \sqrt{[\{\wp(u)-e_1\}\{\wp(u)-e_2\}]}, \sqrt{[\{\wp(u)-e_1\}\{\wp(u)-e_3\}]},$$
$$\sqrt{[\{\wp(u)-e_2\}\{\wp(u)-e_3\}]}, \{\wp(u)+\theta\}. \tag{13}$$

The zeros of solutions of Lamé's equation for other than critical values of B have been considered by other methods by Hodgkinson.†

39. The General Equation of Integral Order‡

Multiplicative Solutions. We saw in § 37 that, if n is a positive integer but B is unrestricted, every solution of

$$\frac{d^2y}{du^2} - [n(n+1)\wp(u) + B]y = 0 \tag{1}$$

is a uniform analytic function of u. When u increases by the period $2\omega_1$, the equation resumes its initial form; hence if $f(u)$ is a solution, then $f(u+2\omega_1), f(u+4\omega_1),\ldots$ are also solutions. If they are not all multiples of the same solution, any three of them must be linearly

---

† J. Hodgkinson, *J. of London Math. Soc.* **5** (1930), 296–306.

‡ C. Hermite, *Œuvres*, iii. 118–22, 266–83, 374–9, 475–8; iv. 8–18; F. Brioschi, *Comptes rendus*, **86** (1878), 313–15.

connected; and so we have an identity

$$f(u+4\omega_1)+2\alpha f(u+2\omega_1)+\beta f(u) = 0. \tag{2}$$

Now we have $\beta = 1$; for when u is changed to $(u+2\omega_1)$, the Wronskian $W[f(u+2\omega_1),f(u)]$ becomes

$$W[f(u+4\omega_1),f(u+2\omega_1)] = \beta W[f(u+2\omega_1),f(u)]. \tag{3}$$

But we know that its value is constant, and so the multiplier $\beta = 1$.

(i) If $\alpha^2 \neq 1$, the equation $(\lambda^2+2\alpha\lambda+1) = 0$ has unequal roots (μ, μ^{-1}); and (2) can be written

$$\left.\begin{array}{l}[f(u+4\omega_1)-\mu^{-1}f(u+2\omega_1)] = \mu[f(u+2\omega_1)-\mu^{-1}f(u)], \\ \text{or} \quad [f(u+4\omega_1)-\mu f(u+2\omega_1)] = \mu^{-1}[f(u+2\omega_1)-\mu f(u)].\end{array}\right\} \tag{4}$$

Thus $[f(u+2\omega_1)-\mu^{-1}f(u)]$ and $[f(u+2\omega_1)-\mu f(u)]$ are linearly independent solutions admitting the multipliers (μ, μ^{-1}) respectively for the period $2\omega_1$. If one is called $\phi(u)$, the other must be a multiple of $\phi(-u)$; for if $\phi(u+2\omega_1) = \mu\phi(u)$, then $\phi(-u-2\omega_1) = \mu^{-1}\phi(-u)$.

Now consider the solution $\phi(u+2\omega_3)$; we must be able to express this in the form

$$\phi(u+2\omega_3) = a\phi(u)+b\phi(-u), \tag{5}$$

and we now have two alternative expressions for

$$\left.\begin{array}{l}\phi(u+2\omega_1+2\omega_3) = a\mu\phi(u)+b\mu^{-1}\phi(-u), \\ = \mu[a\phi(u)+b\phi(-u)].\end{array}\right\} \tag{6}$$

Since this is a single-valued function and $\mu^2 \neq 1$, we must have $b = 0$; hence $\phi(u+2\omega_3) = \nu\phi(u)$ say, and so $\phi(u)$ admits the multipliers (μ, ν), and $\phi(-u)$ the multipliers (μ^{-1}, ν^{-1}) for the periods $(2\omega_1, 2\omega_3)$.

(ii) If $\alpha^2 = 1$, we can find at least *one* solution such that $\phi(u+2\omega_1) = \pm\phi(u)$. If we repeat the argument with $\phi(u)$, $\phi(u+2\omega_3)$, $\phi(u+4\omega_3)$, we obtain *either* two distinct solutions with unequal multipliers (ν, ν^{-1}) for the period $2\omega_3$ and the same multiplier for the period $2\omega_1$, *or* at least one solution with multipliers $(\pm 1, \pm 1)$ for the two periods. This is a doubly-periodic function with periods $(4\omega_1, 4\omega_3)$, in the least favourable case, and so is a *Lamé function*, which we know how to construct.

There cannot be two independent doubly-periodic solutions. For then every solution would admit $(2\omega_1, 2\omega_3)$ as full periods or semi-periods. In particular, the solution belonging to the exponent $(n+1)$ at $u = 0$ would be everywhere bounded, and therefore constant, by Liouville's theorem; and this is absurd.

LAMÉ'S EQUATION

In general $\phi(u)$ is a doubly-periodic function of the second kind, and a method for constructing it was published by Hermite. An easier method, which we here follow, was given in Hermite's lectures and independently published by Brioschi.

Product of Solutions. If (y_1, y_2) are any solutions of (1), their product $Y \equiv y_1 y_2$ is found to be a solution of an equation of the third order

$$\frac{d^3Y}{du^3} - 4[n(n+1)\wp(u) + B]\frac{dY}{du} - 2n(n+1)\wp'(u)Y = 0, \qquad (7)$$

or, if $x = \wp(u)$, $f(x) \equiv 4x^3 - g_2 x - g_3$,

$$f(x)\frac{d^3Y}{dx^3} + \tfrac{3}{2}f'(x)\frac{d^2Y}{dx^2} + \tfrac{1}{2}f''(x)\frac{dY}{dx} -$$
$$- 4[n(n+1)x + B]\frac{dY}{dx} - 2n(n+1)Y = 0. \qquad (8)$$

This has four regular singularities, three with exponents $(0, \tfrac{1}{2}, 1)$ at $x = e_i$ and one with exponents $(n+1, \tfrac{1}{2}, -n)$ at $x = \infty$. Now suppose in particular that Y is the solution $\phi(u)\phi(-u)$; this is an even doubly-periodic function having poles of order $2n$ at the lattice-points $u \equiv 0$, and so it is a polynomial of degree n in $x = \wp(u)$, and must be of the form

$$Y = \sum_{r=0}^{n} c_r (x - e_2)^{n-r}. \qquad (9)$$

We now rewrite (8) in the form

$$4(x-e_2)^3 Y''' + 18(x-e_2)^3 Y'' + [12 - 4n(n+1)](x-e_2)Y' - 2n(n+1)Y +$$
$$+ 12e_2(x-e_2)^2 Y''' + 36e_2(x-e_2)Y'' - 4[B + n(n+1)e_2]Y' +$$
$$+ f'(e_2)[(x-e_2)Y''' + \tfrac{3}{2}Y''] = 0, \qquad (10)$$

and find the recurrence formulae

$$\left.\begin{array}{l}(1-2n)c_1 + [(2n^2-n-3)e_2 - B]c_0 = 0, \\ 2(r+1)(r-2n)(2r-2n+1)c_{r+1} + 4(r-n)[(2n^2-n-3)e_2 + \\ + 3r(r-2n)e_2 - B]c_r + \tfrac{1}{4}f'(e_2)(r-n)(r-n-\tfrac{1}{2})(r-n-1)c_{r-1} = 0.\end{array}\right\} \quad (11)$$

These evidently give $c_{n+1} = c_{n+2} = \ldots = 0$, and we have the required polynomial; on resolving it into factors we have

$$\Phi(u) \equiv \phi(u)\phi(-u) = M \prod_{r=1}^{n} \{\wp(u) - \wp(u_r)\}$$
$$= (-)^n M \prod_{r=1}^{n} \left[\frac{\sigma(u-u_r)\sigma(u+u_r)}{\sigma^2(u)\sigma^2(u_r)}\right]. \qquad (12)$$

The solution $\phi(u)$ of (1) cannot have a double zero without vanishing identically; and $\phi(u)$ and $\phi(-u)$ cannot have a common zero unless they are multiples of one another; each would then have the multipliers $(\pm 1, \pm 1)$ and would be a Lamé function, a case which is here excluded as it has been more simply discussed already.

The Invariant. To complete the solution we require the Wronskian, which is a numerical constant

$$W\{\phi(-u), \phi(u)\} \equiv \phi(-u)\phi'(u) + \phi(u)\phi'(-u) = C. \quad (13)$$

We observe that

$$\Phi'(u) \equiv \phi(-u)\phi'(u) - \phi(u)\phi'(-u), \quad (14)$$

and if we choose the notation in (12) so that $\phi(+u_r) = 0$, $\phi(-u_r) \neq 0$, we find

$$\Phi'(+u_r) = -\Phi'(-u_r) = \phi(-u_r)\phi'(u_r) = C. \quad (15)$$

The $2n$ zeros of $\Phi(u)$ lying in a period-parallelogram can thus be divided into two sets, according as they give $\Phi'(u) = \pm C$; and the zeros of the one set belong to $\phi(u)$ and those of the other to $\phi(-u)$. We now have

$$\frac{\phi'(u)}{\phi(u)} + \frac{\phi'(-u)}{\phi(-u)} = \frac{C}{\Phi(u)}. \quad (16)$$

The expression on the right is an even elliptic function with $2n$ poles in each parallelogram and zeros of order $2n$ at the lattice-points $u \equiv 0$.

From (15) we see that the residues are $(+1, -1)$ at $(u_r, -u_r)$ respectively, and so we must have

$$\frac{\phi'(u)}{\phi(u)} + \frac{\phi'(-u)}{\phi(-u)} = \sum_{r=1}^{n} [\zeta(u-u_r) - \zeta(u+u_r) + 2\zeta(u_r)]. \quad (17)$$

We now integrate and fix the constant of integration by putting $u = 0$, which gives

$$\frac{\phi(u)}{\phi(-u)} = \prod_{r=1}^{n} \left[\frac{\sigma(u-u_r)}{\sigma(u+u_r)} e^{2u\zeta(u_r)} \right]; \quad (18)$$

and on combining this with (12) we have the required solution

$$\phi(u) = (-)^{n/2} M^{\frac{1}{2}} \prod_{r=1}^{n} \left[\frac{\sigma(u-u_r)}{\sigma(u)\sigma(u_r)} e^{u\zeta(u_r)} \right]. \quad (19)$$

40. Equations of Picard's Type†

Commutative Linear Transformations. We now consider more generally an equation

$$\frac{d^n y}{du^n} + p_1(u) \frac{d^{n-1} y}{du^{n-1}} + \ldots + p_n(u) y = 0, \quad (1)$$

† E. Picard, *J. für Math.* **90** (1880), 281–302; *Traité d'analyse*, iii. 437–53; G. Floquet,

whose coefficients are uniform doubly-periodic functions. Suppose that each singularity in a fundamental parallelogram has been examined, and that every solution has been found to be *uniform*. If $\{\phi_i(u)\}$ is any fundamental system of solutions, then

$$\{\phi_i(u+2p\omega_1+2q\omega_3)\}$$

is another, (p, q) being any integers and $(2\omega_1, 2\omega_3)$ a pair of fundamental periods. In terms of the original system we shall have

$$\left. \begin{aligned} (S):\ \phi_i(u+2\omega_1) &= \sum_{j=1}^n a_{ij}\phi_j(u) \quad (i=1,2,...,n), \\ (T):\ \phi_i(u+2\omega_3) &= \sum_{j=1}^n b_{ij}\phi_j(u) \quad (i=1,2,...,n); \end{aligned} \right\} \quad (2)$$

and because $\{\phi_i(u)\}$ are single-valued, the alternative forms of $\{\phi_i(u+2\omega_1+2\omega_3)\}$ given by ST and TS are identical.

Hence the product of the matrices may be written in either of the forms

$$c_{ij} = \sum_{k=1}^n a_{ik}b_{kj} = \sum_{k=1}^n b_{ik}a_{kj} \quad (i,j=1,2,...,n). \quad (3)$$

If we choose a different fundamental system $\psi_i \equiv \sum_{j=1}^n d_{ij}\phi_j$, the matrices corresponding to (a_{ij}) and (b_{ij}) will be $(d_{ij})(a_{ij})(d_{ij})^{-1}$ and $(d_{ij})(b_{ij})(d_{ij})^{-1}$, which are also commutative. If (μ_i) are the roots of the characteristic equation $\Delta(\lambda) \equiv |a_{ij}-\lambda\delta_{ij}| = 0$, we may suppose S reduced to the Jacobian semi-diagonal form

$$\left. \begin{aligned} \phi_1(u+2\omega_1) &= \mu_1\phi_1(u), \\ \phi_2(u+2\omega_1) &= a_{21}\phi_1(u)+\mu_2\phi_2(u), \\ \phi_3(u+2\omega_1) &= a_{31}\phi_1(u)+a_{32}\phi_2(u)+\mu_3\phi_3(u), \\ &\ \cdot\ \cdot\ \cdot\ \cdot\ \cdot\ \cdot\ \cdot\ \cdot\ \cdot\ \cdot\ \cdot\ \cdot\ \cdot \end{aligned} \right\} \quad (4)$$

the notation (a_{ij}) being retained schematically for the new coefficients. If $\lambda = \mu_1$ is a root of multiplicity m, we can arrange the reduction so that

$$\mu_i = \mu_1 \quad (i=1,2,...,m), \qquad \mu_i \neq \mu_1 \quad (i=m+1,m+2,...,n). \quad (5)$$

We shall now show that the m solutions belonging to the characteristic root μ_1 of S are permuted only among themselves by the other transformation T.

Consider first the expression $\phi_1(u+2\omega_3) = \sum_{j=1}^n b_{1j}\phi_j(u)$, and let b_{1s}

be the last coefficient which does not vanish, so that $b_{1s} \neq 0$, $b_{1j} = 0$ $(j > s)$. Then we shall find from (3) that

$$c_{1s} = \mu_1 b_{1s} = b_{1s} \mu_s, \qquad (6)$$

and this would give a contradiction if $s > m$, $\mu_1 \neq \mu_s$. Hence $\phi_1(u+2\omega_3)$ is expressible in terms of the first m solutions. Suppose that this is also true of $\phi_2(u+2\omega_3),...,\phi_k(u+2\omega_3)$; and that in the expression for $\phi_{k+1}(u+2\omega_3)$ the last coefficient which does not vanish is $b_{(k+1)s} \neq 0$, $b_{(k+1)j} = 0$ $(j > s)$. Then we shall have

$$c_{(k+1)s} = \mu_{k+1} b_{(k+1)s} = b_{(k+1)s} \mu_s \quad (k < m), \qquad (7)$$

and again we find a contradiction if $s > m$, $\mu_s \neq \mu_{k+1}$.

We can therefore resolve the fundamental system, first into sets belonging to each distinct characteristic root of S, and then further into sub-sets belonging to each distinct characteristic root of T. The solutions of a sub-set belonging to a particular pair of multipliers (μ, ν) will then be permuted only among themselves in all transformations of the group.

Construction of a Multiplicative Solution. It is not in general necessary to construct *ab initio* a complete system of n distinct solutions; for if $f(u)$ is any solution, other solutions are given by $f(u+2\omega_1), f(u+4\omega_1),...$, and we naturally choose as many of these as are linearly independent as part of our fundamental system. Let $y_i \equiv f\{u+2(i-1)\omega_1\}$ $(i = 1, 2,..., k)$ be linearly independent, then if we have

$$f(u+2k\omega_1) = a_1 f(u) + a_2 f(u+2\omega_1) + ... + a_k f\{u+2(k-1)\omega_1\}, \qquad (8)$$

we begin by reducing to the canonical form the transformation,

$$\left. \begin{array}{l} Y_i = y_{i+1} \quad (i = 1, 2,..., k-1), \\ Y_k = a_1 y_1 + a_2 y_2 + ... + a_k y_k. \end{array} \right\} \qquad (9)$$

We can form at least one combination $g(u)$ of these solutions, such that $g(u+2\omega_1) = \mu_1 g(u)$; by applying the same argument to $g(u+2\omega_3), g(u+4\omega_3),...$, we can then always find one combination $\phi(u)$, such that $\phi(u+2\omega_1) = \mu_1 \phi(u)$ and also $\phi(u+2\omega_3) = \nu_1 \phi(u)$.

It will not be necessary to construct any entirely new solution until all solutions of the type $f(u+2p\omega_1+2q\omega_3)$ have been accounted for and shown to involve less than n independent solutions.

Canonical Sub-sets. It was shown by Jordan that a sub-set belonging to the multipliers (μ, ν) can be so chosen that the matrices are both of Jacobi's type

$$\sigma \equiv \begin{pmatrix} \mu & 0 & 0 & \dots \\ a_{21} & \mu & 0 & \dots \\ a_{31} & a_{32} & \mu & \dots \\ . & . & . & . \end{pmatrix}, \qquad \tau \equiv \begin{pmatrix} \nu & 0 & 0 & \dots \\ b_{21} & \nu & 0 & \dots \\ b_{31} & b_{32} & \nu & \dots \\ . & . & . & . \end{pmatrix}, \qquad (10)$$

where $\sigma\tau \equiv \tau\sigma$.

The theorem is obvious for a sub-set consisting of only one solution. We assume it true for a set of $(k-1)$, and prove it for one of k solutions. By operating within the sub-set as we did for the system as a whole, we can find at least one multiplicative solution $\phi_1(u)$, such that $\phi_1(u+2\omega_1) = \mu\phi_1(u)$ and $\phi_1(u+2\omega_3) = \nu\phi(u)$. We choose $\phi_1(u)$ as the first solution of the canonical sub-set, and retain $(k-1)$ of the others to make up k independent ones altogether. We can now rewrite the transformations in terms of these solutions. If we ignore the first row and column of the matrix, we can reduce the transformation of the last $(k-1)$ to the canonical form, because the theorem was assumed true for $(k-1)$. When we add these relations (with the terms in $\phi_1(u)$ on the right) to those expressing $\phi_1(u+2\omega_1)$ and $\phi_1(u+2\omega_3)$, the k-rowed matrices will be in the required canonical form.

Analytical Form of the Solution. The two multipliers (μ, ν) being never zero, we can reduce both to unity by writing $\phi_i(u) \equiv \chi(u)\psi_i(u)$, where $\chi(u)$ is an auxiliary function of the form

$$\chi(u) \equiv \frac{\sigma(u+a)}{\sigma(u)} e^{bu}. \qquad (11)$$

For the relation $\sigma(u+2\omega_i) = -e^{2(u+\omega_i)\eta_i}\sigma(u)$ gives

$$\chi(u+2\omega_i) = \chi(u)e^{2(a\eta_i+b\omega_i)} \quad (i = 1, 2, 3); \qquad (12)$$

and because $2(\eta_1\omega_3 - \eta_3\omega_1) = \pm i\pi \neq 0$, we can choose (a, b) to make the multipliers corresponding to $(2\omega_1, 2\omega_3)$ have any assigned values (μ, ν). If we write

$$\Delta\psi(u) \equiv \{\psi(u+2\omega_1) - \psi(u)\} \text{ and } \Delta^*\psi(u) \equiv \{\psi(u+2\omega_3) - \psi(u)\},$$

the two transformations become

$$\Delta\psi_1(u) = 0, \qquad\qquad \Delta^*\psi_1(u) = 0,$$
$$\Delta\psi_2(u) = A_{21}\psi_1(u), \qquad \Delta^*\psi_2(u) = B_{21}\psi_1(u),$$
$$\Delta\psi_3(u) = A_{31}\psi_1(u) + A_{32}\psi_2(u), \qquad \Delta^*\psi_3(u) = B_{31}\psi_1(u) + B_{32}\psi_2(u),$$
$$\cdots\cdots\cdots\cdots\cdots\cdots\cdots\cdots\cdots\cdots\cdots\cdots$$

where the matrices (A_{ij}) and (B_{ij}) are commutative. $\qquad (13)$

It is evident that $\psi_1(u)$ is a uniform doubly-periodic function of u; we shall prove by induction that $\psi_k(u)$ is a polynomial of degree $(k-1)$ in $\{u, \zeta(u)\}$, whose coefficients are uniform and doubly-periodic. For this purpose, we take the periods in the order given by $2(\eta_1\omega_3 - \eta_3\omega_1) = +i\pi$, and introduce the expressions

$$\alpha(u) \equiv \frac{\omega_3\zeta(u) - \eta_3 u}{i\pi}, \qquad \beta(u) = \frac{-\omega_1\zeta(u) + \eta_1 u}{i\pi}, \qquad (14)$$

which have the properties

$$\left.\begin{array}{ll} \Delta\alpha(u) = 1, & \Delta\beta(u) = 0, \\ \Delta^*\alpha(u) = 0, & \Delta^*\beta(u) = 1. \end{array}\right\} \qquad (15)$$

Any polynomial of the type in question can then be written in the form

$$f(u) \equiv \sum_{(i)}\sum_{(j)} \Pi_{ij} \alpha_i \beta_j \quad (i+j < k), \qquad (16)$$

where

$$\left.\begin{array}{c} \alpha_k = \dfrac{\alpha(\alpha-1)\ldots(\alpha-k+1)}{k!}, \qquad \beta_k = \dfrac{\beta(\beta-1)\ldots(\beta-k+1)}{k!}, \\ \Delta\alpha_k = \alpha_{k-1}, \qquad \Delta\beta_k = 0, \\ \Delta^*\alpha_k = 0, \qquad \Delta^*\beta_k = \beta_{k-1}. \end{array}\right\} \qquad (17)$$

Suppose expressions of the required form have been found for $\psi_1(u),\ldots,\psi_{k-1}(u)$; at the next stage we are given the expressions

$$\left.\begin{array}{l} \Delta\psi_k(u) = \sum\limits_{(i+j<k-1)}\sum \Pi^{(1)}_{ij}\alpha_i\beta_j, \\ \Delta^*\psi_k(u) = \sum\limits_{(i+j<k-1)}\sum \Pi^{(2)}_{ij}\alpha_i\beta_j. \end{array}\right\} \qquad (18)$$

The condition $(A_{ij})(B_{ij}) = (B_{ij})(A_{ij})$ is equivalent to

$$\Delta^*\Delta\psi_k(u) = \Delta\Delta^*\psi_k(u);$$

and the relations (18) will be compatible if, and only if,

$$\Pi^{(1)}_{i(j+1)} = \Pi^{(2)}_{(i+1)j}. \qquad (19)$$

Accordingly we now write

$$\psi_k(u) = \sum_{(i+j<k)}\sum \Pi_{k,ij}\alpha_i\beta_j, \qquad (20)$$

where $\qquad \Pi_{k,ij} = \Pi^{(1)}_{(i-1)j} \quad or \quad \Pi_{k,ij} = \Pi^{(2)}_{i(j-1)}. \qquad (21)$

All these are given doubly-periodic functions, with the exception of $\Pi_{k,00}$, which is indeterminate. But the relations (18) now reduce to $(\Delta\Pi_{k,00} = 0 = \Delta^*\Pi_{k,00})$, so that this is also a doubly-periodic function. The theorem thus holds for the kth function if it holds for the $(k-1)$th, and the induction is complete.

For the actual construction of the solutions, reference should be made to works on elliptic functions.

EXAMPLES. IX

1. BRIOSCHI'S IDENTITY. If $Y = y_1 y_2$ is the product of two solutions of Lamé's equation, show that the equation of § 39 (7) has a first integral

$$\left(\frac{dY}{du}\right)^2 - 2Y\frac{d^2Y}{du^2} + 4[n(n+1)\wp(u)+B]Y^2 = K.$$

Show also that $\quad K = \left(y_1\dfrac{dy_2}{du} - y_2\dfrac{dy_1}{du}\right)^2 = C^2.$

2. Show that, in Jacobian elliptic functions, Lamé's equation may be written in either of the forms

$$\frac{1}{y}\frac{d^2y}{du^2} = \frac{n(n+1)}{\operatorname{sn}^2 u} + B', \qquad \frac{1}{y}\frac{d^2y}{du^2} = k^2 n(n+1)\operatorname{sn}^2 u + B'.$$

By changing the periods, reduce to Lamé's form the equation

$$\frac{1}{y}\frac{d^2y}{du^2} = \frac{n(n+1)}{\operatorname{sn}^2 u} + k^2 n(n+1)\operatorname{sn}^2 u + B.$$

3. WHITTAKER'S INTEGRAL EQUATION. Show that the Lamé functions of order n which are rational in $\operatorname{sn} u$ satisfy the integral equation

$$y(u) = \lambda \int_0^{4K} P_n(k \operatorname{sn} u \operatorname{sn} v) y(v)\, dv.$$

[E. T. Whittaker, *Proc. London Math. Soc.* (2) **14** (1915), 260–8.]

4. LAMÉ-WANGERIN FUNCTIONS. (i) Show that Laplace's equation $\nabla^2 V = 0$ admits solutions of the type $V = \varpi^{-\frac{1}{2}}\cos m\phi\, W(u,v)$, where (ϖ, z, ϕ) are cylindrical and (u, v, ϕ) curvilinear coordinates, connected by the relation

$$(\varpi + iz) = F(u+iv);$$

and that $\quad\dfrac{1}{W}\left[\dfrac{\partial^2 W}{\partial u^2} + \dfrac{\partial^2 W}{\partial v^2}\right] = \dfrac{(m^2-\frac{1}{4})|F'(u+iv)|^2}{[2RF(u+iv)]^2}.$

(ii) Show that this has normal solutions $W = G(u)H(v)$ if $F(w) \equiv \operatorname{sn} w$, $\operatorname{cn} w$, or $\operatorname{dn} w$; and that $G(u)$ and $H(v)$ then satisfy Lamé's equation of order $(m-\frac{1}{2})$.

(iii) In the most general transformation giving normal solutions, show that

$$\left(\frac{dF}{dw}\right)^2 = AF^4 + BF^3 + CF^2 + DF + E.$$

[A. Wangerin, *Berliner Monatsber.* (1878), 152–66; E. Haentzschel, *Reduktion der Potentialgleichung* (1893).]

5. PENTASPHERICAL COORDINATES. If (S_k) are the powers of any point with respect to five mutually orthogonal spheres of radii (R_k), show that

$$\sum_{k=1}^{5} \frac{S_k^2}{R_k^2} = 0, \qquad \sum_{k=1}^{5} \frac{S_k}{R_k^2} = -2, \qquad \sum_{k=1}^{5} \frac{1}{R_k^2} = 0,$$

$$\sum_{x,y,z}\left(\frac{\partial S_k}{\partial x}\right)^2 = 4S_k+4R_k^2, \qquad \sum_{x,y,z}\left(\frac{\partial S_j}{\partial x}\right)\left(\frac{\partial S_k}{\partial x}\right) = 2S_j+2S_k,$$

$$ds^2 = \frac{1}{4}\sum_{k=1}^{5}\left(\frac{dS_k}{R_k}\right)^2.$$

If V is homogeneous of degree μ in (S_k), show that

$$\nabla^2 V \equiv 4\sum_{k=1}^{5} R_k^2 \frac{\partial^2 V}{\partial S_k^2} + (4\mu+2)\sum_{k=1}^{5}\frac{\partial V}{\partial S_k}.$$

[G. Darboux, *Leçons sur les systèmes orthogonaux* (1910), 287–93; *Principes de géométrie analytique* (1917), 379–404, 462–7.]

6. CONFOCAL CYCLIDES. If $x_k = S_k/R_k$ and if (e_k) are unequal, three surfaces $(\theta = \lambda, \mu, \nu)$ of the system

$$\sum_{k=1}^{5}\frac{x_k^2}{e_k-\theta} = 0, \quad \text{where } \sum_{k=1}^{5} x_k^2 \equiv 0,$$

pass through any point.

If $P \equiv \sum_{k=1}^{5} e_k x_k^2$, $f(\theta) \equiv \prod_{k=1}^{5}(\theta-e_k)$, prove that

$$x_k^2 = \frac{P(e_k-\lambda)(e_k-\mu)(e_k-\nu)}{f'(e_k)},$$

$$ds^2 = \frac{P}{16}\sum_{\lambda,\mu,\nu}\frac{(\lambda-\mu)(\nu-\lambda)\,d\lambda^2}{f(\lambda)},$$

$$\nabla^2 V = \frac{16}{P^{3/2}}\sum_{\lambda,\mu,\nu}\frac{\sqrt{\{f(\lambda)\}}}{(\lambda-\mu)(\nu-\lambda)}\frac{\partial}{\partial\lambda}\left[P^{1/2}\sqrt{\{f(\lambda)\}}\frac{\partial V}{\partial\lambda}\right],$$

$$\nabla^2 P^{-1/4} = 5P^{-9/4}\sum e_k^2 x_k^2 - 2P^{-5/4}\sum e_k,$$
$$= P^{-5/4}[3\sum e_k - 5\lambda - 5\mu - 5\nu),$$

$$\nabla^2[P^{-1/4}\phi] = [3\sum e_k - 5\lambda - 5\mu - 5\nu]P^{-5/4}\phi + 16P^{-5/4}\sum_{\lambda,\mu,\nu}\frac{f(\lambda)\frac{\partial^2\phi}{\partial\lambda^2}+\frac{1}{2}f'(\lambda)\frac{\partial\phi}{\partial\lambda}}{(\lambda-\mu)(\nu-\lambda)}.$$

7. Show that $\nabla^2 V = 0$ is satisfied by $V = P^{-1/4}E_1(\lambda)E_2(\mu)E_3(\nu)$, if $\{E_i(x)\}$ are all solutions of the equation

$$f(x)D^2 y + \tfrac{1}{2}f'(x)Dy + [\tfrac{5}{16}x^3 - \tfrac{3}{16}x^2\sum e_k - Ax - B]y = 0.$$

[A. Wangerin, *J. für Math.* **82** (1876), 145–57; G. Darboux, loc. cit. or *Comptes rendus*, **83** (1876), 1037–40, 1099–1102; M. Bôcher, *Die Reihenentwickelungen der Potentialtheorie* (1894).]

8. THE FLAT RING. Cyclides of revolution of the family

$$\frac{4a^2 z^2}{\theta} + \frac{(a^2-\varpi^2-z^2)^2}{\theta-1} - \frac{(a^2+\varpi^2+z^2)^2}{\theta-(1/k^2)} = 0$$

are given by $\theta = \mathrm{sn}^2 u$, $\theta = \mathrm{sn}^2 iv$, where
$$(\varpi + iz) = (a/k')[\mathrm{dn}(u+iv) - k\,\mathrm{cn}(u+iv)].$$

If $\nabla^2 V = 0$, show that
$$\left[\frac{\partial^2}{\partial u^2} + \frac{\partial^2}{\partial v^2} + k^2(\mathrm{sn}^2 u - \mathrm{sn}^2 iv)\left(\frac{1}{4} + \frac{\partial^2}{\partial \phi^2}\right)\right](\varpi^{\frac{1}{2}} V) = 0,$$

and that there are solutions $V = \varpi^{-\frac{1}{2}}\cos m\phi\, F(u) G(iv)$, where $F(w)$, $G(w)$ satisfy
$$\frac{1}{y}\frac{d^2 y}{dw^2} = (m^2 - \tfrac{1}{4})k^2\,\mathrm{sn}^2 w + B.$$

[E. G. C. Poole, *Proc. London Math. Soc.* (2) **29** (1929), 342–54; **30** (1930), 174–86.]

X
MATHIEU'S EQUATION

41. Nature and Group of Mathieu's Equation†

Introduction. Just as Bessel's equation is a limiting form of Legendre's, so Mathieu's equation (which is now to be discussed) is easily derived from Lamé's. Let the latter be written in the Jacobian form

$$\frac{d^2y}{du^2} - [n(n+1)k^2\text{sn}^2 u + B]y = 0, \tag{1}$$

and let $n(n+1)k^2 = A$ be kept fixed as $k \to 0$ and $n \to \infty$. In the elliptic functions $2K \to \pi$ and $2iK' \to \infty$; the singularities, which are situated at the points $u = \{2pK + (2q+1)iK'\}$ recede to infinity, and we are left with the equation

$$\frac{d^2y}{du^2} - (A\sin^2 u + B)y = 0, \tag{2}$$

which is Mathieu's equation.

In celestial mechanics there occurs the celebrated equation of Hill

$$\frac{d^2y}{d\theta^2} + [\Theta_0 + 2\Theta_1\cos 2\theta + 2\Theta_2\cos 4\theta + ...]y = 0, \tag{3}$$

of which Mathieu's is a very special case. This equation was the occasion for the introduction into analysis of infinite determinants by Hill, and their subsequent justification by Poincaré. In this problem, the coefficients of the equation are given and we have to determine the character of the solution. We here confine our attention to the special case of Mathieu's equation, where the difficulties regarding convergence are trivial and where the determinants are of the special type associated with continued fractions.

The equation arose in a different manner in Mathieu's problem of the vibrations of an elliptical membrane, and in analogous problems regarding the potentials of elliptic cylinders. We transform the two-dimensional wave equation

$$\frac{\partial^2 V}{\partial x^2} + \frac{\partial^2 V}{\partial y^2} - \frac{1}{c^2}\frac{\partial^2 V}{\partial t^2} = 0, \tag{4}$$

---

† E. Mathieu, *Liouville, J. de Math.* (2) **13** (1868), 137–203; G. W. Hill, *Acta Math.* **8** (1886), 1–36; H. Poincaré, *Les méthodes nouvelles de la mécanique céleste (1893)*, ii. 228–80; E. T. Whittaker and G. N. Watson, *Modern Analysis*, ch. xix; P. Humbert, *Mémorial des sciences mathématiques*, x (1926); M. J. O. Strutt, *Ergebnisse der Mathematik*, i, 3 (1932).

to confocal coordinates $(x+iy) = a\cosh(\xi+i\eta)$, and look for normal solutions of the form $V = e^{ikct}F(\xi)G(\eta)$. We then have

$$\frac{F''(\xi)+k^2a^2\cosh^2\xi F(\xi)}{F(\xi)} = \frac{-G''(\eta)+k^2a^2\cos^2\eta G(\eta)}{G(\eta)}, \qquad (5)$$

and both sides of the equation must be equal to a numerical constant p, whose value is unknown. $F(i\theta)$ and $G(\theta)$ will then both satisfy Mathieu's equation

$$\frac{d^2y}{d\theta^2} + (p-k^2a^2\cos^2\theta)y = 0. \qquad (6)$$

In most physical questions, V must be a single-valued function of position, and so p must be chosen so that $G(\eta)$ shall admit the period 2π. These periodic solutions are called *Mathieu functions*, and the attention of writers on mathematical physics has largely been concentrated upon them for evident practical reasons.

Group of the Equation. The equation (6) has no singularities except $\theta = \infty$, so that every solution is an integral function of θ. At the ordinary point $\theta = 0$, two independent solutions are determined by the conditions $(y \neq 0, y' = 0)$ and $(y = 0, y' \neq 0)$. The equation being invariant when we change the sign of θ, one solution will involve only even and the other only odd powers of θ, say

$$f_0(\theta) = f_0(-\theta), \qquad f_1(\theta) = -f_1(-\theta). \qquad (7)$$

Now the coefficients of (6) admit the period π and the solutions are uniform functions of θ; in accordance with Floquet's theory, we express the solutions $f_0(\theta+\pi), f_1(\theta+\pi)$ in the form

$$\left. \begin{array}{l} f_0(\theta+\pi) = \alpha f_0(\theta) + \beta f_1(\theta), \\ f_1(\theta+\pi) = \gamma f_0(\theta) + \delta f_1(\theta). \end{array} \right\} \qquad (8)$$

Since the Wronskian $W\{f_0(\theta), f_1(\theta)\}$ must be a constant, we find on changing θ into $(\theta+\pi)$ that

$$\alpha\delta - \beta\gamma = 1. \qquad (9)$$

We now combine (7) and (8), and obtain the transformation

$$\left. \begin{array}{l} f_0(\pi-\theta) = \alpha f_0(\theta) - \beta f_1(\theta), \\ f_1(\pi-\theta) = \gamma f_0(\theta) - \delta f_1(\theta). \end{array} \right\} \qquad (10)$$

On repeating this transformation, we must have the identical one

$$\begin{pmatrix} \alpha^2-\beta\gamma, & \beta(\delta-\alpha) \\ \gamma(\alpha-\delta), & \delta^2-\beta\gamma \end{pmatrix} \equiv \begin{pmatrix} 1 & 0 \\ 0 & 1 \end{pmatrix}. \qquad (11)$$

(i) If possible, let these relations be satisfied with $(\alpha-\delta) \neq 0$. Then $\beta = \gamma = 0$ and $\alpha = -\delta = \pm 1$; the relations (10) now reduce to

either
$$f_0(\pi-\theta) = f_0(\theta), \quad f_1(\pi-\theta) = f_1(\theta), \tag{12}$$
or
$$f_0(\pi-\theta) = -f_0(\theta), \quad f_1(\pi-\theta) = -f_1(\theta).$$

In the one case, every solution is *even*, in the other *odd* in $\theta' \equiv (\theta-\tfrac{1}{2}\pi)$. But this is impossible; for $\theta = \tfrac{1}{2}\pi$ is an ordinary point of (6), and on taking this as origin we find (by symmetry) that there is always both an odd and an even solution in θ'. In all cases, therefore, we have

$$\alpha = \delta. \tag{13}$$

The relations $(\alpha^2-\beta\gamma) = 1 = (\delta^2-\beta\gamma)$ now follow automatically from (9) and (13).

(ii) The multipliers of the transformation (8) are given by

$$\lambda^2 - 2\alpha\lambda + 1 = 0, \tag{14}$$

and are unequal provided that $\alpha^2 \neq 1$. We write them $(e^{i\pi h}, e^{-i\pi h})$, where $\alpha = \cos\pi h$. The corresponding multiplicative solutions $\{Af_0(\theta)+Bf_1(\theta)\}$ are given by the condition

$$\frac{A}{B} = \frac{\gamma}{\lambda-\alpha} = \frac{\lambda-\delta}{\beta} = \pm\left(\frac{\gamma}{\beta}\right)^{\frac{1}{2}}, \tag{15}$$

since $\alpha = \delta$. We may therefore write them as

$$f(\theta) \equiv \gamma^{\frac{1}{2}}f_0(\theta)+\beta^{\frac{1}{2}}f_1(\theta),$$
$$f(-\theta) \equiv \gamma^{\frac{1}{2}}f_0(\theta)-\beta^{\frac{1}{2}}f_1(\theta). \tag{16}$$

Their product $F(\theta) \equiv f(\theta)f(-\theta)$ is an even function admitting the period π, whose construction is the crucial step in Lindemann's method of solution, based on the Hermite-Brioschi solution of Lamé's equation. We can easily verify that

$$F(\theta) \equiv \gamma f_0^2(\theta) - \beta f_1^2(\theta) \tag{17}$$

is the only independent expression of the type

$$F(\theta) \equiv \{Af_0^2(\theta) + 2Bf_0(\theta)f_1(\theta) + Cf_1^2(\theta)\}$$

such that $F(\theta) = F(-\theta) = F(\pi\pm\theta)$.

In Hill's method of solution we use the property that, if the multiplier $e^{ih\pi}$ belongs to $f(\theta)$, then $\phi(\theta) \equiv e^{-ih\theta}f(\theta)$ is an integral function of θ with the period π. It is therefore a single-valued analytic function of $z \equiv e^{2i\theta}$, whose only singularities are $z = 0, \infty$;

and so it is expressible as a Laurent series $\sum_{n=-\infty}^{\infty} c_n z^n$. We therefore have
$$f(\theta) = \sum_{n=-\infty}^{\infty} c_n e^{i(h+2n)\theta}, \tag{18}$$
where h is at first unknown. But Hill obtained and solved a transcendental equation for h, from the condition that the series (18) should converge.

Periodic Solutions. If the coefficients of Mathieu's equation are real, those of the linear transformation (8) are also real, and the character of the solution depends on the magnitude of α^2.

(i) If $\alpha^2 > 1$, the multipliers (λ, λ^{-1}) are real and unequal; the expression $f(\theta+2m\pi) = \lambda^{2m}f(\theta)$ then tends to infinity or zero for large integers m. Such solutions are called *unstable*.

(ii) If $\alpha^2 < 1$, the multipliers are conjugate imaginaries of modulus unity, and then $|f(\theta+2m\pi)| = |f(\theta)|$ for all integers m; such solutions are called *stable*. The values of h satisfying $\cos \pi h = \alpha$ are real, but not integers (since $\alpha^2 < 1$).

If h has a rational value $h = r/s$, where the fraction is in its lowest terms, the solutions both admit $s\pi$ as a semi-period or full period, according as r is odd or even. For example, if $\alpha = 0$, $h = \frac{1}{2}$, we have
$$f_0(\theta+\pi) = \beta f_1(\theta), \qquad f_1(\theta+\pi) = \gamma f_0(\theta) \quad (\beta\gamma = -1); \tag{19}$$
and then we have, on repeating the operation,
$$f_0(\theta+2\pi) = -f_0(\theta), \qquad f_1(\theta+2\pi) = -f_1(\theta), \tag{20}$$
so that 2π is a *semi-period* of every solution.

(iii) If $\alpha^2 = 1$, the multipliers are given by $(\lambda\pm 1)^2 = 0$. From (9) and (13) we get $\beta\gamma = 0$; hence one at least of the solutions admits π as a semi-period or as a full period. We find four types of periodic solutions
$$\left.\begin{array}{ll} \sum a_{2n} \cos 2n\theta, & \sum b_{2n} \sin 2n\theta, \\ \sum a_{2n+1} \cos(2n+1)\theta, & \sum b_{2n+1} \sin(2n+1)\theta. \end{array}\right\} \tag{21}$$
The four types are distinguished by their parity as even or odd functions of θ and of $(\theta - \frac{1}{2}\pi)$.

Ince's Theorem.[†] It was proved by Ince that there cannot be two independent solutions of period π or 2π. Suppose there are two

---

[†] E. L. Ince, *Proc. Cambridge Phil. Soc.* **21** (1922), 117–20.

solutions, in sines and cosines of even multiples of θ say. Then the recurrence formulae will be

$$(2k^2a^2-4p)a_0+(k^2a^2)a_2 = 0, \\ (k^2a^2)a_{2n-2}+(2k^2a^2-4p+16n^2)a_{2n}+(k^2a^2)a_{2n+2} = 0, \\ b_0 = 0, \\ (k^2a^2)b_{2n-2}+(2k^2a^2-4p+16n^2)b_{2n}+(k^2a^2)b_{2n+2} = 0. \quad (22)$$

If we eliminate the middle coefficient we get

$$\Delta_{2n-2} = \Delta_{2n} = \Delta_{2n+2} = ..., \\ \Delta_{2n} \equiv a_{2n}b_{2n+2}-b_{2n}a_{2n+2}. \quad (23)$$

where

But since both series are to represent integral functions of θ, we must have, as $n \to \infty$,

$$a_{2n} \to 0, \quad b_{2n} \to 0. \quad (24)$$

Hence $\Delta_{2n} = 0$ for all values of n; and in particular $\Delta_0 = a_0 b_2 = 0$. This relation and the relations (23) can only be satisfied if one of the assumed solutions vanishes identically. A similar proof applies to series of odd multiples of θ.

Since β, γ cannot vanish together, the invariant factors of the characteristic matrix are $[(\lambda\pm 1)^2, 1]$. The fundamental solutions can then be reduced to the canonical forms

$$y_1 = f(\theta), \quad y_2 = \theta f(\theta)+g(\theta), \quad (25)$$

where $f(\theta), g(\theta)$ are integral functions of opposite parity, with the same minimum period π or 2π.

42. The Methods of Lindstedt and Hill

Continued Fractions. In considering Lamé's equation, it was mathematically convenient to solve first the cases of greatest physical interest, where there is a periodic Lamé function expressible in finite terms. But when we pass to Mathieu's equation (a confluent Lamé equation of infinite order) there is no simple expression for the interesting solutions of period π or 2π; in the special cases, the periodic solution is about as intractable as the solution in the general case, and the associated non-periodic second solution much more so. We shall therefore begin with the problem as it arises in celestial mechanics, the coefficients of the equation

$$\frac{d^2y}{d\theta^2}+(\Theta_0+2\Theta_1\cos 2\theta)y = 0 \quad (1)$$

being known, but the solution not necessarily periodic.

We write a multiplicative solution in the form
$$y = e^{ih\theta} \sum_{n=-\infty}^{\infty} c_{2n} e^{2in\theta}, \tag{2}$$
and obtain the recurrence formulae
$$\Theta_1 c_{2n-2} + [\Theta_0 - (h+2n)^2] c_{2n} + \Theta_1 c_{2n+2} = 0 \quad (n = 0, \pm 1, \pm 2,...). \tag{3}$$
Following Lindstedt, we use a method applied by Laplace in the theory of the tides on a rotating globe, and greatly developed by Kelvin, Darwin, and Hough.† Let us put for brevity
$$u_{2n+2} = \frac{c_{2n+2}}{c_{2n}}, \qquad L_{2n} = \frac{(h+2n)^2 - \Theta_0}{\Theta_1}; \tag{4}$$
then the recurrence formulae may be written
$$u_{2n+2} = L_{2n} - \frac{1}{u_{2n}}. \tag{5}$$

Let us choose any large value of n; then if u_{2n} is not small, u_{2n+2} will be large and the subsequent ratios will tend to infinity.

But we require a solution (2) which shall converge for all finite values of θ; and this is only possible if, for all large values of n, we have approximately
$$u_{2n} \sim \frac{1}{L_{2n}} \to 0 \quad (n \to \infty). \tag{6}$$

But we can now rewrite (5) in the form
$$u_{2n} = \frac{1}{L_{2n} - u_{2n+2}}, \tag{7}$$
and obtain an expression as an infinite continued fraction
$$u_{2n} = \frac{1}{L_{2n}-}\frac{1}{L_{2n+2}-}\frac{1}{L_{2n+4}-}\cdots. \tag{8}$$

We now apply the same argument to the terms of (2) with negative indices; we now find
$$u_{-2n+2} = L_{-2n} - \frac{1}{u_{-2n}}; \tag{9}$$
and since u_{-2n} must now become *large* for convergence, we get the expansion
$$u_{-2n+2} = L_{-2n} - \frac{1}{L_{-2n-2}-}\frac{1}{L_{-2n-4}-}\cdots. \tag{10}$$

† Cf. H. Lamb, *Hydrodynamics* (5th ed.), 309–35.

Finally, we substitute the expressions (8), (10) for u_2, u_{-2} respectively in the relation

$$L_0 = u_2 + \frac{1}{u_{-2}}, \tag{11}$$

and so obtain the transcendental equation for h

$$L_0 = \left[\frac{1}{L_2^-}\frac{1}{L_4^-}\frac{1}{L_6^-}\cdots\right] + \left[\frac{1}{L_{-2}^-}\frac{1}{L_{-4}^-}\frac{1}{L_{-6}^-}\cdots\right]. \tag{12}$$

Hill's Determinant. The relations (3) can in general be written

$$\left.\begin{array}{l}\alpha_n c_{2n-2} + c_{2n} + \beta_n c_{2n+2} = 0, \\ \alpha_n = \beta_n = -L_{2n}^{-1}, \quad L_{2n} \neq 0,\end{array}\right\} \tag{13}$$

and we obtain an expression for the mth convergent of the continued fraction $u_{2n} = c_{2n}/c_{2n-2}$, by using m successive equations (13) and putting in the last of them $c_{2n+2m} = 0$; on solving the modified system of linear homogeneous equations in (c_{2s}) we get

$$u_{2n}^{(m)} = \frac{-\alpha_n K(n+1, n+m-1)}{K(n, n+m-1)}, \tag{14}$$

where

$$K(n, n') = \begin{vmatrix} 1 & \beta_n & 0 & \cdots & 0 & 0 \\ \alpha_{n+1} & 1 & \beta_{n+1} & \cdots & 0 & 0 \\ \cdot & \cdot & \cdot & \cdot & \cdot & \cdot \\ 0 & 0 & 0 & \cdots & 1 & \beta_{n'-1} \\ 0 & 0 & 0 & \cdots & \alpha_{n'} & 1 \end{vmatrix}. \tag{15}$$

If n remains fixed and $n' \to \infty$, we have an infinite determinant of von Koch's type.† We have

$$K(n, n') = K(n, n'-1) - \alpha_{n'}\beta_{n'-1} K(n, n'-2), \tag{16}$$

and this shows that every term in $K(n, n')$ occurs also in the product $\prod_{r=n}^{n'-1}(1-\alpha_{r+1}\beta_r)$, so that the determinant converges absolutely if $\sum |\alpha_{r+1}\beta_r|$ converges. This condition is satisfied, provided that there is no finite value of n for which $L_{2n} = 0$; for, as $n \to \infty$, we have $\alpha_n = \beta_n = O(n^{-2})$. We may therefore write (8) and the analogous formula (10) in the form

$$\frac{c_{2n}}{c_{2n-2}} = \frac{-\alpha_n K(n+1, \infty)}{K(n, \infty)}, \quad \frac{c_{-2n}}{c_{-2n+2}} = \frac{-\beta_{-n} K(-\infty, -n-1)}{K(-\infty, -n)}, \tag{17}$$

† H. von Koch, *Comptes rendus*, **120** (1895), 144–7.

and we have in general
$$c_{2n} = (-)^n c_0 \alpha_1 \alpha_2 \ldots \alpha_n K(n+1,\infty)/K(1,\infty), \\ c_{-2n} = (-)^n c_0 \beta_{-1} \beta_{-2} \ldots \beta_{-n} K(-\infty,-n-1)/K(-\infty,-1). \quad (18)$$

The condition (11) may be written
$$-\alpha_0 \beta_{-1} K(-\infty,-2) K(1,\infty) + K(-\infty,-1) K(1,\infty) - \\ -\beta_0 \alpha_{-1} K(-\infty,-1) K(2,\infty) = 0, \quad (19)$$

which is the expansion in terms of the elements of the middle row of the doubly infinite determinant
$$\Delta(h) \equiv K(-\infty,\infty) = 0. \quad (20)$$

This is a special case of Hill's determinant
$$\square(h) \equiv |a_{ij}| \quad (i,j = 0, \pm 1, \pm 2, \ldots) \\ a_{nn} = 1, \quad a_{n(n+k)} = a_{n(n-k)} = \Theta_k/[\Theta_0 - (h+2n)^2]. \quad (21)$$

Should one or both of the numbers $(h \pm \Theta_0^{\frac{1}{2}})$ be an even integer, the corresponding $L_{2n} = 0$, and the relation (3) becomes
$$(c_{2n-2} + c_{2n+2}) = 0.$$

In one row of the infinite determinant (or in two at the most) the triad of non-zero elements $(\alpha_n, 1, \beta_n)$ must be replaced by $(1, 0, 1)$. But we can still expand the modified determinant as a finite combination of convergent infinite determinants of von Koch's type.

Evaluation of $\Delta(h)$. Let (m_k), (n_k) be any two sequences of positive integers tending monotonically to infinity, and consider the sequence of rational analytic functions of the complex variable h defined as the determinants
$$f_k(h) = K(-m_k, n_k) \quad (k = 1, 2, \ldots). \quad (22)$$

The poles of these functions are the points
$$h = \pm \Theta_0^{\frac{1}{2}} \pm 2n \quad (n = 0, 1, 2, \ldots); \quad (23)$$

if these are all excluded by small circles of given radius ρ, we can show that the sequence $\{f_k(h)\}$ converges uniformly in the rest of the plane, and so its limit $\Delta(h)$ is an analytic function of h, whose only singularities are poles at the points (23). Corresponding to $\{f_k(h)\}$ we construct as non-analytic functions of h the products of positive terms
$$P_k(h) = \prod_{r=-m_k}^{n_k} \{1 + |\alpha_r| + |\beta_r|\}. \quad (24)$$

Now we have always
$$|f_k(h)| \leqslant P_k(h),$$
$$|f_{k'}(h)-f_k(h)| \leqslant P_{k'}(h)-P_k(h) \quad (k < k'); \qquad (25)$$

for in each case the modulus of every term on the left occurs among the terms on the right, which are all positive. Accordingly the uniform convergence of $\sum_{r=-\infty}^{\infty}\{|\alpha_r|+|\beta_r|\}$ in any domain, which implies that of the sequence $\{P_k(h)\}$, also ensures that of the sequence of analytic functions $\{f_k(h)\}$, whose limit $\Delta(h)$ will (by Weierstrass's theorem) be a uniform analytic function of h. Now in this instance we have

$$|\alpha_r|+|\beta_r| = \frac{2|\Theta_1|}{|(h+2r)^2-\Theta_0|},$$
$$\leqslant \frac{|\Theta_1|}{|h+2r+\Theta_0^{\frac{1}{2}}|^2} + \frac{|\Theta_1|}{|h+2r-\Theta_0^{\frac{1}{2}}|^2}. \qquad (26)$$

Now among the terms $|h+2r+\Theta_0^{\frac{1}{2}}|$ $(r = 0, \pm 1,...)$, there is at most one numerically smaller than unity, which is itself not smaller than ρ outside the system of circles. Accordingly we have

$$\sum_{r=-\infty}^{\infty}\{|\alpha_r|+|\beta_r|\} \leqslant 2|\Theta_1|\left\{\frac{1}{\rho^2}+\sum_{r=1}^{\infty}\frac{2}{(2r-1)^2}\right\},$$
$$\leqslant |\Theta_1|\left\{\frac{2}{\rho^2}+\frac{\pi^2}{2}\right\}, \qquad (27)$$

uniformly in the said region; and so

$$\Delta(h) = \lim_{k\to\infty} f_k(h) \qquad (28)$$

is a single-valued analytic function, whose only singularities are at the points (23). By renumbering or rearranging the rows and columns of the determinant, we now find

$$\Delta(-h) = \Delta(h), \qquad \Delta(h+2) = \Delta(h). \qquad (29)$$

Thus $\Delta(h)$ is an even periodic function, with the period 2; its values everywhere are deducible from those in the strip $0 \leqslant R(h) \leqslant 1$. But in this strip the expression $Z = \cos \pi h$ assumes once every value in the plane of Z, cut from $-\infty$ to -1 and from 1 to ∞. Hence $\Delta(h)$ is a uniform analytic function of $\cos \pi h$, which is also an even analytic function of h with the period 2. The points (23) are those where $\cos \pi h = \cos \pi \Theta_0^{\frac{1}{2}}$; and as h approaches one of these points, one or at most two factors of the product (24) tend to infinity, but in such a manner that $[\cos \pi h - \cos \pi \Theta_0^{\frac{1}{2}}]P_\infty(h)$ remains bounded. *A fortiori*, the expression $[\cos \pi h - \cos \pi \Theta_0^{\frac{1}{2}}]\Delta(h)$ remains bounded also; and so $\Delta(h)$, regarded as a function of $\cos \pi h$, has a pole of the first order only. If K is the residue there, the function

$$\phi(h) \equiv \Delta(h) - \frac{K}{[\cos \pi h - \cos \pi \Theta_0^{\frac{1}{2}}]}, \tag{30}$$

remains bounded for all finite values of h. But if $h \to i\infty$, $\Delta(h) \to 1$ and the second term in (30) tends to zero; and since $\phi(h+2) = \phi(h)$, we see that $\phi(h)$ is everywhere bounded, and therefore is a constant, by Liouville's theorem.

Hence we have finally

$$\Delta(h) \equiv 1 + \frac{K}{[\cos \pi h - \cos \pi \Theta_0^{\frac{1}{2}}]}. \tag{31}$$

The coefficient K is determined by substituting any convenient value of h, such as $h = 0$, which gives

$$K = [\Delta(0) - 1][1 - \cos \pi \Theta_0^{\frac{1}{2}}]. \tag{32}$$

The equation $\Delta(h) = 0$ is thus equivalent to

$$\sin^2 \tfrac{1}{2}\pi h = \Delta(0) \sin^2 \tfrac{1}{2}\pi \Theta_0^{\frac{1}{2}}. \tag{33}$$

This equation may be read in two ways. In the astronomical problem, where the coefficients of the differential equation are given, it serves to determine the multipliers $e^{\pm i\pi h}$ of the solutions. On the other hand, if Θ_0 is not given, it may be read as a transcendental equation to determine Θ_0, when the solution has assigned multipliers, for example, when it admits the period 4π.

43. Mathieu Functions†

Mathieu's Method. We must now examine more particularly the solutions of period π or 2π, and we begin by remarking that, when $\Theta_1 = 0$, the critical numbers become $\Theta_0 = n^2$ and give the solutions $\cos n\theta$, $\sin n\theta$. By the methods of Sturm and Liouville we know also that, for any given real Θ_1, the critical numbers giving the even and odd Mathieu functions with n zeros in an interval of length π must satisfy the inequality

$$|\Theta_0 - n^2| \leqslant 2|\Theta_1|. \tag{1}$$

† See (in addition to previous references) E. Heine, *Kugelfunktionen* (1878), i. 401–15; E. L. Ince, *Proc. Roy. Soc. Edinburgh*, **46** (1925–6), 20–9, 316–22; **47** (1927), 294–301; S. Goldstein, *Trans. Cambridge Phil. Soc.* **23** (1927), 303–36; *Proc. Cambridge Phil. Soc.* **24** (1928), 223–30.

Accordingly Mathieu assumes that the solution and its corresponding critical number can be expanded as power-series in Θ_1, and gives rules for evaluating those series term by term. We write in Mathieu's equation (§ 42 (1)) a solution of the type

$$\left.\begin{aligned}\operatorname{ce}_n(\theta) &= \cos n\theta + \Theta_1 U_1(\theta) + \Theta_1^2 U_2(\theta) + \ldots, \\ \Theta_0 &= n^2 + A_1 \Theta_1 + A_2 \Theta_1^2 + \ldots,\end{aligned}\right\} \quad (2)$$

or,
$$\left.\begin{aligned}\operatorname{se}_n(\theta) &= \sin n\theta + \Theta_1 V_1(\theta) + \Theta_1^2 V_2(\theta) + \ldots, \\ \Theta_0 &= n^2 + B_1 \Theta_1 + B_2 \Theta_1^2 + \ldots,\end{aligned}\right\} \quad (3)$$

where n is zero or a positive integer. We now equate to zero the coefficients of each power of Θ_1, which gives for the even Mathieu functions the set of conditions

$$\left.\begin{aligned}U_1'' + n^2 U_1 + (A_1 + 2\cos 2\theta)\cos n\theta &= 0, \\ U_2'' + n^2 U_2 + (A_1 + 2\cos 2\theta)U_1 + A_2 \cos n\theta &= 0, \\ U_3'' + n^2 U_3 + (A_1 + 2\cos 2\theta)U_2 + A_2 U_1 + A_3 \cos n\theta &= 0, \\ \cdot \quad \cdot \quad \cdot \quad \cdot \quad \cdot \quad \cdot \quad \cdot \quad \cdot \quad \cdot \quad \cdot \quad \cdot \quad \cdot \quad \cdot \quad \cdot\end{aligned}\right\} \quad (4)$$

At each stage we first determine A_k by the condition that $U_k(\theta)$ shall be periodic; and then we determine $U_k(\theta)$ uniquely by the condition that $\cos n\theta$, $\sin n\theta$ must appear only in the leading term. Now a particular integral of the first equation (4) is

$$\left.\begin{aligned}U_1(\theta) &= \frac{\cos(n+2)\theta}{4(n+1)} - \frac{\cos(n-2)\theta}{4(n-1)} - \frac{A_1 \theta \sin n\theta}{2n} &\quad (n > 1), \\ \text{or,} \quad &= \tfrac{1}{8}\cos 3\theta - \tfrac{1}{2}(1+A_1)\theta\sin\theta &\quad (n = 1), \\ \text{or,} \quad &= \tfrac{1}{2}\cos 2\theta - \tfrac{1}{2}A_1 \theta^2 &\quad (n = 0),\end{aligned}\right\} \quad (5)$$

which is not periodic unless

$$A_1 = 0 \quad (n \neq 1), \quad \text{or} \quad A_1 = -1 \quad (n = 1). \quad (6)$$

The complementary function $C\cos n(\theta - \theta_0)$ need not be added, in accordance with our rule, because all such multiples of θ are to appear only in the first term. We have therefore definitely

$$\left.\begin{aligned}U_1(\theta) &= \frac{\cos(n+2)\theta}{4(n+1)} - \frac{\cos(n-2)\theta}{4(n-1)} &\quad (n \neq 1) \\ &= \tfrac{1}{8}\cos 3\theta &\quad (n = 1).\end{aligned}\right\} \quad (7)$$

At the next stage we have in general (if $n \geqslant 2$)

$$U_2'' + n^2 U_2 + \frac{\cos(n+4)\theta}{4(n+1)} - \frac{\cos(n-4)\theta}{4(n-1)} + \left[A_2 - \frac{1}{2(n^2-1)}\right]\cos n\theta = 0, \quad (8)$$

and the condition for a periodic solution gives
$$A_2 = \frac{1}{2(n^2-1)} \quad (n > 2), \qquad A_2 = \frac{5}{12} \quad (n = 2); \tag{9}$$
and we then have a unique expression
$$U_2(\theta) = \frac{\cos(n+4)\theta}{4.8(n+1)(n+2)} + \frac{\cos(n-4)\theta}{4.8(n-1)(n-2)} \quad (n > 2). \tag{10}$$
The second term is dropped if $n = 2$; the modified forms corresponding to $n = 0, 1$ are left to the reader. It is evident that the formal process can be continued indefinitely, with proper precautions from the nth stage when terms in $\cos(-n\theta)$ first appear. The same method is applicable to the odd Mathieu functions. It was pointed out by Mathieu that the series giving Θ_0 for $ce_n(\theta)$ and $se_n(\theta)$ begin alike, but differ after n terms. The coefficients of the trigonometrical polynomials $\{U_k(\theta)\}$ and $\{V_k(\theta)\}$ begin to differ as soon as negative multiples of θ make their appearance. Various improvements in the method have been made by Whittaker, and it has been shown by Watson that the series converge if Θ_1 is sufficiently small.

As Θ_1 increases, however, the method breaks down from one of two causes. The true connexion between Θ_0 and Θ_1 is found by putting $h = 0$ or $h = 1$ in Hill's determinantal equation § 42 (33). This is a transcendental equation for Θ_0, whose roots can be expanded as convergent power-series in Θ_1, if and only if Θ_1 is sufficiently small. Again, in constructing the nth even Mathieu function, we divided by the coefficient of $\cos n\theta$. But, for certain values of Θ_1, the coefficient in question vanishes; and, if we try to equate it to unity, the other coefficients become infinite. These limitations of Mathieu's method have been more or less clearly realized by mathematicians since Heine, who applied the method of continued fractions; this method was employed by Lindstedt in the associated astronomical problem, as we have already seen, and was also employed in connexion with the solutions of period $2s\pi$ ($s > 1$) by the writer. The most recent and complete results relating to the Mathieu functions of period π or 2π have been based on the use of infinite determinants or infinite continued fractions. The former were used by Ince to calculate the critical values of Θ_0, and the latter by Goldstein for expanding the Mathieu functions and also the associated second solutions.

Continued Fractions. To construct the functions of period π, we put $h = 0$ in the analysis of § 42. It is easily verified that, by symmetry, we have

$$L_{2n} = L_{-2n},$$
$$\alpha_n = \beta_n = \alpha_{-n} = \beta_{-n},$$
$$\frac{c_{2n+2}}{c_{2n}} = \frac{c_{-2n-2}}{c_{-2n}}, \quad (11)$$

and this gives either

$$c_0 \neq 0, \quad c_{2n} = c_{-2n},$$
$$\text{or} \quad c_0 = 0, \quad c_{2n} = -c_{-2n}. \quad (12)$$

For the even solution we get

$$-L_0 c_0 + 2c_2 = 0,$$
$$c_{2n-2} - L_{2n} c_{2n} + c_{2n+2} = 0 \quad (n \geqslant 1); \quad (13)$$

and so § 42 (12) is replaced by the condition

$$\tfrac{1}{2} L_0 = \frac{1}{L_2 -}\frac{1}{L_4 -}\frac{1}{L_6 -}\cdots, \quad (14)$$

or § 42 (20) by the simply-infinite determinant

$$\begin{vmatrix} 1, & 2\alpha_0, & 0, & 0, & \ldots \\ \alpha_1, & 1, & \beta_1, & 0, & \ldots \\ 0, & \alpha_2, & 1, & \beta_2, & \ldots \\ \cdot & \cdot & \cdot & \cdot & \cdot \end{vmatrix} = 0. \quad (15)$$

For the odd solution, we have in the same way

$$-L_2 c_2 + c_4 = 0,$$
$$c_{2n-2} - L_{2n} c_{2n} + c_{2n+2} = 0 \quad (n \geqslant 2); \quad (16)$$

which give

$$L_2 = \frac{1}{L_4 -}\frac{1}{L_6 -}\cdots, \quad (17)$$

or

$$\begin{vmatrix} 1, & \alpha_1, & 0, & 0, & \ldots \\ \alpha_2, & 1, & \beta_2, & 0, & \ldots \\ 0, & \alpha_3, & 1, & \beta_3, & \ldots \\ \cdot & \cdot & \cdot & \cdot & \cdot \end{vmatrix} = 0. \quad (18)$$

For the solutions of period 2π we put $h = 1$ and adopt the more convenient notation

$$y = \sum_{n=-\infty}^{\infty} c_{2n+1} e^{(2n+1)i\theta},$$
$$L_{2n+1} = L_{-2n-1} = [(2n+1)^2 - \Theta_0]/\Theta_1. \quad (19)$$

We then have, by symmetry

$$\frac{c_{2n+1}}{c_{2n-1}} = \frac{c_{-2n-1}}{c_{-2n+1}}, \tag{20}$$

and so in particular

$$\frac{c_1}{c_{-1}} = \frac{c_{-1}}{c_1}, \qquad c_{2n+1} = \pm c_{-2n-1}. \tag{21}$$

For the even solution we get now

$$\left.\begin{array}{l}(1-L_1)c_1+c_3 = 0,\\ c_{2n-1}-L_{2n+1}c_{2n+1}+c_{2n+3} = 0 \quad (n \geqslant 1),\end{array}\right\} \tag{22}$$

giving the condition

$$1 = L_1 - \frac{1}{L_3-}\frac{1}{L_5-}\cdots. \tag{23}$$

For the odd solution we get

$$\left.\begin{array}{l}-(1+L_1)c_1+c_3 = 0,\\ c_{2n-1}-L_{2n+1}c_{2n+1}+c_{2n+3} = 0 \quad (n \geqslant 1),\end{array}\right\} \tag{24}$$

giving the condition

$$0 = 1+L_1 - \frac{1}{L_3-}\frac{1}{L_5-}\cdots. \tag{25}$$

The critical values are calculated by the method of successive approximations, following Hough's procedure in the theory of the tides. By the methods of Sturm and Liouville, we know that one of the roots of (14), for example, is given to a first approximation by $L_{2n} = 0$. To evaluate this root more exactly, we write (14) in the equivalent form

$$L_{2n} = \left[\frac{1}{L_{2n-2}-}\frac{1}{L_{2n-4}-}\cdots\frac{1}{L_2-}\frac{2}{L_0}\right] + \left[\frac{1}{L_{2n+2}-}\frac{1}{L_{2n+4}-}\frac{1}{L_{2n+6}-}\cdots\right]. \tag{26}$$

The value of Θ_0 given by $L_{2n} = 0$ is now introduced in the two expressions on the right, and a second approximation is then obtained, and so on.

The Second Solution. When one solution is a known Mathieu function, the other can be expressed by a quadrature; but a more convenient construction based on the method of infinite continued fractions has been given by Goldstein. As a typical case, suppose we know the Mathieu function

$$\mathrm{ce}_{2s+1}(\theta) = \sum_{n=0}^{\infty} a_{2n+1}\cos(2n+1)\theta, \tag{27}$$

and the appropriate value of Θ_0. According to § 41 (25), the second solution will be of the form

$$\text{in}_{2s+1}(\theta) = \theta \text{ce}_{2s+1}(\theta) + \sum_{n=0}^{\infty} a'_{2n+1} \sin(2n+1)\theta, \qquad (28)$$

and the recurrence formulae are

$$\left.\begin{aligned}
(1-L_1)a_1 + a_3 &= 0, \\
a_{2n-1} - L_{2n+1}a_{2n+1} + a_{2n+3} &= 0, \\
-(1+L_1)a'_1 + a'_3 &= 2a_1/\Theta_1, \\
a'_{2n-1} - L_{2n+1}a'_{2n+1} + a'_{2n+3} &= 2(2n+1)a_{2n+1}/\Theta_1.
\end{aligned}\right\} \qquad (29)$$

We now introduce an auxiliary sequence (A_{2n+1}), defined by the relations

$$\left.\begin{aligned}
-(1+L_1)A_1 + A_3 &= 0, \\
A_{2n-1} - L_{2n+1}A_{2n+1} + A_{2n+3} &= 0.
\end{aligned}\right\} \qquad (30)$$

This must diverge to infinity; for otherwise we should have both an odd and an even Mathieu function for the same value of Θ_0. As in § 41 (23), we find that

$$\left.\begin{aligned}
a_{2n+1}A_{2n+3} - a_{2n+3}A_{2n+1} &= a_{2n-1}A_{2n+1} - a_{2n+1}A_{2n-1} \\
&= \ldots \ldots \\
&= 2a_1 A_1 \neq 0.
\end{aligned}\right\} \qquad (31)$$

Following Goldstein, we write in (29)

$$a'_{2n+1} = \sum_{r=0}^{\infty} a^{(2r+1)}_{2n+1} \quad (n = 0, 1, 2, \ldots), \qquad (32)$$

where the $(r+1)$th set of terms is defined by the relations

$$\left.\begin{aligned}
-(1+L_1)a^{(2r+1)}_1 + a^{(2r+1)}_3 &= 0, \\
a^{(2r+1)}_{2n-1} - L_{2n+1}a^{(2r+1)}_{2n+1} + a^{(2r+1)}_{2n+3} &= 0 \qquad (n \neq r) \\
&= 2(2r+1)a_{2r+1}/\Theta_1 \quad (n = r).
\end{aligned}\right\} \qquad (33)$$

To satisfy these conditions by a sequence of values tending rapidly to zero, we put

$$\left.\begin{aligned}
a^{(2r+1)}_{2n+1} &= \lambda_{2r+1} A_{2n+1} a_{2r+1} \quad (n \leqslant r) \\
&= \lambda_{2r+1} A_{2r+1} a_{2n+1} \quad (n \geqslant r),
\end{aligned}\right\} \qquad (34)$$

where λ_{2r+1} is determined by the condition

$$[A_{2r-1}a_{2r+1} - L_{2r+1}A_{2r+1}a_{2r+1} + A_{2r+1}a_{2r+3}]\Theta_1 \lambda_{2r+1} = (4r+2)a_{2r+1}. \qquad (35)$$

By using (30) and then (31), this is reduced to

$$A_1 a_1 \Theta_1 \lambda_{2r+1} = -(2r+1)a_{2r+1}; \qquad (36)$$

hence we have
$$a_{2n+1}^{(2r+1)} = -(2r+1)\lambda A_{2n+1} a_{2r+1}^2 \qquad (n \leqslant r),$$
$$\qquad\qquad = -(2r+1)\lambda A_{2r+1} a_{2r+1} a_{2n+1} \qquad (n \geqslant r), \quad (37)$$

where $\lambda = (A_1 a_1 \Theta_1)^{-1}$. The expression (32) now becomes

$$a'_{2n+1} = -\lambda a_{2n+1}\left[\sum_{r=0}^{n-1}(2r+1)A_{2r+1}a_{2r+1}\right] - \lambda A_{2n+1}\left[\sum_{r=n}^{\infty}(2r+1)a_{2r+1}^2\right]. \quad (38)$$

Now for large values of n we have

$$\frac{a_{2n-1}}{a_{2n+1}} \sim \frac{A_{2n+3}}{A_{2n+1}} \sim L_{2n+1} \to \infty, \qquad (39)$$

$$\lim_{n\to\infty}\frac{(2n+3)A_{2n+3}a_{2n+3}}{(2n+1)A_{2n+1}a_{2n+1}} = \lim_{n\to\infty}\frac{(2n+3)L_{2n+1}}{(2n+1)L_{2n+3}} = 1. \qquad (40)$$

For any fixed positive α, we can choose M so that
$$|(2n+1)A_{2n+1}a_{2n+1}| < M(1+\alpha)^{2n+1}; \qquad (41)$$
and so the first series in (38) has the upper bound

$$\sum_{r=0}^{n-1}|(2r+1)A_{2r+1}a_{2r+1}| < M\sum_{r=0}^{n-1}(1+\alpha)^{2r+1}$$
$$< M(1+\alpha)^{2n-1}\sum_{s=0}^{\infty}(1+\alpha)^{-2s}$$
$$< \frac{M(1+\alpha)^{2n+1}}{(1+\alpha)^2-1} = M'(1+\alpha)^{2n+1}. \qquad (42)$$

On the other hand, the second series in (38) converges with extreme rapidity and (except for those values of n, limited in number, for which $a_{2n+1} = 0$) its sum is nearly equal to the first term, for $n > n_0$ say. Hence

$$\left|A_{2n+1}\sum_{r=n}^{\infty}(2r+1)a_{2n+1}^2\right| = O[(2n+1)|A_{2n+1}a_{2n+1}^2|]$$
$$= O[(1+\alpha)^{2n+1}|a_{2n+1}|]. \qquad (43)$$

From (38), (42), and (43) we now have
$$a'_{2n+1} = O[|a_{2n+1}|(1+\alpha)^{2n+1}] \quad (n > n_0); \qquad (44)$$
and since (a_{2n+1}) are the coefficients of an integral function, the series $\sum a'_{2n+1}\sin(2n+1)\theta$ is also convergent for all finite values of θ.

44. The Methods of Lindemann and Stieltjes

Product of Solutions. We conclude our account of Mathieu's equation with the solution of Lindemann and Stieltjes, modelled on the Hermite-Brioschi treatment of Lamé's equation. It is convenient to revert to the notation of § 41 (6), namely

$$\frac{d^2y}{d\theta^2}+(p-k^2a^2\cos^2\theta)y = 0. \tag{1}$$

The product of any two solutions, $Y = y_1 y_2$, satisfies the equation

$$\frac{d^3Y}{d\theta^3}+4(p-k^2a^2\cos^2\theta)\frac{dY}{d\theta}+4k^2a^2\cos\theta\sin\theta\, Y = 0; \tag{2}$$

and on multiplying this by $-2Y$ and integrating, we have the Brioschi identity

$$\left(\frac{dY}{d\theta}\right)^2 - 2Y\frac{d^2Y}{d\theta^2} - 4(p-k^2a^2\cos^2\theta)Y^2 = C^2. \tag{3}$$

If we put $Y = y_1 y_2$ and use (1), the left-hand side reduces to the square of the Wronskian; when Y is known, we can calculate C^2 from (3) and we then obtain the numerical value of the Wronskian

$$y_1\frac{dy_2}{d\theta}-y_2\frac{dy_1}{d\theta} = \pm C, \tag{4}$$

which will enable us to complete the solution by a quadrature.

Now whether the equation (1) has or has not a solution of period π or 2π, it is easily verified that (2) has always one and only one solution which is an even function of θ with the period π. This solution is either the square of a Mathieu function, or else the product $F(\theta) \equiv f(\theta)f(-\theta)$ of the two principal multiplicative solutions of (1). We could expand $F(\theta)$ either in the form $\sum A_{2n}\cos 2n\theta$, or as an even power-series in $\cos\theta$ or $\sin\theta$. The latter forms are more convenient in connexion with the relation (3). For to calculate C we insert some particular value $\theta = 0$ or $\tfrac{1}{2}\pi$; and in the one case Y and its derivatives are expressed as infinite series, and in the other in a finite form. We therefore put $z = \cos^2\theta$ and obtain

$$z(1-z)Y'''+\tfrac{3}{2}(1-2z)Y''+(p-1-k^2a^2z)Y'-\tfrac{1}{2}k^2a^2Y = 0, \tag{2*}$$

$$z(1-z)(Y'^2-2YY'')-(1-2z)YY'-(p-k^2a^2z)Y^2 = \tfrac{1}{4}C^2, \tag{3*}$$

$$z^{\frac{1}{2}}(1-z)^{\frac{1}{2}}(y_1y_2'-y_2y_1') = \pm\tfrac{1}{2}C, \tag{4*}$$

where primes denote differentiation with respect to z. Now the equation (2*) has an irregular singularity at $z = \infty$, and two regular singularities free from logarithms, with exponents $(0, \tfrac{1}{2}, 1)$, at $z = 0, 1$. We can express each of the principal branches at $z = 0$ in terms of the three principal branches at $z = 1$; and we must construct a combination of the two branches holomorphic at $z = 0$, which shall *not* involve the branch belonging to the exponent $\frac{1}{2}$ at $z = 1$. This solution will be holomorphic for all finite values of z, and so it will be the required integral function. Now every solution $Y = \sum C_n z^n$, which is holomorphic at $z = 0$, satisfies the recurrence formulae

$$n(n+\tfrac{1}{2})(n+1)C_{n+1} = n(n^2-p)C_n + k^2 a^2 (n-\tfrac{1}{2})C_{n-1} \quad (n=1,2,\ldots), \quad (5)$$

which may be written

$$u_{n+1} = \frac{(n^2-p)}{(n+\tfrac{1}{2})(n+1)} + \frac{k^2 a^2 (n-\tfrac{1}{2})}{n(n+\tfrac{1}{2})(n+1) u_n} \quad \left(u_{n+1} = \frac{C_{n+1}}{C_n} \right). \quad (6)$$

In general, if u_n is not small for some fairly large value of n, the sequence (u_n) tends to the limit unity. If the ratio u_1 is arbitrarily chosen, the solution converges when $|z| < 1$, but has a singularity at $z = 1$. But, for a certain properly chosen value of u_1, the sequence tends to zero and we have approximately

$$u_n \sim \frac{-k^2 a^2 (n-\tfrac{1}{2})}{n(n^2-p)} \to 0. \quad (7)$$

We write (5) in the form

$$\left. \begin{array}{c} \alpha_n C_{n-1} + C_n + \beta_n C_{n+1} = 0, \\ \alpha_n = \dfrac{k^2 a^2 (n-\tfrac{1}{2})}{n(n^2-p)}, \quad \beta_n = -\dfrac{(n+\tfrac{1}{2})(n+1)}{(n^2-p)}, \end{array} \right\} \quad (8)$$

which gives in general

$$u_n = -\frac{\alpha_n}{1 + \beta_n u_{n+1}},$$

$$= -\frac{\alpha_n}{1-} \frac{\alpha_{n+1} \beta_n}{1-} \frac{\alpha_{n+2} \beta_{n+1}}{1-} \ldots. \quad (9)$$

This infinite continued fraction is equivalent to the quotient of two infinite determinants

$$u_n = -\frac{\alpha_n K(n+1, \infty)}{K(n, \infty)}, \quad (10)$$

where

$$K(n, \infty) = \begin{vmatrix} 1 & \beta_n & 0 & 0 & \ldots \\ \alpha_{n+1} & 1 & \beta_{n+1} & 0 & \ldots \\ \cdot & \cdot & \cdot & \cdot & \cdot \end{vmatrix}. \quad (11)$$

This converges by von Koch's rule, because $\sum \alpha_{n+1} \beta_n$ is absolutely convergent. The coefficients of the solution may now be written

$$C_n = (-)^n C_0 \alpha_1 \alpha_2 \ldots \alpha_n K(n+1, \infty)/K(1, \infty), \quad (12)$$

and the series $\sum C_n z^n$ converges for all finite values of z.

If p is the square of an integer ($p = m^2$), we have exceptionally, for that one value of the suffix,

$$u_m = \frac{k^2 a^2 (m-\tfrac{1}{2})}{m(m+\tfrac{1}{2})(m+1) u_{m+1}}. \tag{13}$$

The expressions (12) as they stand are indeterminate of the type ∞/∞; but the correct values are found by omitting α_m, whenever it occurs as a factor of the numerator, and at the same time replacing the elements $(\alpha_m, 1, \beta_m)$ by $\left(1, 0, \dfrac{-m(m+\tfrac{1}{2})(m+1)}{k^2 a^2 (m-\tfrac{1}{2})}\right)$ in the determinants $K(n, \infty)$, where $n \leqslant m$.

The Invariant. The series $Y = \sum C_n z^n$ being now known, we substitute it in the identity (3*) and put $z = 0$. This gives the value of the constant

$$C^2 = -4[C_0 C_1 + p C_0^2]. \tag{14}$$

If there is a Mathieu function of period π or 2π, the constant C vanishes and $f(\theta) = \pm f(-\theta)$; the solution is then more conveniently constructed by other methods. We accordingly suppose $C^2 \neq 0$; it is immaterial which sign is given to C, as this merely changes the notation by interchanging $f(\theta)$ and $f(-\theta)$. We now have

$$\left. \begin{array}{l} F(\theta) = f(\theta) f(-\theta) = \sum\limits_{n=0}^{\infty} C_n \cos 2n\theta, \\[4pt] \dfrac{f'(\theta)}{f(\theta)} + \dfrac{f'(-\theta)}{f(-\theta)} = \dfrac{C}{F(\theta)} \quad (C \neq 0). \end{array} \right\} \tag{15}$$

Hence
$$\frac{f(\theta)}{f(-\theta)} = \exp\left[\int_0^\theta \frac{C\, d\theta}{F(\theta)}\right], \tag{16}$$

both sides reducing to unity as $\theta \to 0$, since $f(0) \neq 0$ when there is no solution of period π or 2π. If we put $\theta = \pi$ in (16) and use the multipliers, we have Stieltjes's formula

$$2i\pi h = \int_0^\pi \frac{C\, d\theta}{F(\theta)}. \tag{17}$$

Finally, we obtain the solutions

$$\left. \begin{array}{l} f(\theta) = \{F(\theta)\}^{\frac{1}{2}} \exp\left[\dfrac{1}{2} \displaystyle\int_0^\theta \dfrac{C\, d\theta}{F(\theta)}\right], \\[10pt] f(-\theta) = \{F(\theta)\}^{\frac{1}{2}} \exp\left[-\dfrac{1}{2} \displaystyle\int_0^\theta \dfrac{C\, d\theta}{F(\theta)}\right]. \end{array} \right\} \tag{18}$$

EXAMPLES. X

1. The four types of Mathieu functions are the solutions of the equation
$$\frac{d^2y}{d\theta^2} + (p - k^2a^2\cos^2\theta)y = 0$$
determined by the boundary conditions $y = 0$ or $\frac{dy}{d\theta} = 0$, at the two points $\theta = 0$ and $\theta = \frac{1}{2}\pi$.

2. Mathieu functions belonging to different critical values p_1, p_2 satisfy the orthogonal relation
$$\int_0^{2\pi} f_1(\theta) f_2(\theta)\, d\theta = 0.$$
If $p_1 > p_2$ and k^2a^2 is real, the zeros of $f_1(\theta)$ separate those of $f_2(\theta)$ along the axis of real values of θ.

3. If w is a solution of the auxiliary equation
$$\frac{d^2w}{d\theta^2} + n^2w = 0,$$
show that
$$\int_a^b [n^2 - p + k^2a^2\cos^2\theta] w(\theta) y(\theta)\, d\theta = [w(\theta) y'(\theta) - y(\theta) w'(\theta)]_a^b.$$
If $n_1^2 = p$, $n_2^2 = p - k^2a^2$ ($k^2a^2 > 0$), the real zeros of $y(\theta)$ separate those of $\cos n_2(\theta - \theta_0)$ and are themselves separated by those of $\cos n_1(\theta - \theta_0)$.

[The zeros in the complex plane are examined by E. Hille, *Proc. London Math. Soc.* (2) **23** (1924), 185–237.]

4. INTEGRAL EQUATIONS. (i) If $x = \cos\theta$, show that Mathieu's equation may be written in the form
$$(1 - x^2) D^2 y - x\, Dy + (p - k^2a^2x^2) y = 0.$$

(ii) Show that this can be solved by Laplace's transformation
$$y = \int_C e^{ikaxt} \phi(t)\, dt,$$
provided that
$$(1 - t^2) \phi''(t) - t\phi'(t) + (p - k^2a^2t^2) \phi(t) = 0,$$
and
$$[e^{ikaxt}\{ikax(1 - t^2)\phi(t) + t\phi(t) - (1 - t^2)\phi'(t)\}]_C = 0.$$

(iii) By making $\phi(\cos\theta)$ a Mathieu function, obtain Whittaker's equations of the form
$$\mathrm{ce}_{2n}(\theta) = \lambda \int_0^\pi \cos(ka\cos\theta\cos\theta')\, \mathrm{ce}_{2n}(\theta')\, d\theta',$$
$$\mathrm{ce}_{2n+1}(\theta) = \lambda \int_0^\pi \sin(ka\cos\theta\cos\theta')\, \mathrm{ce}_{2n+1}(\theta')\, d\theta'.$$

(iv) By taking $\sin\theta$ as variable, obtain similarly Whittaker's equations
$$\mathrm{ce}_{2n}(\theta) = \lambda \int_0^\pi \cosh(ka\sin\theta\sin\theta')\, \mathrm{ce}_{2n}(\theta')\, d\theta',$$

$$\mathrm{se}_{2n+1}(\theta) = \lambda \int_0^\pi \sinh(ka\sin\theta\sin\theta')\mathrm{se}_{2n+1}(\theta')\,d\theta'.$$

[E. T. Whittaker and G. N. Watson, *Modern Analysis*, 400–2.]

5. By expanding as a series of Mathieu functions $\sum A_n F_n(\xi)G_n(\eta)$ the solutions of the equation of wave-motion

$$e^{ikx}, \qquad e^{iky}, \qquad ye^{ikx}, \qquad xe^{iky},$$

obtain Whittaker's equations and the following:

$$\mathrm{se}_{2n+1}(\theta) = \lambda \int_0^\pi \sin\theta\sin\theta'\cos(ka\cos\theta\cos\theta')\mathrm{se}_{2n+1}(\theta')\,d\theta',$$

$$\mathrm{se}_{2n}(\theta) = \lambda \int_0^\pi \sin\theta\sin\theta'\sin(ka\cos\theta\cos\theta')\mathrm{se}_{2n}(\theta')\,d\theta',$$

$$\mathrm{ce}_{2n+1}(\theta) = \lambda \int_0^\pi \cos\theta\cos\theta'\cosh(ka\sin\theta\sin\theta')\mathrm{ce}_{2n+1}(\theta')\,d\theta',$$

$$\mathrm{se}_{2n}(\theta) = \lambda \int_0^\pi \cos\theta\cos\theta'\sinh(ka\sin\theta\sin\theta')\mathrm{se}_{2n}(\theta')\,d\theta'.$$

6. If $\mathrm{ce}_{m+\frac{1}{2}}(\theta)$, $\mathrm{se}_{m+\frac{1}{2}}(\theta)$ are the even and odd solutions of period 4π, whose limiting forms are $\cos(m+\tfrac{1}{2})\theta$, $\sin(m+\tfrac{1}{2})\theta$, show that

$$\mathrm{ce}_{m+\frac{1}{2}}(\theta) = \lambda \int_0^{2\pi} K(\theta,\theta')\mathrm{ce}_{m+\frac{1}{2}}(\theta')\,d\theta',$$

$$\mathrm{se}_{m+\frac{1}{2}}(\theta) = \lambda \int_0^{2\pi} K(\theta+\pi,\theta'+\pi)\mathrm{se}_{m+\frac{1}{2}}(\theta')\,d\theta'.$$

where
$$K(\theta,\theta') = e^{ika\cos\theta\cos\theta'} \int_0^{2\cos\frac{1}{2}\theta\cos\frac{1}{2}\theta'} e^{-ikat^2}\,dt.$$

[Transform to parabolic cylindrical coordinates

$$(x+a+iy) = \tfrac{1}{2}a(\xi+i\eta)^2$$

the equation $\dfrac{\partial^2 V}{\partial x^2} + \dfrac{\partial^2 V}{\partial y^2} - k^2 V = 0$, and expand in a series of products of Mathieu functions of period 4π the particular solution

$$V = e^{ika(\xi^2-\eta^2)} \int_0^{i\eta} e^{-ikat^2}\,dt.$$

E. G. C. Poole, *Proc. London Math. Soc.* (2) **20** (1922), 374–88.]

7. By means of Whittaker's integral equation, show that

$$\mathrm{ce}_{2n}(\theta) = \sum_{r=-\infty}^{\infty} a_{2r}\cos 2r\theta = \lambda\pi \sum_{r=-\infty}^{\infty} (-)^r a_{2r} J_{2r}(ka\cos\theta).$$

[Heine; Whittaker; Goldstein, loc. cit.]

8. The solutions $c_i(\theta)$, $s_i(\theta)$ of Mathieu's equation with the parameter p_i are determined by the conditions

$$c_i(0) = 1, \qquad c_i'(0) = 0, \qquad s_i(0) = 0, \qquad s_i'(0) = 1.$$

Show that
$$s_2(\theta) = s_1(\theta)+(p_1-p_2)\int_0^\theta [s_1(\theta)c_1(\theta')-c_1(\theta)s_1(\theta')]s_2(\theta')\,d\theta',$$

$$c_2(\theta) = c_1(\theta)+(p_1-p_2)\int_0^\theta [s_1(\theta)c_1(\theta')-c_1(\theta)s_1(\theta')]c_2(\theta')\,d\theta'.$$

[Z. Marković, *Proc. London Math. Soc.* (2) **31** (1930), 417–38.]

9. RECURRENCE FORMULAE. (i) If $(x+iy) = a\cosh(\xi+i\eta)$, show that
$$\frac{\partial V}{\partial x} = \frac{1}{a(\cosh^2\xi-\cos^2\eta)}\left[\sinh\xi\cos\eta\frac{\partial V}{\partial \xi}-\cosh\xi\sin\eta\frac{\partial V}{\partial \eta}\right].$$

(ii) If $V \equiv \mathrm{ce}_n(i\xi)\mathrm{ce}_n(\eta)$, show by expanding $\partial V/\partial x$ as a series of Mathieu functions that

$$i\sinh\xi\,\mathrm{ce}_n'(i\xi)\int_{-\pi}^{\pi}\frac{\cos\eta\,\mathrm{ce}_n(\eta)\mathrm{ce}_{n-1}(\eta)}{\cosh^2\xi-\cos^2\eta}\,d\eta - \cosh\xi\,\mathrm{ce}_n(i\xi)\int_{-\pi}^{\pi}\frac{\sin\eta\,\mathrm{ce}_n'(\eta)\mathrm{ce}_{n-1}(\eta)}{\cosh^2\xi-\cos^2\eta}\,d\eta$$

$$= A_{n-1}\,\mathrm{ce}_{n-1}(i\xi)\int_{-\pi}^{\pi}\mathrm{ce}_{n-1}^2(\eta)\,d\eta.$$

[E. T. Whittaker, *J. of London Math. Soc.* **4** (1929), 88–96.]

SHORT BIBLIOGRAPHY†

A. GENERAL TREATISES IN ENGLISH

H. BATEMAN, *Differential Equations* (Longmans, 1918).
A. R. FORSYTH, *A Treatise on Differential Equations* (6th ed., Macmillan, 1929).
A. R. FORSYTH, *Theory of Differential Equations*, Part III, Volume 4 (Cambridge, 1902).
E. L. INCE, *Ordinary Differential Equations* (Longmans, 1927).
F. R. MOULTON, *Differential Equations* (Macmillan, New York, 1930).
H. T. H. PIAGGIO, *Differential Equations* (Bell, 1920).
E. T. WHITTAKER and G. N. WATSON, *A Course of Modern Analysis* (4th ed., Cambridge, 1927).

B. FOREIGN GENERAL TREATISES

E. GOURSAT, *Cours d'analyse mathématique* (5th ed., Paris, 1927).
J. HADAMARD, *Cours d'analyse professé à l'École Polytechnique* (Paris, 1927).
L. HEFFTER, *Einleitung in die Theorie der linearen Differentialgleichungen* (Leipzig, 1894).
C. JORDAN, *Cours d'analyse de l'École Polytechnique* (3rd ed., Paris, 1909).
E. KAMKE, *Differentialgleichungen reeller Funktionen* (Leipzig, 1930).
E. PICARD, *Traité d'analyse* (3rd ed., Paris, 1922).
L. SCHLESINGER, *Handbuch der Theorie der linearen Differentialgleichungen* (Leipzig, 1895).
L. SCHLESINGER, *Vorlesungen über lineare Differentialgleichungen* (Leipzig, 1908).
CH. J. DE LA VALLÉE POUSSIN, *Cours d'analyse infinitésimale* (4th ed., Louvain and Paris, 1921).

C. WORKS OF REFERENCE

E. HILB, 'Lineare Differentialgleichungen im komplexen Gebiet' (1916), *Encyklopädie der math. Wiss.* II, B (5), 471–562.
P. HUMBERT, 'Fonctions de Lamé et fonctions de Mathieu' (1926), *Mémorial des sc. math.* x.
L. SCHLESINGER, 'Bericht über die Entwickelung der Theorie der linearen Differentialgleichungen seit 1865' (1909). *Jahresbericht der deutschen Mathematikervereinigung*, XVIII.
M. J. O. STRUTT, 'Lamé-sche, Mathieu-sche und verwandte Funktionen in Physik und Technik' (1932), *Ergebnisse der Mathematik*, I (iii).
E. VESSIOT, 'Gewöhnliche Differentialgleichungen: elementare Integrationsmethoden' (1900), *Encyklopädie der math. Wiss.* II, A (4 b), 230–93 (especially 260–74).
A. WANGERIN, 'Theorie der Kugelfunktionen und der verwandten Funktionen' (1904), *Encyklopädie der math. Wiss.* II, A (10), 695–759.

† Further references to memoirs and treatises of a more special character are given in the footnotes.

CATALOGUE OF

CONCERNING THE NATURE OF THINGS, Sir William Bragg. Christmas lectures at Royal Society by Nobel laureate, dealing with atoms, gases, liquids, and various types of crystals. No scientific background is needed to understand this remarkably clear introduction to basic processes and aspects of modern science. "More interesting than any bestseller," London Morning Post. 32pp. of photos. 57 figures. xii + 232pp. 5⅜ x 8. **T31 Paperbound $1.35**

THE RISE OF THE NEW PHYSICS, A. d'Abro. Half million word exposition, formerly titled "The Decline of Mechanism," for readers not versed in higher mathematics. Only thorough explanation in everyday language of core of modern mathematical physical theory, treating both classical, modern views. Scientifically impeccable coverage of thought from Newtonian system through theories of Dirac, Heisenberg, Fermi's statistics. Combines history, exposition; broad but unified, detailed view, with constant comparison of classical, modern views. "A must for anyone doing serious study in the physical sciences," J. of the Franklin Inst. "Extraordinary faculty . . . to explain ideas and theories . . . in language of everyday life," Isis. Part I of set: philosophy of science, from practice of Newton, Maxwell, Poincaré, Einstein, etc. Modes of thought, experiment, causality, etc. Part II: 100 pp. on grammar, vocabulary of mathematics, discussions of functions, groups, series, Fourier series, etc. Remainder treats concrete, detailed coverage of both classical, quantum physics: analytic mechanics, Hamilton's principle, electromagnetic waves, thermodynamics, Brownian movement, special relativity, Bohr's atom, de Broglie's wave mechanics, Heisenberg's uncertainty, scores of other important topics. Covers discoveries, theories of d'Alembert, Born, Cantor, Debye, Euler, Foucault, Galois, Gauss, Hadamard, Kelvin, Kepler Laplace, Maxwell, Pauli, Rayleigh Volterra, Weyl, more than 180 others. 97 illustrations. ix + 982pp. 5⅜ x 8.
T3 Vol. I Paperbound $2.00
T4 Vol. II Paperbound $2.00

SPINNING TOPS AND GYROSCOPIC MOTION, John Perry. Well-known classic of science still unsurpassed for lucid, accurate, delightful exposition. How quasi-rigidity is induced in flexible, fluid bodies by rapid motions; why gyrostat falls, top rises; nature, effect of internal fluidity on rotating bodies; etc. Appendixes describe practical use of gyroscopes in ships, compasses, monorail transportation. 62 figures. 128pp. 5⅜ x 8.
T416 Paperbound $1.00

FOUNDATIONS OF PHYSICS, R. B. Lindsay, H. Margenau. Excellent bridge between semipopular and technical writings. Discussion of methods of physical description, construction of theory; valuable to physicist with elementary calculus. Gives meaning to data, tools of modern physics. Contents: symbolism, mathematical equations; space and time; foundations of mechanics; probability; physics, continua; electron theory; relativity; quantum mechanics; causality; etc. "Thorough and yet not overdetailed. Unreservedly recommended," Nature. Unabridged corrected edition. 35 illustrations. xi + 537pp. 5⅜ x 8. **S377 Paperbound $2.45**

FADS AND FALLACIES IN THE NAME OF SCIENCE, Martin Gardner. Formerly entitled "In the Name of Science," the standard account of various cults, quack systems, delusions which have masqueraded as science: hollow earth fanatics, orgone sex energy, dianetics, Atlantis, Forteanism, flying saucers, medical fallacies like zone therapy, etc. New chapter on Bridey Murphy, psionics, other recent manifestations. A fair reasoned appraisal of eccentric theory which provides excellent innoculation. "Should be read by everyone, scientist or nonscientist alike," R. T. Birge, Prof. Emeritus of Physics, Univ. of Calif; Former Pres. Amer. Physical Soc. x + 365pp. 5⅜ x 8. **T394 Paperbound $1.50**

ON MATHEMATICS AND MATHEMATICIANS, R. E. Moritz. A 10 year labor of love by discerning, discriminating Prof. Moritz, this collection conveys the full sense of mathematics and personalities of great mathematicians. Anecdotes, aphorisms, reminiscences, philosophies, definitions, speculations, biographical insights, etc. by great mathematicians, writers: Descartes, Mill, Locke, Kant, Coleridge, Whitehead, etc. Glimpses into lives of great mathematicians, from Archimedes to Euler, Gauss, Weierstrass. To mathematicians, a superb browsing-book. To laymen, exciting revelation of fullness of mathematics. Extensive cross index. 410pp. 5⅜ x 8. **T489 Paperbound $1.95**

GUIDE TO THE LITERATURE OF MATHEMATICS AND PHYSICS, N. G. Parke III. Over 5000 entries under approximately 120 major subject headings, of selected most important books, monographs, periodicals, articles in English, plus important works in German, French, Italian, Spanish, Russian. (many recently available works). Covers every branch of physics, math, related engineering. Includes author, title, edition, publisher, place, date, number of volumes, number of pages. 40 page introduction on basic problems of research, study provides useful information on organization, use of libraries, psychology of learning, etc. Will save you hours of time. 2nd revised edition. Indices of authors, subjects. 464pp. 5⅜ x 8. **S447 Paperbound $2.49**

THE STRANGE STORY OF THE QUANTUM, An Account for the General Reader of the Growth of Ideas Underlying Our Present Atomic Knowledge, B. Hoffmann. Presents lucidly, expertly, with barest amount of mathematics, problems and theories which led to modern quantum physics. Begins with late 1800's when discrepancies were noticed; with illuminating analogies, examples, goes through concepts of Planck, Einstein, Pauli, Schroedinger, Dirac, Sommerfield, Feynman, etc. New postscript through 1958. "Of the books attempting an account of the history and contents of modern atomic physics which have come to my attention, this is the best," H. Margenau, Yale U., in Amer. J. of Physics. 2nd edition. 32 tables, illustrations. 275pp. 5⅜ x 8. **T518 Paperbound $1.45**

INDEX OF NAMES

Abel, 12, 13, 102.
Airy, 140.
Aitken, 27, 60.
d'Alembert, 16, 70.
Appell, 44.

Baker, 2.
Barnes, 104.
Bateman, 139, 153.
Berg, 18.
Bessel, 79, 142, 147–51.
Birkhoff, 57.
Bôcher, 27, 36, 60, 166, 176.
Boole, 18.
Borel, 32.
Brassine, 34.
Brioschi, 163, 167, 169, 175, 180, 194.
Burnside, 132, 133.

Caqué, 2.
Carson, 18.
Cauchy, 2, 4, 17, 18, 32.
Cayley, 120, 122.
Chaundy, 109.
Christoffel, 134, 159.
Courant, 95, 152, 153.
Crawford, 156, 163, 164.

Darboux, 36, 42, 43, 113, 176.
Darwin, 183.
Dixon, 104.

Elliot, 113.
Euler, 26, 53, 104, 107.

Ferrar, 117.
Floquet, 170.
Forsyth, 132.
Fricke, 133.
Frobenius, 42, 63, 70–4.
Fuchs, 4, 6, 16, 55, 74–9.

Gauss, 92, 101, 109.
Goldstein, 187, 189, 191, 192, 198.
Goursat, 97, 106.
Green, 164.

Haentzschel, 175.
Halphen, 154, 162, 164.
Hamburger, 53–5, 65.
Hankel, 79, 138, 148, 149.
Haupt, 124, 132.
Heaviside, 18–22, 30–2.

Heffter, 63, 70.
Heine, 158, 187, 189, 198.
Hermite, 32, 140, 152, 167, 169, 180, 194.
Hilbert, 95, 152, 153.
Hill, 178, 180, 182–7.
Hille, 197.
Hilton, 60.
Hobson, 98, 115, 154.
Hodgkinson, 132, 167.
Hough, 183, 191.
Humbert, 18, 154, 178.

Ince, 181, 187, 189.

Jacobi, 7, 43, 51, 59, 94, 106, 111, 152, 171, 173.
Jeffreys, 18.
Jordan, 27, 51, 104, 106, 136, 139, 160, 171, 172.

Kelvin, 183.
Klein, 104, 124, 132, 133, 158, 166.
von Koch, 184.
Kronecker, 8.
Kummer, 80, 87–8, 97, 102–4, 109, 112, 120, 139, 140, 143–7.

Lagrange, 37, 120, 153.
Laguerre, 153.
Lamb, 82, 183.
Lamé, 154–77.
Laplace, 82, 136–53, 183.
Legendre, 81, 95, 96, 98, 110, 114, 115, 116, 142.
Liapounoff, 57.
Lindemann, 180, 193–6.
Lindstedt, 182–7, 189.
Liouville, 2, 12–13, 44, 156, 166, 187, 191.
Love, 82, 134.
Lowry, 31, 32.

Markoff, 113.
Marković, 199.
Mathieu, 178–99.
Mellin, 104.
Molk, 116.

Papperitz, 86.
Peano, 2, 11.
Perron, 136.

INDEX OF NAMES

Picard, 2, 44, 136, 170–5.
Pidduck, 36.
Pincherle, 104.
Pochhammer, 106.
Poincaré, 46, 149, 178.
van der Pol, 32.
Pólya, 95, 153, 158.

Riemann, 46, 83–7, 88–92, 98, 100, 104, 107, 118, 120, 131.
Rodrigues, 111.

Schendel, 111.
Scherk, 152.
Schläfli, 114.
Schlesinger, 2, 44, 104.
Schrödinger, 153.
Schwarz, 118–22, 125, 128, 132, 134, 135, 159.
Smith, 28.
Sonine, 143, 145, 153.
Steiner, 127.

Stieltjes, 158, 193–6.
Strutt, 154, 178.
Sturm, 156, 163, 165, 187, 191.
Szegö, 95, 153, 158.

Tannery, 116.
Titchmarsh, 125.
Tschebyscheff, 95, 111.
Turnbull, 27, 60.

van Vleck, 113.

Wangerin, 175, 176.
Watson, 104, 139, 147, 152, 154, 178, 189, 198.
Weber, 133, 140.
Weierstrass, 51, 159, 186.
Whipple, 98.
Whittaker, 104, 139, 141, 152, 154, 175, 178, 189, 197, 198, 199.
Wirtinger, 104, 108, 124.
Wronski, 10–12, 41.

Catalogue of Dover
SCIENCE BOOKS

BOOKS THAT EXPLAIN SCIENCE

THE NATURE OF LIGHT AND COLOUR IN THE OPEN AIR, M. Minnaert. Why is falling snow sometimes black? What causes mirages, the fata morgana, multiple suns and moons in the sky; how are shadows formed? Prof. Minnaert of U. of Utrecht answers these and similar questions in optics, light, colour, for non-specialists. Particularly valuable to nature science students, painters, photographers. "Can best be described in one word—fascinating," Physics Today. Translated by H. M. Kremer-Priest, K. Jay. 202 illustrations, including photos. xvi + 362pp. 5⅜ x 8. T196 Paperbound $1

THE RESTLESS UNIVERSE, Max Born. New enlarged version of this remarkably readable account by a Nobel laureate. Moving from sub-atomic particles to universe, the author explains in very simple terms the latest theories of wave mechanics. Partial contents: and its relatives, electrons and ions, waves and particles, electronic structure of atom, nuclear physics. Nearly 1000 illustrations, including 7 animated sequences. 6 x 9. T412 Paperbound

MATTER AND LIGHT, THE NEW PHYSICS, L. de Broglie. Non-technical papers by a laureate explain electromagnetic theory, relativity, matter, light, radiation, wave me quantum physics, philosophy of science. Einstein, Planck, Bohr, others explained s that no mathematical training is needed for all but 2 of the 21 chapters. "Easy s and lucidity . . . should make this source-book of modern physcis available to public," Saturday Review. Unabridged. 300pp. 5⅜ x 8. T35 Paperbo

THE COMMON SENSE OF THE EXACT SCIENCES, W. K. Clifford. Introduction by J man, edited by Karl Pearson. For 70 years this has been a guide to classical mathematical thought. Explains with unusual clarity basic concepts such as e meaning of symbols, characteristics of surface boundaries, properties of pl vectors, Cartesian method of determining position, etc. Long preface by Bertr Bibliography of Clifford. Corrected. 130 diagrams redrawn. 249pp. 5⅜ x 8. T61 Pape

THE EVOLUTION OF SCIENTIFIC THOUGHT FROM NEWTON TO EINSTEIN, A. d'A special, general theories of relativity, with historical implications, analyzed i terms. Excellent accounts of contributions of Newton, Riemann, Weyl, Pla Maxwell, Lorentz, etc., are treated in terms of space, time, equations of e finiteness of universe, methodology of science. "Has become a standard wo diagrams. 482pp. 5⅜ x 8. T2 P

BRIDGES AND THEIR BUILDERS, D. Steinman, S. R. Watson. Engineers, hi ever fascinated by great spans will find this an endless source of informa Dr. Steinman, recent recipient of Louis Levy Medal, is one of the great engineers of all time. His analysis of great bridges of history is both easily followed. Greek, Roman, medieval, oriental bridges; modern work Bridge, Golden Gate Bridge, etc. described in terms of history, const artistry, function. Most comprehensive, accurate semi-popular history of English. New, greatly revised, enlarged edition. 23 photographs, 26 li 401pp. 5⅜ x 8.

DOVER SCIENCE BOOKS

HISTORY OF SCIENCE
AND PHILOSOPHY OF SCIENCE

THE VALUE OF SCIENCE, Henri Poincaré. Many of most mature ideas of "last scientific universalist" for both beginning, advanced workers. Nature of scientific truth, whether order is innate in universe or imposed by man, logical thought vs. intuition (relating to Weierstrass, Lie, Riemann, etc), time and space (relativity, psychological time, simultaneity), Herz's concept of force, values within disciplines of Maxwell, Carnot, Mayer, Newton, Lorentz, etc. iii + 147pp. 5⅜ x 8. S469 Paperbound **$1.35**

PHILOSOPHY AND THE PHYSICISTS, L. S. Stebbing. Philosophical aspects of modern science examined in terms of lively critical attack on ideas of Jeans, Eddington. Tasks of science, causality, determinism, probability, relation of world physics to that of everyday experience, philosophical significance of Planck-Bohr concept of discontinuous energy levels, inferences to be drawn from Uncertainty Principle, implications of "becoming" involved in 2nd law of thermodynamics, other problems posed by discarding of Laplacean determinism. 285pp. 5⅜ x 8. T480 Paperbound **$1.65**

THE PRINCIPLES OF SCIENCE, A TREATISE ON LOGIC AND THE SCIENTIFIC METHOD, W. S. Jevons. Milestone in development of symbolic logic remains stimulating contribution to investigation of inferential validity in sciences. Treats inductive, deductive logic, theory of number, probability, limits of scientific method; significantly advances Boole's logic, contains detailed introduction to nature and methods of probability in physics, astronomy, everyday affairs, etc. In introduction, Ernest Nagel of Columbia U. says,"[Jevons] continues to be of interest as an attempt to articulate the logic of scientific inquiry." liii + 786pp. 5⅜ x 8. S446 Paperbound **$2.98**

A HISTORY OF ASTRONOMY FROM THALES TO KEPLER, J. L. E. Dreyer. Only work in English to give complete history of cosmological views from prehistoric times to Kepler. Partial contents: Near Eastern astronomical systems, Early Greeks, Homocentric spheres of Euxodus, Epicycles, Ptolemaic system, Medieval cosmology, Copernicus, Kepler, much more. "Especially useful to teachers and students of the history of science . . . unsurpassed in its field," Isis. Formerly "A History of Planetary Systems from Thales to Kepler." Revised foreword by W. H. Stahl. xvii + 430pp. 5⅜ x 8. S79 Paperbound **$1.98**

A CONCISE HISTORY OF MATHEMATICS, D. Struik. Lucid study of development of ideas, techniques, from Ancient Near East, Greece, Islamic science, Middle Ages, Renaissance, modern times. Important mathematicians described in detail. Treatment not anecdotal, but analytical development of ideas. Non-technical—no math training needed. "Rich in content, thoughtful in interpretations," U.S. Quarterly Booklist. 60 illustrations including Greek, Egyptian manuscripts, portraits of 31 mathematicians. 2nd edition. xix + 299pp. 5⅜ x 8. S255 Paperbound **$1.75**

THE PHILOSOPHICAL WRITINGS OF PEIRCE, edited by Justus Buchler. A carefully balanced expositon of Peirce's complete system, written by Peirce himself. It covers such matters as scientific method, pure chance vs. law, symbolic logic, theory of signs, pragmatism, experiment, and other topics. "Excellent selection . . . gives more than adequate evidence of the range and greatness," Personalist. Formerly entitled "The Philosophy of Peirce." xvi + 368pp. T217 Paperbound **$1.95**

SCIENCE AND METHOD, Henri Poincaré. Procedure of scientific discovery, methodology, experiment, idea-germination—processes by which discoveries come into being. Most significant and interesting aspects of development, application of ideas. Chapters cover selection of facts, chance, mathematical reasoning, mathematics and logic; Whitehead, Russell, Cantor, the new mechanics, etc. 288pp. 5⅜ x 8. S222 Paperbound **$1.35**

SCIENCE AND HYPOTHESIS, Henri Poincaré. Creative psychology in science. How such concepts as number, magnitude, space, force, classical mechanics developed, how modern scientist uses them in his thought. Hypothesis in physics, theories of modern physics. Introduction by Sir James Larmor. "Few mathematicians have had the breadth of vision of Poincaré, and none is his superior in the gift of clear exposition," E. T. Bell. 272pp. 5⅜ x 8. S221 Paperbound **$1.35**

ESSAYS IN EXPERIMENTAL LOGIC, John Dewey. Stimulating series of essays by one of most influential minds in American philosophy presents some of his most mature thoughts on wide range of subjects. Partial contents: Relationship between inquiry and experience; dependence of knowledge upon thought; character logic; judgments of practice, data, and meanings; stimuli of thought, etc. viii + 444pp. 5⅜ x 8. T73 Paperbound **$1.95**

WHAT IS SCIENCE, Norman Campbell. Excellent introduction explains scientific method, role of mathematics, types of scientific laws. Contents: 2 aspects of science, science and nature, laws of chance, discovery of laws, explanation of laws, measurement and numerical laws, applications of science. 192pp. 5⅜ x 8. S43 Paperbound **$1.25**

CATALOGUE OF

FROM EUCLID TO EDDINGTON: A STUDY OF THE CONCEPTIONS OF THE EXTERNAL WORLD, Sir Edmund Whittaker. Foremost British scientist traces development of theories of natural philosophy from western rediscovery of Euclid to Eddington, Einstein, Dirac, etc. 5 major divisions: Space, Time and Movement; Concepts of Classical Physics; Concepts of Quantum Mechanics; Eddington Universe. Contrasts inadequacy of classical physics to understand physical world with present day attempts of relativity, non-Euclidean geometry, space curvature, etc. 212pp. 5⅜ x 8.
T491 Paperbound **$1.35**

THE ANALYSIS OF MATTER, Bertrand Russell. How do our senses accord with the new physics? This volume covers such topics as logical analysis of physics, prerelativity physics, causality, scientific inference, physics and perception, special and general relativity, Weyl's theory, tensors, invariants and their physical interpretation, periodicity and qualitative series. "The most thorough treatment of the subject that has yet been published," The Nation. Introduction by L. E. Denonn. 422pp. 5⅜ x 8.
T231 Paperbound **$1.95**

LANGUAGE, TRUTH, AND LOGIC, A. Ayer. A clear introduction to the Vienna and Cambridge schools of Logical Positivism. Specific tests to evaluate validity of ideas, etc. Contents: function of philosophy, elimination of metaphysics, nature of analysis, a priori, truth and probability, etc. 10th printing. "I should like to have written it myself," Bertrand Russell. 160pp. 5⅜ x 8.
T10 Paperbound **$1.25**

THE PSYCHOLOGY OF INVENTION IN THE MATHEMATICAL FIELD, J. Hadamard. Where do ideas come from? What role does the unconscious play? Are ideas best developed by mathematical reasoning, word reasoning, visualization? What are the methods used by Einstein, Poincaré, Galton, Riemann? How can these techniques be applied by others? One of the world's leading mathematicians discusses these and other questions. xiii + 145pp. 5⅜ x 8.
T107 Paperbound **$1.25**

GUIDE TO PHILOSOPHY, C. E. M. Joad. By one of the ablest expositors of all time, this is not simply a history or a typological survey, but an examination of central problems in terms of answers afforded by the greatest thinkers: Plato, Aristotle, Scholastics, Leibniz, Kant, Whitehead, Russell, and many others. Especially valuable to persons in the physical sciences; over 100 pages devoted to Jeans, Eddington, and others, the philosophy of modern physics, scientific materialism, pragmatism, etc. Classified bibliography. 592pp. 5⅜ x 8.
T50 Paperbound **$2.00**

SUBSTANCE AND FUNCTION, and EINSTEIN'S THEORY OF RELATIVITY, Ernst Cassirer. Two books bound as one. Cassirer establishes a philosophy of the exact sciences that takes into consideration new developments in mathematics, shows historical connections. Partial contents: Aristotelian logic, Mill's analysis, Helmholtz and Kronecker, Russell and cardinal numbers, Euclidean vs. non-Euclidean geometry, Einstein's relativity. Bibliography. Index. xxi + 464pp. 5⅜ x 8.
T50 Paperbound **$2.00**

FOUNDATIONS OF GEOMETRY, Bertrand Russell. Nobel laureate analyzes basic problems in the overlap area between mathematics and philosophy: the nature of geometrical knowledge, the nature of geometry, and the applications of geometry to space. Covers history of non-Euclidean geometry, philosophic interpretations of geometry, especially Kant, projective and metrical geometry. Most interesting as the solution offered in 1897 by a great mind to a problem still current. New introduction by Prof. Morris Kline, N.Y. University. "Admirably clear, precise, and elegantly reasoned analysis," International Math. News. xii + 201pp. 5⅜ x 8.
S233 Paperbound **$1.60**

THE NATURE OF PHYSICAL THEORY, P. W. Bridgman. How modern physics looks to a highly unorthodox physicist—a Nobel laureate. Pointing out many absurdities of science, demonstrating inadequacies of various physical theories, weighs and analyzes contributions of Einstein, Bohr, Heisenberg, many others. A non-technical consideration of correlation of science and reality. xi + 138pp. 5⅜ x 8.
S33 Paperbound **$1.25**

EXPERIMENT AND THEORY IN PHYSICS, Max Born. A Nobel laureate examines the nature and value of the counterclaims of experiment and theory in physics. Synthetic versus analytical scientific advances are analyzed in works of Einstein, Bohr, Heisenberg, Planck, Eddington, Milne, others, by a fellow scientist. 44pp. 5⅜ x 8.
S308 Paperbound **60¢**

A SHORT HISTORY OF ANATOMY AND PHYSIOLOGY FROM THE GREEKS TO HARVEY, Charles Singer. Corrected edition of "The Evolution of Anatomy." Classic traces anatomy, physiology from prescientific times through Greek, Roman periods, dark ages, Renaissance, to beginning of modern concepts. Centers on individuals, movements, that definitely advanced anatomical knowledge. Plato, Diocles, Erasistratus, Galen, da Vinci, etc. Special section on Vesalius. 20 plates. 270 extremely interesting illustrations of ancient, Medieval, Renaissance, Oriental origin. xii + 209pp. 5⅜ x 8.
T389 Paperbound **$1.75**

SPACE - TIME - MATTER, Hermann Weyl. "The standard treatise on the general theory of relativity," (Nature), by world renowned scientist. Deep, clear discussion of logical coherence of general theory, introducing all needed tools: Maxwell, analytical geometry, non-Euclidean geometry, tensor calculus, etc. Basis is classical space-time, before absorption of relativity. Contents: Euclidean space, mathematical form, metrical continuum, general theory, etc. 15 diagrams. xviii + 330pp. 5⅜ x 8.
S267 Paperbound **$1.75**

DOVER SCIENCE BOOKS

MATTER AND MOTION, James Clerk Maxwell. Excellent exposition begins with simple particles, proceeds gradually to physical systems beyond complete analysis; motion, force, properties of centre of mass of material system; work, energy, gravitation, etc. Written with all Maxwell's original insights and clarity. Notes by E. Larmor. 17 diagrams. 178pp. 5⅜ x 8.
S188 Paperbound **$1.25**

PRINCIPLES OF MECHANICS, Heinrich Hertz. Last work by the great 19th century physicist is not only a classic, but of great interest in the logic of science. Creating a new system of mechanics based upon space, time, and mass, it returns to axiomatic analysis, understanding of the formal or structural aspects of science, taking into account logic, observation, a priori elements. Of great historical importance to Poincaré, Carnap, Einstein, Milne. A 20 page introduction by R. S. Cohen, Wesleyan University, analyzes the implications of Hertz's thought and the logic of science. 13 page introduction by Helmholtz. xlii + 274pp. 5⅜ x 8.
S316 Clothbound **$3.50**
S317 Paperbound **$1.75**

FROM MAGIC TO SCIENCE, Charles Singer. A great historian examines aspects of science from Roman Empire through Renaissance. Includes perhaps best discussion of early herbals, penetrating physiological interpretation of "The Visions of Hildegarde of Bingen." Also examines Arabian, Galenic influences; Pythagoras' sphere, Paracelsus; reawakening of science under Leonardo da Vinci, Vesalius; Lorica of Gildas the Briton; etc. Frequent quotations with translations from contemporary manuscripts. Unabridged, corrected edition. 158 unusual illustrations from Classical, Medieval sources. xxvii + 365pp. 5⅜ x 8.
T390 Paperbound **$2.00**

A HISTORY OF THE CALCULUS, AND ITS CONCEPTUAL DEVELOPMENT, Carl B. Boyer. Provides laymen, mathematicians a detailed history of the development of the calculus, from beginnings in antiquity to final elaboration as mathematical abstraction. Gives a sense of mathematics not as technique, but as habit of mind, in progression of ideas of Zeno, Plato, Pythagoras, Eudoxus, Arabic and Scholastic mathematicians, Newton, Leibniz, Taylor, Descartes, Euler, Lagrange, Cantor, Weierstrass, and others. This first comprehensive, critical history of the calculus was originally entitled "The Concepts of the Calculus." Foreword by R. Courant. 22 figures. 25 page bibliography. v + 364pp. 5⅜ x 8.
S509 Paperbound **$2.00**

A DIDEROT PICTORIAL ENCYCLOPEDIA OF TRADES AND INDUSTRY, Manufacturing and the Technical Arts in Plates Selected from "L'Encyclopédie ou Dictionnaire Raisonné des Sciences, des Arts, et des Métiers" of Denis Diderot. Edited with text by C. Gillispie. First modern selection of plates from high-point of 18th century French engraving. Storehouse of technological information to historian of arts and science. Over 2,000 illustrations on 485 full page plates, most of them original size, show trades, industries of fascinating era in such great detail that modern reconstructions might be made of them. Plates teem with men, women, children performing thousands of operations; show sequence, general operations, closeups, details of machinery. Illustrates such important, interesting trades, industries as sowing, harvesting, beekeeping, tobacco processing, fishing, arts of war, mining, smelting, casting iron, extracting mercury, making gunpowder, cannons, bells, shoeing horses, tanning, papermaking, printing, dying, over 45 more categories. Professor Gillispie of Princeton supplies full commentary on all plates, identifies operations, tools, processes, etc. Material is presented in lively, lucid fashion. Of great interest to all studying history of science, technology. Heavy library cloth. 920pp. 9 x 12.
T421 2 volume set **$18.50**

DE MAGNETE, William Gilbert. Classic work on magnetism, founded new science. Gilbert was first to use word "electricity," to recognize mass as distinct from weight, to discover effect of heat on magnetic bodies; invented an electroscope, differentiated between static electricity and magnetism, conceived of earth as magnet. This lively work, by first great experimental scientist, is not only a valuable historical landmark, but a delightfully easy to follow record of a searching, ingenious mind. Translated by P. F. Mottelay. 25 page biographical memoir. 90 figures. lix + 368pp. 5⅜ x 8.
S470 Paperbound **$2.00**

HISTORY OF MATHEMATICS, D. E. Smith. Most comprehensive, non-technical history of math in English. Discusses lives and works of over a thousand major, minor figures, with footnotes giving technical information outside book's scheme, and indicating disputed matters. Vol. I: A chronological examination, from primitive concepts through Egypt, Babylonia, Greece, the Orient, Rome, the Middle Ages, The Renaissance, and to 1900. Vol. II: The development of ideas in specific fields and problems, up through elementary calculus. "Marks an epoch . . . will modify the entire teaching of the history of science," George Sarton. 2 volumes, total of 510 illustrations, 1355pp. 5⅜ x 8. Set boxed in attractive container.
T429, 430 Paperbound, the set **$5.00**

THE PHILOSOPHY OF SPACE AND TIME, H. Reichenbach. An important landmark in development of empiricist conception of geometry, covering foundations of geometry, time theory, consequences of Einstein's relativity, including: relations between theory and observations; coordinate definitions; relations between topological and metrical properties of space; psychological problem of visual intuition of non-Euclidean structures; many more topics important to modern science and philosophy. Majority of ideas require only knowledge of intermediate math. "Still the best book in the field," Rudolf Carnap. Introduction by R. Carnap. 49 figures. xviii + 296pp. 5⅜ x 8.
S443 Paperbound **$2.00**

CATALOGUE OF

FOUNDATIONS OF SCIENCE: THE PHILOSOPHY OF THEORY AND EXPERIMENT, N. Campbell. A critique of the most fundamental concepts of science, particularly physics. Examines why certain propositions are accepted without question, demarcates science from philosophy, etc. Part I analyzes presuppositions of scientific thought: existence of material world, nature of laws, probability, etc; part 2 covers nature of experiment and applications of mathematics: conditions for measurement, relations between numerical laws and theories, error, etc. An appendix covers problems arising from relativity, force, motion, space, time. A classic in its field. "A real grasp of what science is," Higher Educational Journal. xiii + 565pp. 5⅝ x 8⅜.
S372 Paperbound **$2.95**

THE STUDY OF THE HISTORY OF MATHEMATICS and THE STUDY OF THE HISTORY OF SCIENCE, G. Sarton. Excellent introductions, orientation, for beginning or mature worker. Describes duty of mathematical historian, incessant efforts and genius of previous generations. Explains how today's discipline differs from previous methods. 200 item bibliography with critical evaluations, best available biographies of modern mathematicians, best treatises on historical methods is especially valuable. 10 illustrations. 2 volumes bound as one. 113pp. + 75pp. 5⅜ x 8.
T240 Paperbound **$1.25**

MATHEMATICAL PUZZLES

MATHEMATICAL PUZZLES OF SAM LOYD, selected and edited by **Martin Gardner.** 117 choice puzzles by greatest American puzzle creator and innovator, from his famous "Cyclopedia of Puzzles." All unique style, historical flavor of originals. Based on arithmetic, algebra, probability, game theory, route tracing, topology, sliding block, operations research, geometrical dissection. Includes famous "14-15" puzzle which was national craze, "Horse of a Different Color" which sold millions of copies. 120 line drawings, diagrams. Solutions. xx + 167pp. 5⅜ x 8.
T498 Paperbound **$1.00**

SYMBOLIC LOGIC and THE GAME OF LOGIC, Lewis Carroll. "Symbolic Logic" is not concerned with modern symbolic logic, but is instead a collection of over 380 problems posed with charm and imagination, using the syllogism, and a fascinating diagrammatic method of drawing conclusions. In "The Game of Logic" Carroll's whimsical imagination devises a logical game played with 2 diagrams and counters (included) to manipulate hundreds of tricky syllogisms. The final section, "Hit or Miss" is a lagniappe of 101 additional puzzles in the delightful Carroll manner. Until this reprint edition, both of these books were rarities costing up to $15 each. Symbolic Logic: Index. xxxi + 199pp. The Game of Logic: 96pp. 2 vols. bound as one. 5⅜ x 8.
T492 Paperbound **$1.50**

PILLOW PROBLEMS and A TANGLED TALE, Lewis Carroll. One of the rarest of all Carroll's works, "Pillow Problems" contains 72 original math puzzles, all typically ingenious. Particularly fascinating are Carroll's answers which remain exactly as he thought them out, reflecting his actual mental process. The problems in "A Tangled Tale" are in story form, originally appearing as a monthly magazine serial. Carroll not only gives the solutions, but uses answers sent in by readers to discuss wrong approaches and misleading paths, and grades them for insight. Both of these books were rarities until this edition, "Pillow Problems" costing up to $25, and "A Tangled Tale" $15. Pillow Problems: Preface and Introduction by Lewis Carroll. xx + 109pp. A Tangled Tale: 6 illustrations. 152pp. Two vols. bound as one. 5⅜ x 8.
T493 Paperbound **$1.50**

NEW WORD PUZZLES, G. L. Kaufman. 100 brand new challenging puzzles on words, combinations, never before published. Most are new types invented by author, for beginners and experts both. Squares of letters follow chess moves to build words; symmetrical designs made of synonyms; rhymed crostics; double word squares; syllable puzzles where you fill in missing syllables instead of missing letter; many other types, all new. Solutions. "Excellent," Recreation. 100 puzzles. 196 figures. vi + 122pp. 5⅜ x 8.
T344 Paperbound **$1.00**

MATHEMATICAL EXCURSIONS, H. A. Merrill. Fun, recreation, insights into elementary problem solving. Math expert guides you on by-paths not generally travelled in elementary math courses—divide by inspection, Russian peasant multiplication; memory systems for pi; odd, even magic squares; dyadic systems; square roots by geometry; Tchebichev's machine; dozens more. Solutions to more difficult ones. "Brain stirring stuff . . . a classic," Genie. 50 illustrations. 145pp. 5⅜ x 8.
T350 Paperbound **$1.00**

THE BOOK OF MODERN PUZZLES, G. L. Kaufman. Over 150 puzzles, absolutely all new material based on same appeal as crosswords, deduction puzzles, but with different principles, techniques. 2-minute teasers, word labyrinths, design, pattern, logic, observation puzzles, puzzles testing ability to apply general knowledge to peculiar situations, many others. Solutions. 116 illustrations. 192pp. 5⅜ x 8.
T143 Paperbound **$1.00**

MATHEMAGIC, MAGIC PUZZLES, AND GAMES WITH NUMBERS, R. V. Heath. Over 60 puzzles, stunts, on properties of numbers. Easy techniques for multiplying large numbers mentally, identifying unknown numbers, finding date of any day in any year. Includes The Lost Digit, 3 Acrobats, Psychic Bridge, magic squares, triangles, cubes, others not easily found elsewhere. Edited by J. S. Meyer. 76 illustrations. 128pp. 5⅜ x 8.
T110 Paperbound **$1.00**

DOVER SCIENCE BOOKS

PUZZLE QUIZ AND STUNT FUN, J. Meyer. 238 high-priority puzzles, stunts, tricks—math puzzles like The Clever Carpenter, Atom Bomb, Please Help Alice; mysteries, deductions like The Bridge of Sighs, Secret Code; observation puzzlers like The American Flag, Playing Cards, Telephone Dial; over 200 others with magic squares, tongue twisters, puns, anagrams. Solutions. Revised, enlarged edition of "Fun-To-Do." Over 100 illustrations. 238 puzzles, stunts, tricks. 256pp. 5⅜ x 8. T337 Paperbound **$1.00**

101 PUZZLES IN THOUGHT AND LOGIC, C. R. Wylie, Jr. For readers who enjoy challenge, stimulation of logical puzzles without specialized math or scientific knowledge. Problems entirely new, range from relatively easy to brainteasers for hours of subtle entertainment. Detective puzzles, find the lying fisherman, how a blind man identifies color by logic, many more. Easy-to-understand introduction to logic of puzzle solving and general scientific method. 128pp. 5⅜ x 8. T367 Paperbound **$1.00**

CRYPTANALYSIS, H. F. Gaines. Standard elementary, intermediate text for serious students. Not just old material, but much not generally known, except to experts. Concealment, Transposition, Substitution ciphers; Vigenere, Kasiski, Playfair, multafid, dozens of other techniques. Formerly "Elementary Cryptanalysis." Appendix with sequence charts, letter frequencies in English, 5 other languages, English word frequencies. Bibliography. 167 codes. New to this edition: solutions to codes. vi + 230pp. 5⅜ x 8⅜. T97 Paperbound **$1.95**

CRYPTOGRAPY, L. D. Smith. Excellent elementary introduction to enciphering, deciphering secret writing. Explains transposition, substitution ciphers; codes; solutions; geometrical patterns, route transcription, columnar transposition, other methods. Mixed cipher systems; single, polyalphabetical substitutions; mechanical devices; Vigenere; etc. Enciphering Japanese; explanation of Baconian biliteral cipher; frequency tables. Over 150 problems. Bibliography. Index. 164pp. 5⅜ x 8. T247 Paperbound **$1.00**

MATHEMATICS, MAGIC AND MYSTERY, M. Gardner. Card tricks, metal mathematics, stage mind-reading, other "magic" explained as applications of probability, sets, number theory, etc. Creative examination of laws, applications. Scores of new tricks, insights. 115 sections on cards, dice, coins; vanishing tricks, many others. No sleight of hand—math guarantees success. "Could hardly get more entertainment . . . easy to follow," Mathematics Teacher. 115 illustrations. xii + 174pp. 5⅜ x 8. T335 Paperbound **$1.00**

AMUSEMENTS IN MATHEMATICS, H. E. Dudeney. Foremost British originator of math puzzles, always witty, intriguing, paradoxical in this classic. One of largest collections. More than 430 puzzles, problems, paradoxes. Mazes, games, problems on number manipulations, unicursal, other route problems, puzzles on measuring, weighing, packing, age, kinship, chessboards, joiners', crossing river, plane figure dissection, many others. Solutions. More than 450 illustrations. viii + 258pp. 5⅜ x 8. T473 Paperbound **$1.25**

THE CANTERBURY PUZZLES H. E. Dudeney. Chaucer's pilgrims set one another problems in story form. Also Adventures of the Puzzle Club, the Strange Escape of the King's Jester, the Monks of Riddlewell, the Squire's Christmas Puzzle Party, others. All puzzles are original, based on dissecting plane figures, arithmetic, algebra, elementary calculus, other branches of mathematics, and purely logical ingenuity. "The limit of ingenuity and intricacy," The Observer. Over 110 puzzles, full solutions. 150 illustrations. viii + 225 pp. 5⅜ x 8. T474 Paperbound **$1.25**

MATHEMATICAL PUZZLES FOR BEGINNERS AND ENTHUSIASTS, G. Mott-Smith. 188 puzzles to test mental agility. Inference, interpretation, algebra, dissection of plane figures, geometry, properties of numbers, decimation, permutations, probability, all are in these delightful problems. Includes the Odic Force, How to Draw an Ellipse, Spider's Cousin, more than 180 others. Detailed solutions. Appendix with square roots, triangular numbers, primes, etc. 135 illustrations. 2nd revised edition. 248pp. 5⅜ x 8. T198 Paperbound **$1.00**

MATHEMATICAL RECREATIONS, M. Kraitchik. Some 250 puzzles, problems, demonstrations of recreation mathematics on relatively advanced level. Unusual historical problems from Greek, Medieval, Arabic, Hindu sources; modern problems on "mathematics without numbers," geometry, topology, arithmetic, etc. Pastimes derived from figurative, Mersenne, Fermat numbers: fairy chess; latruncles: reversi; etc. Full solutions. Excellent insights into special fields of math. "Strongly recommended to all who are interested in the lighter side of mathematics," Mathematical Gaz. 181 illustrations. 330pp. 5⅜ x 8. T163 Paperbound **$1.75**

FICTION

FLATLAND, E. A. Abbott. A perennially popular science-fiction classic about life in a 2-dimensional world, and the impingement of higher dimensions. Political, satiric, humorous, moral overtones. This land where women are straight lines and the lowest and most dangerous classes are isosceles triangles with 3° vertices conveys brilliantly a feeling for many concepts of modern science. 7th edition. New introduction by Banesh Hoffmann. 128pp. 5⅜ x 8. T1 Paperbound **$1.00**

CATALOGUE OF

SEVEN SCIENCE FICTION NOVELS OF H. G. WELLS. Complete texts, unabridged, of seven of Wells' greatest novels: The War of the Worlds, The Invisible Man, The Island of Dr. Moreau, The Food of the Gods, First Men in the Moon, In the Days of the Comet, The Time Machine. Still considered by many experts to be the best science-fiction ever written, they will offer amusements and instruction to the scientific minded reader. "The great master," Sky and Telescope. 1051pp. 5⅜ x 8.
T264 Clothbound **$3.95**

28 SCIENCE FICTION STORIES OF H. G. WELLS. Unabridged! This enormous omnibus contains 2 full length novels—Men Like Gods, Star Begotten—plus 26 short stories of space, time, invention, biology, etc. The Crystal Egg, The Country of the Blind, Empire of the Ants, The Man Who Could Work Miracles, Aepyornis Island, A Story of the Days to Come, and 20 others "A master . . . not surpassed by . . . writers of today," The English Journal. 915pp. 5⅜ x 8.
T265 Clothbound **$3.95**

FIVE ADVENTURE NOVELS OF H. RIDER HAGGARD. All the mystery and adventure of darkest Africa captured accurately by a man who lived among Zulus for years, who knew African ethnology, folkways as did few of his contemporaries. They have been regarded as examples of the very best high adventure by such critics as Orwell, Andrew Lang, Kipling. Contents: She, King Solomon's Mines, Allan Quatermain, Allan's Wife, Maiwa's Revenge. "Could spin a yarn so full of suspense and color that you couldn't put the story down," Sat. Review. 821pp. 5⅜ x 8.
T108 Clothbound **$3.95**

CHESS AND CHECKERS

LEARN CHESS FROM THE MASTERS, Fred Reinfeld. Easiest, most instructive way to improve your game—play 10 games against such masters as Marshall, Znosko-Borovsky, Bronstein, Najdorf, etc., with each move graded by easy system. Includes ratings for alternate moves possible. Games selected for interest, clarity, easily isolated principles. Covers Ruy Lopez, Dutch Defense, Vienna Game openings; subtle, intricate middle game variations; all-important end game. Full annotations. Formerly "Chess by Yourself." 91 diagrams. viii + 144pp. 5⅜ x 8.
T362 Paperbound **$1.00**

REINFELD ON THE END GAME IN CHESS, Fred Reinfeld. Analyzes 62 end games by Alekhine, Flohr, Tarrasch, Morphy, Capablanca, Rubinstein, Lasker, Reshevsky, other masters. Only 1st rate book with extensive coverage of error—tell exactly what is wrong with each move you might have made. Centers around transitions from middle play to end play. King and pawn, minor pieces, queen endings; blockage, weak, passed pawns, etc. "Excellent . . . a boon," Chess Life. Formerly "Practical End Play." 62 figures. vi + 177pp. 5⅜ x 8.
T417 Paperbound **$1.25**

HYPERMODERN CHESS as developed in the games of its greatest exponent, ARON NIMZOVICH, edited by Fred Reinfeld. An intensely original player, analyst, Nimzovich's approaches startled, often angered the chess world. This volume, designed for the average player, shows how his iconoclastic methods won him victories over Alekhine, Lasker, Marshall, Rubinstein, Spielmann, others, and infused new life into the game. Use his methods to startle opponents, invigorate play. "Annotations and introductions to each game . . . are excellent," Times (London). 180 diagrams. viii + 220pp. 5⅜ x 8.
T448 Paperbound **$1.35**

THE ADVENTURE OF CHESS, Edward Lasker. Lively reader, by one of America's finest chess masters, including: history of chess, from ancient Indian 4-handed game of Chaturanga to great players of today; such delights and oddities as Maelzel's chess-playing automaton that beat Napoleon 3 times; etc. One of most valuable features is author's personal recollections of men he has played against—Nimzovich, Emanuel Lasker, Capablanca, Alekhine, etc. Discussion of chess-playing machines (newly revised). 5 page chess primer. 11 illustrations. 53 diagrams. 296pp. 5⅜ x 8.
S510 Paperbound **$1.45**

THE ART OF CHESS, James Mason. Unabridged reprinting of latest revised edition of most famous general study ever written. Mason, early 20th century master, teaches beginning, intermediate player over 90 openings; middle game, end game, to see more moves ahead, to plan purposefully, attack, sacrifice, defend, exchange, govern general strategy. "Classic . . . one of the clearest and best developed studies," Publishers Weekly. Also included, a complete supplement by F. Reinfeld, "How Do You Play Chess?", invaluable to beginners for its lively question-and-answer method. 448 diagrams. 1947 Reinfeld-Bernstein text. Bibliography. xvi + 340pp. 5⅜ x 8.
T463 Paperbound **$1.85**

MORPHY'S GAMES OF CHESS, edited by P. W. Sergeant. Put boldness into your game by flowing brilliant, forceful moves of the greatest chess player of all time. 300 of Morphy's best games, carefully annotated to reveal principles. 54 classics against masters like Andersson, Harrwitz, Bird, Paulsen, and others. 52 games at odds; 54 blindfold games; plus over 100 others. Follow his interpretation of Dutch Defense, Evans Gambit, Giuoco Piano, Ruy Lopez, many more. Unabridged reissue of latest revised edition. New introduction by F. Reinfeld. Annotations, introduction by Sergeant. 235 diagrams. x + 352pp. 5⅜ x 8.
T386 Paperbound **$1.75**

DOVER SCIENCE BOOKS

WIN AT CHECKERS, M. Hopper. (Formerly "Checkers.") Former World's Unrestricted Checker Champion discusses principles of game, expert's shots, traps, problems for beginner, standard openings, locating best move, end game, opening "blitzkrieg" moves to draw when behind, etc. Over 100 detailed questions, answers anticipate problems. Appendix. 75 problems with solutions, diagrams. 79 figures. xi + 107pp. 5⅜ x 8. T363 Paperbound **$1.00**

HOW TO FORCE CHECKMATE, Fred Reinfeld. If you have trouble finishing off your opponent, here is a collection of lightning strokes and combinations from actual tournament play. Starts with 1-move checkmates, works up to 3-move mates. Develops ability to look ahead, gain new insights into combinations, complex or deceptive positions; ways to estimate weaknesses, strengths of you and your opponent. "A good deal of amusement and instruction," Times, (London). 300 diagrams. Solutions to all positions. Formerly "Challenge to Chess Players." 111pp. 5⅜ x 8. T417 Paperbound **$1.25**

A TREASURY OF CHESS LORE, edited by Fred Reinfeld. Delightful collection of anecdotes, short stories, aphorisms by, about masters; poems, accounts of games, tournaments, photographs; hundreds of humorous, pithy, satirical, wise, historical episodes, comments, word portraits. Fascinating "must" for chess players; revealing and perhaps seductive to those who wonder what their friends see in game. 49 photographs (14 full page plates). 12 diagrams. xi + 306pp. 5⅜ x 8. T458 Paperbound **$1.75**

WIN AT CHESS, Fred Reinfeld. 300 practical chess situations, to sharpen your eye, test skill against masters. Start with simple examples, progress at own pace to complexities. This selected series of crucial moments in chess will stimulate imagination, develop stronger, more versatile game. Simple grading system enables you to judge progress. "Extensive use of diagrams is a great attraction," Chess. 300 diagrams. Notes, solutions to every situation. Formerly "Chess Quiz." vi + 120pp. 5⅜ x 8. T433 Paperbound **$1.00**

MATHEMATICS:
ELEMENTARY TO INTERMEDIATE

HOW TO CALCULATE QUICKLY, H. Sticker. Tried and true method to help mathematics of everyday life. Awakens "number sense"—ability to see relationships between numbers as whole quantities. A serious course of over 9000 problems and their solutions through techniques not taught in schools: left-to-right multiplications, new fast division, etc. 10 minutes a day will double or triple calculation speed. Excellent for scientist at home in higher math, but dissatisfied with speed and accuracy in lower math. 256pp. 5 x 7¼. Paperbound **$1.00**

FAMOUS PROBLEMS OF ELEMENTARY GEOMETRY, Felix Klein. Expanded version of 1894 Easter lectures at Göttingen. 3 problems of classical geometry: squaring the circle, trisecting angle, doubling cube, considered with full modern implications: transcendental numbers, pi, etc. "A modern classic . . . no knowledge of higher mathematics is required," Scientia. Notes by R. Archibald. 16 figures. xi + 92pp. 5⅜ x 8. T298 Paperbound **$1.00**

HIGHER MATHEMATICS FOR STUDENTS OF CHEMISTRY AND PHYSICS, J. W. Mellor. Practical, not abstract, building problems out of familiar laboratory material. Covers differential calculus, coordinate, analytical geometry, functions, integral calculus, infinite series, numerical equations, differential equations, Fourier's theorem probability, theory of errors, calculus of variations, determinants. "If the reader is not familiar with this book, it will repay him to examine it," Chem. and Engineering News. 800 problems. 189 figures. xxi + 641pp. 5⅜ x 8. S193 Paperbound **$2.25**

TRIGONOMETRY REFRESHER FOR TECHNICAL MEN, A. A. Klaf. 913 detailed questions, answers cover most important aspects of plane, spherical trigonometry—particularly useful in clearing up difficulties in special areas. Part I: plane trig, angles, quadrants, functions, graphical representation, interpolation, equations, logs, solution of triangle, use of slide rule, etc. Next 188 pages discuss applications to navigation, surveying, elasticity, architecture, other special fields. Part 3: spherical trig, applications to terrestrial, astronomical problems. Methods of time-saving, simplification of principal angles, make book most useful. 913 questions answered. 1738 problems, answers to odd numbers. 494 figures. 24 pages of formulas, functions. x + 629pp. 5⅜ x 8. T371 Paperbound **$2.00**

CALCULUS REFRESHER FOR TECHNICAL MEN, A. A. Klaf. 756 questions examine most important aspects of integral, differential calculus. Part I: simple differential calculus, constants, variables, functions, increments, logs, curves, etc. Part 2: fundamental ideas of integrations, inspection, substitution, areas, volumes, mean value, double, triple integration, etc. Practical aspects stressed. 50 pages illustrate applications to specific problems of civil, nautical engineering, electricity, stress, strain, elasticity, similar fields. 756 questions answered. 566 problems, mostly answered. 36pp. of useful constants, formulas. v + 431pp. 5⅜ x 8. T370 Paperbound **$2.00**

CATALOGUE OF

MONOGRAPHS ON TOPICS OF MODERN MATHEMATICS, edited by J. W. A. Young. Advanced mathematics for persons who have forgotten, or not gone beyond, high school algebra. 9 monographs on foundation of geometry, modern pure geometry, non-Euclidean geometry, fundamental propositions of algebra, algebraic equations, functions, calculus, theory of numbers, etc. Each monograph gives proofs of important results, and descriptions of leading methods, to provide wide coverage. "Of high merit," Scientific American. New introduction by Prof. M. Kline, N.Y. Univ. 100 diagrams. xvi + 416pp. 6⅛ x 9¼.
S289 Paperbound **$2.00**

MATHEMATICS IN ACTION, O. G. Sutton. Excellent middle level application of mathematics to study of universe, demonstrates how math is applied to ballistics, theory of computing machines, waves, wave-like phenomena, theory of fluid flow, meteorological problems, statistics, flight, similar phenomena. No knowledge of advanced math required. Differential equations, Fourier series, group concepts, Eigenfunctions, Planck's constant, airfoil theory, and similar topics explained so clearly in everyday language that almost anyone can derive benefit from reading this even if much of high-school math is forgotten. 2nd edition. 88 figures. viii + 236pp. 5⅜ x 8.
T450 Clothbound **$3.50**

ELEMENTARY MATHEMATICS FROM AN ADVANCED STANDPOINT, Felix Klein. Classic text, an outgrowth of Klein's famous integration and survey course at Göttingen. Using one field to interpret, adjust another, it covers basic topics in each area, with extensive analysis. Especially valuable in areas of modern mathematics. "A great mathematician, inspiring teacher, . . . deep insight," Bul., Amer. Math Soc.

Vol. I. ARITHMETIC, ALGEBRA, ANALYSIS. Introduces concept of function immediately, enlivens discussion with graphical, geometric methods. Partial contents: natural numbers, special properties, complex numbers. Real equations with real unknowns, complex quantities. Logarithmic, exponential functions, infinitesimal calculus. Transcendence of e and pi, theory of assemblages. Index. 125 figures. ix + 274pp. 5⅜ x 8.
S151 Paperbound **$1.75**

Vol. II. GEOMETRY. Comprehensive view, accompanies space perception inherent in geometry with analytic formulas which facilitate precise formulation. Partial contents: Simplest geometric manifold; line segments, Grassman determinant principles, classication of configurations of space. Geometric transformations: affine, projective, higher point transformations, theory of the imaginary. Systematic discussion of geometry and its foundations. 141 illustrations. ix + 214pp. 5⅜ x 8.
S151 Paperbound **$1.75**

A TREATISE ON PLANE AND ADVANCED TRIGONOMETRY, E. W. Hobson. Extraordinarily wide coverage, going beyond usual college level, one of few works covering advanced trig in full detail. By a great expositor with unerring anticipation of potentially difficult points. Includes circular functions; expansion of functions of multiple angle; trig tables; relations between sides, angles of triangles; complex numbers; etc. Many problems fully solved. "The best work on the subject," Nature. Formerly entitled "A Treatise on Plane Trigonometry." 689 examples. 66 figures. xvi + 383pp. 5⅜ x 8.
S353 Paperbound **$1.95**

NON-EUCLIDEAN GEOMETRY, Roberto Bonola. The standard coverage of non-Euclidean geometry. Examines from both a historical and mathematical point of view geometries which have arisen from a study of Euclid's 5th postulate on parallel lines. Also included are complete texts, translated, of Bolyai's "Theory of Absolute Space," Lobachevsky's "Theory of Parallels." 180 diagrams. 431pp. 5⅜ x 8.
S27 Paperbound **$1.95**

GEOMETRY OF FOUR DIMENSIONS, H. P. Manning. Unique in English as a clear, concise introduction. Treatment is synthetic, mostly Euclidean, though in hyperplanes and hyperspheres at infinity, non-Euclidean geometry is used. Historical introduction. Foundations of 4-dimensional geometry. Perpendicularity, simple angles. Angles of planes, higher order. Symmetry, order, motion; hyperpyramids, hypercones, hyperspheres; figures with parallel elements; volume, hypervolume in space; regular polyhedroids. Glossary. 78 figures. ix + 348pp. 5⅜ x 8.
S182 Paperbound **$1.95**

MATHEMATICS: INTERMEDIATE TO ADVANCED

GEOMETRY (EUCLIDEAN AND NON-EUCLIDEAN)

THE GEOMETRY OF RENÉ DESCARTES. With this book, Descartes founded analytical geometry. Original French text, with Descartes's own diagrams, and excellent Smith-Latham translation. Contains: Problems the Construction of Which Requires only Straight Lines and Circles; On the Nature of Curved Lines; On the Construction of Solid or Supersolid Problems. Diagrams. 258pp. 5⅜ x 8.
S68 Paperbound **$1.50**

DOVER SCIENCE BOOKS

THE WORKS OF ARCHIMEDES, edited by T. L. Heath. All the known works of the great Greek mathematician, including the recently discovered Method of Archimedes. Contains: On Sphere and Cylinder, Measurement of a Circle, Spirals, Conoids, Spheroids, etc. Definitive edition of greatest mathematical intellect of ancient world. 186 page study by Heath discusses Archimedes and history of Greek mathematics. 563pp. 5⅜ x 8. S9 Paperbound **$2.00**

COLLECTED WORKS OF BERNARD RIEMANN. Important sourcebook, first to contain complete text of 1892 "Werke" and the 1902 supplement, unabridged. 31 monographs, 3 complete lecture courses, 15 miscellaneous papers which have been of enormous importance in relativity, topology, theory of complex variables, other areas of mathematics. Edited by R. Dedekind, H. Weber, M. Noether, W. Wirtinger. German text; English introduction by Hans Lewy. 690pp. 5⅜ x 8. S226 Paperbound **$2.85**

THE THIRTEEN BOOKS OF EUCLID'S ELEMENTS, edited by Sir Thomas Heath. Definitive edition of one of very greatest classics of Western world. Complete translation of Heiberg text, plus spurious Book XIV. 150 page introduction on Greek, Medieval mathematics, Euclid, texts, commentators, etc. Elaborate critical apparatus parallels text, analyzing each definition, postulate, proposition, covering textual matters, refutations, supports, extrapolations, etc. This is the full Euclid. Unabridged reproduction of Cambridge U. 2nd edition. 3 volumes. 995 figures. 1426pp. 5⅜ x 8. S88, 89, 90, 3 volume set, paperbound **$6.00**

AN INTRODUCTION TO GEOMETRY OF N DIMENSIONS, D. M. Y. Sommerville. Presupposes no previous knowledge of field. Only book in English devoted exclusively to higher dimensional geometry. Discusses fundamental ideas of incidence, parallelism, perpendicularity, angles between linear space, enumerative geometry, analytical geometry from projective and metric views, polytopes, elementary ideas in analysis situs, content of hyperspacial figures. 60 diagrams. 196pp. 5⅜ x 8. S494 Paperbound **$1.50**

ELEMENTS OF NON-EUCLIDEAN GEOMETRY, D. M. Y. Sommerville. Unique in proceeding step-by-step. Requires only good knowledge of high-school geometry and algebra, to grasp elementary hyperbolic, elliptic, analytic non-Euclidean Geometries; space curvature and its implications; radical axes; homopethic centres and systems of circles; parataxy and parallelism; Gauss' proof of defect area theorem; much more, with exceptional clarity. 126 problems at chapter ends. 133 figures. xvi + 274pp. 5⅜ x 8. S460 Paperbound **$1.50**

THE FOUNDATIONS OF EUCLIDEAN GEOMETRY, H. G. Forder. First connected, rigorous account in light of modern analysis, establishing propositions without recourse to empiricism, without multiplying hypotheses. Based on tools of 19th and 20th century mathematicians, who made it possible to remedy gaps and complexities, recognize problems not earlier discerned. Begins with important relationship of number systems in geometrical figures. Considers classes, relations, linear order, natural numbers, axioms for magnitudes, groups, quasi-fields, fields, non-Archimedian systems, the axiom system (at length), particular axioms (two chapters on the Parallel Axioms), constructions, congruence, similarity, etc. Lists: axioms employed, constructions, symbols in frequent use. 295pp. 5⅜ x 8.
S481 Paperbound **$2.00**

CALCULUS, FUNCTION THEORY (REAL AND COMPLEX), FOURIER THEORY

FIVE VOLUME "THEORY OF FUNCTIONS" SET BY KONRAD KNOPP. Provides complete, readily followed account of theory of functions. Proofs given concisely, yet without sacrifice of completeness or rigor. These volumes used as texts by such universities as M.I.T., Chicago, N.Y. City College, many others. "Excellent introduction . . . remarkably readable, concise, clear, rigorous," J. of the American Statistical Association.

ELEMENTS OF THE THEORY OF FUNCTIONS, Konrad Knopp. Provides background for further volumes in this set, or texts on similar level. Partial contents: Foundations, system of complex numbers and Gaussian plane of numbers, Riemann sphere of numbers, mapping by linear functions, normal forms, the logarithm, cyclometric functions, binomial series. "Not only for the young student, but also for the student who knows all about what is in it," Mathematical Journal. 140pp. 5⅜ x 8. S154 Paperbound **$1.35**

THEORY OF FUNCTIONS, PART I, Konrad Knopp. With volume II, provides coverage of basic concepts and theorems. Partial contents: numbers and points, functions of a complex variable, integral of a continuous function, Cauchy's intergral theorem, Cauchy's integral formulae, series with variable terms, expansion and analytic function in a power series, analytic continuation and complete definition of analytic functions, Laurent expansion, types of singularities. vii + 146pp. 5⅜ x 8. S156 Paperbound **$1.35**

THEORY OF FUNCTIONS, PART II, Konrad Knopp. Application and further development of general theory, special topics. Single valued functions, entire, Weierstrass. Meromorphic functions: Mittag-Leffler. Periodic functions. Multiple valued functions. Riemann surfaces. Algebraic functions. Analytical configurations, Riemann surface. x + 150pp. 5⅜ x 8.
S157 Paperbound **$1.35**

CATALOGUE OF

PROBLEM BOOK IN THE THEORY OF FUNCTIONS, VOLUME I, Konrad Knopp. Problems in elementary theory, for use with Knopp's "Theory of Functions," or any other text. Arranged according to increasing difficulty. Fundamental concepts, sequences of numbers and infinite series, complex variable, integral theorems, development in series, conformal mapping. Answers. viii + 126pp. 5⅜ x 8.
S 158 **Paperbound $1.35**

PROBLEM BOOK IN THE THEORY OF FUNCTIONS, VOLUME II, Konrad Knopp. Advanced theory of functions, to be used with Knopp's "Theory of Functions," or comparable text. Singularities, entire and meromorphic functions, periodic, analytic, continuation, multiple-valued functions, Riemann surfaces, conformal mapping. Includes section of elementary problems. "The difficult task of selecting . . . problems just within the reach of the beginner is here masterfully accomplished," AM. MATH. SOC. Answers. 138pp. 5⅜ x 8.
S159 Paperbound **$1.35**

ADVANCED CALCULUS, E. B. Wilson. Still recognized as one of most comprehensive, useful texts. Immense amount of well-represented, fundamental material, including chapters on vector functions, ordinary differential equations, special functions, calculus of variations, etc., which are excellent introductions to these areas. Requires only one year of calculus. Over 1300 exercises cover both pure math and applications to engineering and physical problems. Ideal reference, refresher. 54 page introductory review. ix + 566pp. 5⅜ x 8.
S504 Paperbound **$2.45**

LECTURES ON THE THEORY OF ELLIPTIC FUNCTIONS, H. Hancock. Reissue of only book in English with so extensive a coverage, especially of Abel, Jacobi, Legendre, Weierstrass, Hermite, Liouville, and Riemann. Unusual fullness of treatment, plus applications as well as theory in discussing universe of elliptic integrals, originating in works of Abel and Jacobi. Use is made of Riemann to provide most general theory. 40-page table of formulas. 76 figures. xxiii + 498pp. 5⅜ x 8.
S483 Paperbound **$2.55**

THEORY OF FUNCTIONALS AND OF INTEGRAL AND INTEGRO-DIFFERENTIAL EQUATIONS, Vito Volterra. Unabridged republication of only English translation. General theory of functions depending on continuous set of values of another function. Based on author's concept of transition from finite number of variables to a continually infinite number. Includes much material on calculus of variations. Begins with fundamentals, examines generalization of analytic functions, functional derivative equations, applications, other directions of theory, etc. New introduction by G. C. Evans. Biography, criticism of Volterra's work by E. Whittaker. xxxx + 226pp. 5⅜ x 8.
S502 Paperbound **$1.75**

AN INTRODUCTION TO FOURIER METHODS AND THE LAPLACE TRANSFORMATION, Philip Franklin. Concentrates on essentials, gives broad view, suitable for most applications. Requires only knowledge of calculus. Covers complex qualities with methods of computing elementary functions for complex values of argument and finding approximations by charts; Fourier series; harmonic anaylsis; much more. Methods are related to physical problems of heat flow, vibrations, electrical transmission, electromagnetic radiation, etc. 828 problems, answers. Formerly entitled "Fourier Methods." x + 289pp. 5⅜ x 8.
S452 Paperbound **$1.75**

THE ANALYTICAL THEORY OF HEAT, Joseph Fourier. This book, which revolutionized mathematical physics, has been used by generations of mathematicians and physicists interested in heat or application of Fourier integral. Covers cause and reflection of rays of heat, radiant heating, heating of closed spaces, use of trigonometric series in theory of heat, Fourier integral, etc. Translated by Alexander Freeman. 20 figures. xxii + 466pp. 5⅜ x 8.
S93 Paperbound **$2.00**

ELLIPTIC INTEGRALS, H. Hancock. Invaluable in work involving differential equations with cubics, quatrics under root sign, where elementary calculus methods are inadequate. Practical solutions to problems in mathematics, engineering, physics; differential equations requiring integration of Lamé's, Briot's, or Bouquet's equations; determination of arc of ellipse, hyperbola, lemiscate; solutions of problems in elastics; motion of a projectile under resistance varying as the cube of the velocity; pendulums; more. Exposition in accordance with Legendre-Jacobi theory. Rigorous discussion of Legendre transformations. 20 figures. 5 place table. 104pp. 5⅜ x 8.
S484 Paperbound **$1.25**

THE TAYLOR SERIES, AN INTRODUCTION TO THE THEORY OF FUNCTIONS OF A COMPLEX VARIABLE, P. Dienes. Uses Taylor series to approach theory of functions, using ordinary calculus only, except in last 2 chapters. Starts with introduction to real variable and complex algebra, derives properties of infinite series, complex differentiation, integration, etc. Covers biuniform mapping, overconvergence and gap theorems, Taylor series on its circle of convergence, etc. Unabridged corrected reissue of first edition. 186 examples, many fully worked out. 67 figures. xii + 555pp. 5⅜ x 8.
S391 Paperbound **$2.75**

LINEAR INTEGRAL EQUATIONS, W. V. Lovitt. Systematic survey of general theory, with some application to differential equations, calculus of variations, problems of math, physics. Includes: integral equation of 2nd kind by successive substitutions; Fredholm's equation as ratio of 2 integral series in lambda, applications of the Fredholm theory, Hilbert-Schmidt theory of symmetric kernels, application, etc. Neumann, Dirichlet, vibratory problems. ix + 253pp. 5⅜ x 8.
S175 Clothbound **$3.50**
S176 Paperbound **$1.60**

12

DOVER SCIENCE BOOKS

DICTIONARY OF CONFORMAL REPRESENTATIONS, H. Kober. Developed by British Admiralty to solve Laplace's equation in 2 dimensions. Scores of geometrical forms and transformations for electrical engineers, Joukowski aerofoil for aerodynamics, Schwartz-Christoffel transformations for hydro-dynamics, transcendental functions. Contents classified according to analytical functions describing transformations with corresponding regions. Glossary. Topological index. 447 diagrams. 6⅛ x 9¼. .S160 Paperbound **$2.00**

ELEMENTS OF THE THEORY OF REAL FUNCTIONS, J. E. Littlewood. Based on lectures at Trinity College, Cambridge, this book has proved extremely successful in introducing graduate students to modern theory of functions. Offers full and concise coverage of classes and cardinal numbers, well ordered series, other types of series, and elements of the theory of sets of points. 3rd revised edition. vii + 71pp. 5⅜ x 8. S171 Clothbound **$2.85**
S172 Paperbound **$1.25**

INFINITE SEQUENCES AND SERIES, Konrad Knopp. 1st publication in any language. Excellent introduction to 2 topics of modern mathematics, designed to give student background to penetrate further alone. Sequences and sets, real and complex numbers, etc. Functions of a real and complex variable. Sequences and series. Infinite series. Convergent power series. Expansion of elementary functions. Numerical evaluation of series. v + 186pp. 5⅜ x 8.
S152 Clothbound **$3.50**
S153 Paperbound **$1.75**

THE THEORY AND FUNCTIONS OF A REAL VARIABLE AND THE THEORY OF FOURIER'S SERIES, E. W. Hobson. One of the best introductions to set theory and various aspects of functions and Fourier's series. Requires only a good background in calculus. Exhaustive coverage of: metric and descriptive properties of sets of points; transfinite numbers and order types; functions of a real variable; the Riemann and Lebesgue integrals; sequences and series of numbers; power-series; functions representable by series sequences of continuous functions; trigonometrical series; representation of functions by Fourier's series; and much more. "The best possible guide," Nature. Vol. I: 88 detailed examples, 10 figures. Index. xv + 736pp. Vol. II: 117 detailed examples, 13 figures. x + 780pp. 6⅛ x 9¼.
Vol. I: S387 Paperbound **$3.00**
Vol. II: S388 Paperbound **$3.00**

ALMOST PERIODIC FUNCTIONS, A. S. Besicovitch. Unique and important summary by a well known mathematician covers in detail the two stages of development in Bohr's theory of almost periodic functions: (1) as a generalization of pure periodicity, with results and proofs; (2) the work done by Stepanof, Wiener, Weyl, and Bohr in generalizing the theory. xi + 180pp. 5⅜ x 8. S18 Paperbound **$1.75**

INTRODUCTION TO THE THEORY OF FOURIER'S SERIES AND INTEGRALS, H. S. Carslaw. 3rd revised edition, an outgrowth of author's courses at Cambridge. Historical introduction, rational, irrational numbers, infinite sequences and series, functions of a single variable, definite integral, Fourier series, and similar topics. Appendices discuss practical harmonic analysis, periodogram analysis, Lebesgue's theory. 84 examples. xiii + 368pp. 5⅜ x 8.
S48 Paperbound **$2.00**

SYMBOLIC LOGIC

THE ELEMENTS OF MATHEMATICAL LOGIC, Paul Rosenbloom. First publication in any language. For mathematically mature readers with no training in symbolic logic. Development of lectures given at Lund Univ., Sweden, 1948. Partial contents: Logic of classes, fundamental theorems, Boolean algebra, logic of propositions, of propositional functions, expressive languages, combinatory logics, development of math within an object language, paradoxes, theorems of Post, Goedel, Church, and similar topics. iv + 214pp. 5⅜ x 8.
S227 Paperbound **$1.45**

INTRODUCTION TO SYMBOLIC LOGIC AND ITS APPLICATION, R. Carnap. Clear, comprehensive, rigorous, by perhaps greatest living master. Symbolic languages analyzed, one constructed. Applications to math (axiom systems for set theory, real, natural numbers), topology (Dedekind, Cantor continuity explanations), physics (general analysis of determination, causality, space-time topology), biology (axiom system for basic concepts). "A masterpiece," Zentralblatt für Mathematik und Ihre Grenzgebiete. Over 300 exercises. 5 figures. xvi + 241pp. 5⅜ x 8. S453 Paperbound **$1.85**

AN INTRODUCTION TO SYMBOLIC LOGIC, Susanne K. Langer. Probably clearest book for the philosopher, scientist, layman—no special knowledge of math required. Starts with simplest symbols, goes on to give masterful grasp of Boole-Schroeder, Russell-Whitehead systems, clearly, quickly. Partial Contents: Forms, Generalization, Classes, Deductive System of Classes, Algebra of Logic, Assumptions of Principia Mathematica, Logistics, Proofs of Theorems, etc. "Clearest . . . simplest introduction . . . the intelligent non-mathematician should have no difficulty," MATHEMATICS GAZETTE. Revised, expanded 2nd edition. Truth-value tables. 368pp. 5⅜ 8. S164 Paperbound **$1.75**

CATALOGUE OF

TRIGONOMETRICAL SERIES, Antoni Zygmund. On modern advanced level. Contains carefully organized analyses of trigonometric, orthogonal, Fourier systems of functions, with clear adequate descriptions of summability of Fourier series, proximation theory, conjugate series, convergence, divergence of Fourier series. Especially valuable for Russian, Eastern European coverage. 329pp. 5⅜ x 8.
S290 Paperbound **$1.50**

THE LAWS OF THOUGHT, George Boole. This book founded symbolic logic some 100 years ago. It is the 1st significant attempt to apply logic to all aspects of human endeavour. Partial contents: derivation of laws, signs and laws, interpretations, eliminations, conditions of a perfect method, analysis, Aristotelian logic, probability, and similar topics. xvii + 424pp. 5⅜ x 8.
S28 Paperbound **$2.00**

SYMBOLIC LOGIC, C. I. Lewis, C. H. Langford. 2nd revised edition of probably most cited book in symbolic logic. Wide coverage of entire field; one of fullest treatments of paradoxes; plus much material not available elsewhere. Basic to volume is distinction between logic of extensions and intensions. Considerable emphasis on converse substitution, while matrix system presents supposition of variety of non-Aristotelian logics. Especially valuable sections on strict limitations, existence theorems. Partial contents: Boole-Schroeder algebra; truth value systems, the matrix method; implication and deductibility; general theory of propositions; etc. "Most valuable," Times, London. 506pp. 5⅜ x 8. S170 Paperbound **$2.00**

GROUP THEORY AND LINEAR ALGEBRA, SETS, ETC.

LECTURES ON THE ICOSAHEDRON AND THE SOLUTION OF EQUATIONS OF THE FIFTH DEGREE, Felix Klein. Solution of quintics in terms of rotations of regular icosahedron around its axes of symmetry. A classic, indispensable source for those interested in higher algebra, geometry, crystallography. Considerable explanatory material included. 230 footnotes, mostly bibliography. "Classical monograph . . . detailed, readable book," Math. Gazette. 2nd edition. xvi + 289pp. 5⅜ x 8.
S314 Paperbound **$1.85**

INTRODUCTION TO THE THEORY OF GROUPS OF FINITE ORDER, R. Carmichael. Examines fundamental theorems and their applications. Beginning with sets, systems, permutations, etc., progresses in easy stages through important types of groups: Abelian, prime power, permutation, etc. Except 1 chapter where matrices are desirable, no higher math is needed. 783 exercises, problems. xvi + 447pp. 5⅜ x 8.
S299 Clothbound **$3.95**
S300 Paperbound **$2.00**

THEORY OF GROUPS OF FINITE ORDER, W. Burnside. First published some 40 years ago, still one of clearest introductions. Partial contents: permutations, groups independent of representation, composition series of a group, isomorphism of a group with itself, Abelian groups, prime power groups, permutation groups, invariants of groups of linear substitution, graphical representation, etc. "Clear and detailed discussion . . . numerous problems which are instructive," Design News. xxiv + 512pp. 5⅜ x 8.
S38 Paperbound **$2.45**

COMPUTATIONAL METHODS OF LINEAR ALGEBRA, V. N. Faddeeva, translated by C. D. Benster. 1st English translation of unique, valuable work, only one in English presenting systematic exposition of most important methods of linear algebra—classical, contemporary. Details of deriving numerical solutions of problems in mathematical physics. Theory and practice. Includes survey of necessary background, most important methods of solution, for exact, iterative groups. One of most valuable features is 23 tables, triple checked for accuracy, unavailable elsewhere. Translator's note. x + 252pp. 5⅜ x 8. S424 Paperbound **$1.95**

THE CONTINUUM AND OTHER TYPES OF SERIAL ORDER, E. V. Huntington. This famous book gives a systematic elementary account of the modern theory of the continuum as a type of serial order. Based on the Cantor-Dedekind ordinal theory, which requires no technical knowledge of higher mathematics, it offers an easily followed analysis of ordered classes, discrete and dense series, continuous series, Cantor's transfinite numbers. "Admirable introduction to the rigorous theory of the continuum . . . reading easy," Science Progress. 2nd edition. viii + 82pp. 5⅜ x 8.
S129 Clothbound **$2.75**
S130 Paperbound **$1.00**

THEORY OF SETS, E. Kamke. Clearest, amplest introduction in English, well suited for independent study. Subdivisions of main theory, such as theory of sets of points, are discussed, but emphasis is on general theory. Partial contents: rudiments of set theory, arbitrary sets, their cardinal numbers, ordered sets, their order types, well-ordered sets, their cardinal numbers. vii + 144pp. 5⅜ x 8.
S141 Paperbound **$1.35**

CONTRIBUTIONS TO THE FOUNDING OF THE THEORY OF TRANSFINITE NUMBERS, Georg Cantor. These papers founded a new branch of mathematics. The famous articles of 1895-7 are translated, with an 82-page introduction by P. E. B. Jourdain dealing with Cantor, the background of his discoveries, their results, future possibilities. ix + 211pp. 5⅜ x 8.
S45 Paperbound **$1.25**

DOVER SCIENCE BOOKS

NUMERICAL AND GRAPHICAL METHODS, TABLES

JACOBIAN ELLIPTIC FUNCTION TABLES, L. M. Milne-Thomson. Easy-to-follow, practical, not only useful numerical tables, but complete elementary sketch of application of elliptic functions. Covers description of principle properties; complete elliptic integrals; Fourier series, expansions; periods, zeros, poles, residues, formulas for special values of argument; cubic, quartic polynomials; pendulum problem; etc. Tables, graphs form body of book: Graph, 5 figure table of elliptic function sn (u m); cn (u m); dn (u m). 8 figure table of complete elliptic integrals K, K', E, E', nome q. 7 figure table of Jacobian zeta-function Z(u). 3 figures. xi + 123pp. 5⅜ x 8. S194 Paperbound **$1.35**

TABLES OF FUNCTIONS WITH FORMULAE AND CURVES, E. Jahnke, F. Emde. Most comprehensive 1-volume English text collection of tables, formulae, curves of transcendent functions. 4th corrected edition, new 76-page section giving tables, formulae for elementary functions not in other English editions. Partial contents: sine, cosine, logarithmic integral; error integral; elliptic integrals; theta functions; Legendre, Bessel, Riemann, Mathieu, hypergeometric functions; etc. "Out-of-the-way functions for which we know no other source," Scientific Computing Service, Ltd. 212 figures. 400pp. 5⅝ x 8⅜. S133 Paperbound **$2.00**

MATHEMATICAL TABLES, H. B. Dwight. Covers in one volume almost every function of importance in applied mathematics, engineering, physical sciences. Three extremely fine tables of the three trig functions, inverses, to 1000th of radian; natural, common logs; squares, cubes; hyperbolic functions, inverses; $(a^2 + b^2)$ exp. ½a; complete elliptical integrals of 1st, 2nd kind; sine, cosine integrals; exponential integrals; Ei(x) and Ei($-x$); binomial coefficients; factorials to 250; surface zonal harmonics, first derivatives; Bernoulli, Euler numbers, their logs to base of 10; Gamma function; normal probability integral; over 60pp. Bessel functions; Riemann zeta function. Each table with formulae generally used, sources of more extensive tables, interpolation data, etc. Over half have columns of differences, to facilitate interpolation. viii + 231pp. 5⅜ x 8. S445 Paperbound **$1.75**

PRACTICAL ANALYSIS, GRAPHICAL AND NUMERICAL METHODS, F. A. Willers. Immensely practical hand-book for engineers. How to interpolate, use various methods of numerical differentiation and integration, determine roots of a single algebraic equation, system of linear equations, use empirical formulas, integrate differential equations, etc. Hundreds of shortcuts for arriving at numerical solutions. Special section on American calculating machines, by T. W. Simpson. Translation by R. T. Beyer. 132 illustrations. 422pp. 5⅜ x 8.
S273 Paperbound **$2.00**

NUMERICAL SOLUTIONS OF DIFFERENTIAL EQUATIONS, H. Levy, E. A. Baggott. Comprehensive collection of methods for solving ordinary differential equations of first and higher order. 2 requirements: practical, easy to grasp; more rapid than school methods. Partial contents: graphical integration of differential equations, graphical methods for detailed solution. Numerical solution. Simultaneous equations and equations of 2nd and higher orders. "Should be in the hands of all in research and applied mathematics, teaching," Nature. 21 figures. viii + 238pp. 5⅜ x 8. S168 Paperbound **$1.75**

NUMERICAL INTEGRATION OF DIFFERENTIAL EQUATIONS, Bennet, Milne, Bateman. Unabridged republication of original prepared for National Research Council. New methods of integration by 3 leading mathematicians: "The Interpolational Polynomial," "Successive Approximation," A. A. Bennett, "Step-by-step Methods of Integration," W. W. Milne. "Methods for Partial Differential Equations," H. Bateman. Methods for partial differential equations, solution of differential equations to non-integral values of a parameter will interest mathematicians, physicists. 288 footnotes, mostly bibliographical. 235 item classified bibliography. 108pp. 5⅜ x 8. S305 Paperbound **$1.35**

Write for free catalogs!
Indicate your field of interest. Dover publishes books on physics, earth sciences, mathematics, engineering, chemistry, astronomy, anthropology, biology, psychology, philosophy, religion, history, literature, mathematical recreations, languages, crafts, art, graphic arts, etc.

Science A

Write to Dept. catr
Dover Publications, Inc.
180 Varick St., N. Y. 14, N. Y.